ATMOSPHERIC TURBULENCE AND MESOSCALE METEOROLOGY

Bringing together leading researchers, this volume surveys recent developments in the fields of atmospheric turbulence and mesoscale meteorology, with particular emphasis on the areas pioneered by Douglas K. Lilly.

Numerical simulation is an increasingly important tool for improving our understanding of a wide range of atmospheric phenomena. The first part of this book looks at the development of theoretical and computational analyses of atmospheric turbulent flows, and reviews current research advances in this area. Chapters in the second part look at various aspects of mesoscale weather phenomena: from the numerical forecasting of individual thunderstorms to understanding how mountains affect local weather and climate.

Researchers and graduate students will find the book to be an excellent resource summarizing the development of techniques as well as current and future work in the fields of atmospheric turbulence and mesoscale meteorology.

EVGENI FEDOROVICH specialises in the dynamics of atmospheric boundary layers. He has authored or co-authored numerous articles in meteorology, fluid dynamics, and geophysics. In 1998 he co-edited *Buoyant Convection in Geophysical Flows* (Kluwer) having organised a NATO Advanced Study Institute meeting with the same name.

RICHARD ROTUNNO has spent most of the last 25 years at the National Center for Atmospheric Research in Boulder, Colorado, where he has been Senior Scientist since 1989. For his work directed at the understanding needed to make progress in the forecasting of mesoscale weather phenomena he received the Jule G. Charney award of the American Meteorological Society in 2004.

With a background in electrical engineering, BJORN STEVENS has extensively studied the dynamics of shallow cumulus convection. In recognition of his contributions to the field, he was presented with the Clarence Leroy Meisinger award by the American Meteorological Society in 2002.

ATMOSPHERIC TURBULENCE AND MESOSCALE METEOROLOGY

Scientific Research Inspired by Doug Lilly

Edited by

EVGENI FEDOROVICH

School of Meteorology, University of Oklahoma, Norman, USA

RICHARD ROTUNNO

Mesoscale and Microscale Meteorology Division, National Center for Atmospheric Research, Boulder, USA

BJORN STEVENS

Department of Atmospheric and Oceanic Sciences, University of California, Los Angeles, USA

CAMBRIDGE
UNIVERSITY PRESS

phys

₀l3S7095x

PUBLISHED BY THE PRESS SYNDICATE OF THE UNIVERSITY OF CAMBRIDGE
The Pitt Building, Trumpington Street, Cambridge, United Kingdom

CAMBRIDGE UNIVERSITY PRESS
The Edinburgh Building, Cambridge CB2 2RU, UK
40 West 20th Street, New York, NY 10011–4211, USA
477 Williamstown Road, Port Melbourne, VIC 3207, Australia
Ruiz de Alarcón 13, 28014 Madrid, Spain
Dock House, The Waterfront, Cape Town 8001, South Africa

http://www.cambridge.org

First published 2004

Printed in the United Kingdom at the University Press, Cambridge

Typeface Times 11/14 pt.   *System* LaTeX 2_ε [TB]

A catalog record for this book is available from the British Library

ISBN 0 521 83588 7 hardback

Contents

The color plates are situated between pages 118 and 119.

Preface

Nature is not particularly generous when it comes to producing individuals with both great intellectual and humanitarian qualities. Douglas K. Lilly, or simply Doug, is a rare example of a person who essentially combines these two virtues.

As will become apparent from the scientific articles collected in this volume, Doug has earned an outstanding reputation worldwide for the very high caliber of his contributions to the fields of meteorology and geophysical fluid dynamics. Less evident, but not less striking, is the dignity of his character, his modesty, and his dedication to truth. Of pioneering stock, Doug still embodies the best of the pioneering spirit: vision, individualism, fearlessness, and obliviousness of authority. His fairness of judgements co-existing with his friendliness to colleagues and dedication to students has become almost legendary. Doug has made many friends throughout the years at the various places where he has worked, and is respected and admired by students and prominent scientists alike.

This collective volume, dedicated to Doug on the occasion of his 75th birthday, begins by focusing on Doug the man. His biography, written by K. Kanak, a recent Ph.D. student of Doug, traces his scientific evolution by incorporating recollections of several people, who worked with Doug, beginning with those of J. Smagorinsky, Doug's post-doc advisor. Doug's fundamental work on the numerical simulation of turbulence dates back to his interactions with Smagorinsky and K. Bryan while at the Geophysical Fluid Dynamics Laboratory (GFDL) in the late 1950s and early 1960s. In the latter part of the 1960s, Doug went to the National Center for Atmospheric Research (NCAR), where he continued to develop his interests in numerical simulation in collaboration with J. Deardorff. During that period of his career, Doug began nurturing interest also in observational techniques for measuring atmospheric turbulent flows. D. Lenschow, who worked closely with Doug during those years, reflects upon this interest in his recollection note.

In the early 1980s, Doug left NCAR to become Professor of Meteorology at the University of Oklahoma (OU). While at OU, Doug developed strong ties to the

National Severe Storms Laboratory (NSSL) of the National Oceanic and Atmospheric Administration (NOAA). He directed the OU/NOAA Cooperative Institute for Mesoscale Meteorological Studies (CIMMS) and was a co-founder of OU's Center for the Analysis and Prediction of Storms (CAPS). These various endeavors of Doug are highlighted in the biography through the recollections of P. Lamb, K. Droegemeier, and J. Kimpel, who worked closely with Doug while he was at OU.

The bulk of this book, however, focuses on the impact of Doug's scientific contributions. Individual chapters in Part I: *Atmospheric turbulence* and Part II: *Mesoscale meteorology* represent invited contributions from renowned experts in these two areas, whose careers, either directly or indirectly, were touched by interactions with Doug. Consequently, both the contents of and works cited in each chapter serve as testaments to Doug's scientific contributions. At the same time, the chapter sequence defines a course in atmospheric dynamics on scales ranging from the micro to the meso, which should be of interest to graduate students or beginning scientists who perhaps have never heard of Doug, as well as to seasoned practitioners interested in the latest assessment of advances in the fields of atmospheric turbulence and mesoscale meteorology.

For instance, it should be readily apparent from the *Atmospheric turbulence* chapters that the thread of Doug's ideas is deeply woven into the fabric of modern research on the topic. These chapters serve both to introduce and to reflect upon the birth of an entirely new scientific methodology – the numerical simulation of atmospheric turbulent flows. The birth of this new methodology raises novel questions, such as:

- How does one rationalize simulation?
- How should simulations be optimally constructed, and then used?
- In what ways can simulations be integrated with established methodologies such as experiments, observations, and more traditional theoretical work?

These questions form the subtext of nearly a half-century of research by Doug Lilly. They should continue to interest current and future generations of students, researchers, and practitioners involved in atmospheric turbulence research and its applications.

This idea of Doug that numerical simulation represents a new frontier for explorations of turbulence is eloquently articulated in the overview chapter by J. Wyngaard, wherein he discusses Doug's strategy for numerical simulation, which he calls a bold "three-phase plan of attack." With one exception, the other chapters in Part I (and many chapters in Part II, especially the contribution by J. Klemp and W. Skamarock) help to exemplify implementation of this plan. In particular, the use of Direct Numerical Simulation (DNS) to study fundamental problems in turbulence research, which J. Wyngaard credits with revitalizing the field, is exemplified in the article on two-dimensional and stratified turbulence by J. McWilliams. The continuing efforts to rationalize Large Eddy Simulation (LES), which is the most active area of turbulence research, and the contributions of Doug in this respect are reviewed and extended in the chapter by J. Wyngaard and that by C. Higgins, C. Meneveau, and M. Parlange. As another example of the implementation of Doug's strategy, the chapter by C.-H. Moeng, B. Stevens, and P. Sullivan shows how LES is being used to study basic problems pertaining to cloud-topped mixed layers. The only chapter to deviate from this three-phase strategy is illustrative in that it focuses on the history of a simple theoretical framework for studying a rich and complex geophysical problem: the stratocumulus mixed layer. This framework, known as mixed-layer theory, was developed by Doug and is reviewed in the chapter by D. Randall and W. Schubert.

However, it remains the centerpiece of efforts to use numerical simulation to understand stratocumulus-topped boundary layers. As such, it reminds us that Doug's contributions to an emerging scientific methodology are probably no more than the byproduct of a brilliant scientist searching to understand and explain phenomena.

Introducing the chapters in Part II, it is worth recalling that Doug did his Ph.D. work on buoyant convection in 1959, when mesoscale meteorology had yet to become a common term describing research on phenomena occurring on spatial scales of less than approximately 1000 kilometers and time scales of less than about a day [see Fujita, T. (1963). Analytical mesometeorology: a review. In *Severe Local Storms*, Meteorological Monographs No. 27, **5**, American Meteorological Society, 77–125, for the history of the term "mesometeorology"]. In non-technical terms, mesoscale meteorology is the science of the weather phenomena that are directly experienced by human beings. The examples of such phenomena are thunderstorms, cold fronts, strong local winds, fog and rain, etc. Doug's deep interest in and enthusiasm for these weather phenomena have had such a great influence that it is probably no exaggeration to say that mesoscale meteorology, as we know it today, is largely his creation.

The science of mesoscale meteorology, as developed by Doug Lilly, was based on the pursuit of innovative observational technologies, recognition of the vast potential of computer simulations to aid in the interpretation and forecasting of the small-scale weather systems and, finally, a skilled use of analytical theory. As both scientist and scientific manager, Doug pursued all of these areas with great vigor and, consequently, many of the things we take for granted today, such as computer models of thunderstorms and aircraft measurements of mesoscale air motions, are a direct consequence of his leadership.

In Part II we have included a broad selection of topics, to which Doug has made fundamental contributions. He both pioneered simulations of atmospheric convection through his early work and fostered its later development through to the experimental forecasting of individual thunderstorms. The chapter by J. Klemp and W. Skamarock and that by J. Sun address recent advances in these respective areas. Doug also made fundamental contributions through theory, modeling, and novel use of observations to the understanding of how mountains affect weather and climate; the chapter by R. Smith likewise addresses some of the newer findings in that field. In mesoscale meteorology, Doug's curiosity knew no bound, and he tried his hand at just about everything in this area at one time or another. The chapter by K. Emanuel tells the interesting tale of Doug Lilly's influence on Emanuel's tropical cyclone research. Finally, Doug had the knack of recognizing a problem before the rest of us knew it was a problem. Such was the case with explaining the atmospheric energy spectra over a very large range of scales. The chapter by K. Gage chronicles the evolution of this field, in which Doug's original contribution continues to play an important role.

As John Wyngaard stated in his response to the invitation to contribute to the book, "Doug is not just another excellent scientist; he has changed the face of modern small-scale meteorology." Therefore, we expect that this book will be of significant interest to the meteorological community all over the world. Furthermore, since the book includes a variety of papers on fundamental aspects of turbulence and convection, we also anticipate interest from engineers, fluid dynamicists, oceanographers, and environmental scientists. We hope that the audience for the book will include researchers, university instructors, and first- or second-year graduate students in the fields of atmospheric dynamics, turbulence modeling and simulation, boundary-layer meteorology, air-pollution meteorology, convective storms, mountain meteorology, ocean dynamics, and computational fluid dynamics. Some operational meteorologists may be interested in reading the chapters on downslope windstorms and convective storm modeling. The book may also be used as a supplementary textbook for graduate courses in atmospheric turbulence and mesoscale meteorology, as well as a research compendium in these two areas.

The editors of this book gratefully acknowledge the assistance of Katharine Kanak and Bob Conzemius in editing the chapters and preparing figures for the book, and the help of Mark Laufersweiler in managing electronic versions of the book components. They are indebted to Katharine Kanak for the compilation of the Appendices.

Douglas K. Lilly: a biography

Katharine M. Kanak

with recollections from
K. Bryan, J. Deardorff, K. Droegemeier, J. Kimpel, P. Lamb,
D. Lenschow, and J. Smagorinsky

Douglas (Doug) Lilly was born on June 16, 1929 in San Francisco, California. He grew up on the San Francisco peninsula where, as he describes it, "there is not much weather!" He states that he was interested in weather and the atmosphere starting from his years in high school in California. The predominant cloud features there were stratus decks that would come into the bay area, stay for a while, and eventually break up. He used to borrow the family car to drive up hills to observe these stratus decks. One might say this was his high school hobby.

Doug attended Stanford University and completed a Bachelor of Science degree in Physics in 1950. At Stanford, he was a member of the rowing crew and of the Naval Reserve Officers Training Corps (ROTC) program. From 1950 to 1953, during the Korean War, he was on active duty in the Navy. He was stationed for a while in Hawaii, and then later on a minesweeper off the coast of Korea.

After completing his military service Doug decided to pursue a graduate degree in Meteorology. It was early in his graduate studies at Florida State University (FSU) that Doug first met Judith (Judy) Anne Schuh, who would later become his wife. She was pursuing a degree in Education with a minor in German. They dated for one year and married on August 12, 1954 (the year Judy graduated) in her home town Jacksonville, Florida. Their first child, Kathryn Elizabeth, was born July 19, 1955 in Tallahassee, Florida. In this same year, Doug completed his Master of Science degree in Meteorology at FSU. During their time living in Florida, and driving back and forth between Jacksonville and Tallahassee, Doug was fascinated with the tropical convection and spent a good deal of the rides with his head out of the car window!

In 1956, Doug took a job with Radio Free Europe and the family moved to Munich, Germany. His responsibilities there included prediction of wind directions and weather conditions for the purposes of launching balloons with news pamphlets into Eastern Europe during the early years of the cold war. This was a nice opportunity for Judy to perfect her German in which she had earned a minor in college. Doug also learned German there and later also some French. During this year they lived in the apartment of a retired opera singer.

In the summer of 1957, the family returned to Redwood City, California, where their second child, Donald Roger, was born on July 31, 1957. The following year, they returned to Tallahassee for one more year in order for Doug to complete his Ph.D. in 1958 with Seymour L. Hess as his major advisor. The title of his dissertation was "On the Theory of Disturbances in a Conditionally Unstable Atmosphere." This novel theoretical work was later published under the same title in 1960 in *Monthly Weather Review* (Lilly, 1960).

After completing his Ph.D., Doug took a position as a Research Meteorologist at the US Weather Bureau's General Circulation Laboratory (GCL), a division of the National Oceanic and Atmospheric Administration (NOAA) in Washington, DC (predecessor to the NOAA Geophysical Fluid Dynamics Laboratory, GFDL). In Washington, DC, Doug and Judy's third child, Carol Susan, was born on August 18, 1959.

It was at GCL that Doug met and worked with Joseph Smagorinsky. During that time, Doug contributed to some of the very earliest efforts towards numerical simulation of atmospheric convection. He developed a series of numerical techniques and methods that are still used today. He also worked on laboratory studies of vortices and convection, and, of course, on the theory of atmospheric convection. To this day, Doug considers Joe Smagorinsky as his most esteemed mentor and respected friend Joe writes:

I first met Doug Lilly in the late 1950s. I was a very young laboratory director looking for talent. Bob Simpson and Werner Baum talked to me about a brilliant, though brash, graduate student at Florida State University who had not yet finished his thesis work. Doug Lilly's interests were in tropical meteorology and convection. This coincided with my intentions to begin a modeling and simulation activity at the GCL.

I went to Florida to talk to Doug. It was one of the hottest spells of the year and I can remember my introduction, the following morning, to grits. The scientific precedents to my objectives were the work of Joanne Simpson of Woods Hole Oceanographic Institution and Georg Witt of the University of Stockholm. The modeling of atmospheric convection at that time was virtually non-existent, as it was in other domains of geophysical fluid dynamics, such as climate, oceanography, hurricanes, mesoscale meteorology, and extended-range weather prediction.

Doug agreed to take the job and came to GCL in 1959. He and Syukuru (Suki) Manabe arrived within a few days of each other: two nascent superstars. Doug started by modeling the boundary layer in today's Large Eddy Simulation (LES) sense and then derived appropriate parameterizations – something that hasn't yet been done properly 40 years later. But Doug did invent the essence of LES along the way!

Considering that Doug did his undergraduate work at Stanford, it was not too surprising that he got a hankering to return to the West. Doug applied for a sabbatical to NCAR. The country, replete with horses, was irresistible to Doug and his family. This was the beginning of the end of the GCL phase of his career. Doug decided to make his stay at Boulder more permanent. We missed his presence at GCL.

I can't help thinking "this lanky kid came a long way, and some of it started here."

(*Joseph Smagorinsky*, GFDL, Princeton)

Another colleague of Doug's from his time at the GCL, Kirk Bryan, remembers:

I was one of a very small group of scientists who worked in GCL from 1950 to 1965. For a brief period, Doug Lilly, Syukuru Manabe, and I all shared the same office. I remember those years as one of the most intellectually stimulating periods of my life. At that time Doug was working on the simulation of convection, but his interests extended far beyond that. He had completely mastered what was then known about numerical methods and worked out many original approaches to numerical simulation on his own. What little I have learned about the subject is through him.

Lengthy scientific discussions during the day left little time for actual work at the office. That was done after hours. Almost every day Doug would come in the office, looking somewhat haggard, and we would spend two or three hours at the blackboard going over the problem he had "solved" or "almost solved" the previous night.

We would spend hours dissecting each lecture given at the Laboratory. Doug was very quick to penetrate to the essential ideas. This was not easy to do in those days, because geophysical fluid dynamics was still in a very rudimentary state. This is what has made him an inspiring teacher and mentor to many coming into the field. In a sense, I consider myself one of his first students.

(*Kirk Bryan*, GFDL, Princeton)

In 1964, Doug took a position as a Senior Scientist at the National Center for Atmospheric Research (NCAR) in Boulder, Colorado. Not long after the family moved to Boulder, construction was started on the NCAR Mesa Laboratory, which was completed in 1966. Doug and his colleagues referred to the Mesa Laboratory as "Mount Olympus: Home of the Gods," which seemed appropriate due to the building's impressive nature.

In 1966, the Lilly family moved to a home east of Boulder. A few years later they had an anemometer mounted outside the house with the recorder indoors in order to measure the wind speeds at the house. Doug's daughter Carol recounts that the instrument had its own graph and red ink, and that three kids loved to watch it recording wind speeds during the severe wind storms, which were common in the lee of the Rocky Mountains.

At some point, the kids became aware that their father was becoming renowned as a meteorologist. He would appear on television occasionally, but he was always "low key," just as he is today – modest about his achievements. At one time the interviewer asked Doug what the wind was going to do this year, and his simple chuckling response was: "Blow!"

Doug loved his work and he worked all the time. After coming home from the office, he would eat dinner and then sit in an easy chair, writing on his lap from about eight until midnight. Even today, this is his preferred method of working. Anyone who knows him can visualize him with his simple thin spiral notebook (nothing fancy for elegant equations!) and a pencil. His constant working was a source of amusement to the children and their friends. They would come in the house, all pass by him and say "Hello, Dr. Lilly" one after another, and then giggle upstairs because he never even looked up! Carol says he was also a wonderful role model in finding work that he loved so much.

Doug's career is distinguished by his broad range of interests and abilities, but it might be said that stratocumulus clouds were, and still are, his favorite topic. While at NCAR in 1968, he published a seminal paper on stratocumulus clouds in the *Quarterly Journal of the Royal Meteorological Society* (Lilly, 1968). It is remarkable that much later, in the 2000s, Doug has again returned to stratocumulus (Lilly, 2002a, b; Stevens *et al.*, 2003) and is writing a book on the topic.

Whilst at NCAR, Doug also began to conduct laboratory and observational studies with Jim Deardorff and Don Lenschow. He did work on turbulence in the atmospheric boundary layer and in the stratosphere, and also became

Doug Lilly in the mid 1960s at NCAR speaking about two-dimensional turbulence.

interested in predictability and numerical simulation of turbulence. Jim Deardorff
writes:

Around the time that Doug joined NCAR, in the early 1960s, I was getting started in
the numerical simulation of two-dimensional thermal convection. Doug showed me how
to finite-difference the vorticity and thermal-diffusion equations so as to conserve kinetic
energy and temperature variance in the absence of sources and sinks, and helped direct me
to references on the subject, which had already started to appear in the literature. His papers
in *Tellus* (Lilly, 1962) and in *Journal of the Atmospheric Sciences* (Lilly, 1964) were of
great help to me.

After I started to realize the advantages of utilizing the equations of motion directly, along
with the thermal-diffusion equation, it was Doug who showed me how to use a "staggered"
grid system for the velocity components in a manner that would conserve kinetic energy
and avoid non-linear computational instability. We in Doug's group (I think NCAR had
developed a small-scale group by then which he had agreed to head), could count on him
to keep up on the latest developments in our fields of interest, and he had acquired an
understanding of Arakawa's (1963) pioneering work along these lines at an early stage.

In the 1960s, Glen Willis and I were busy pursuing turbulent thermal convection in the
laboratory, in which Doug was quite interested, and we were making frequent observations
of the rate of increase in the height of the turbulent layer through the process of penetrative

convection and entrainment. Doug already had a decent understanding of this process, through his studies of the work of Turner and others, and could explain it to us along with the expectation that the magnitude of the downward heat flux in the entrainment region would be some fraction of the upward heat flux at the surface. His input here was too valuable for him to avoid being a co-author of our paper on laboratory investigation of non-steady penetrative convection (Deardorff *et al.*, 1969).

After computing power had increased to the point where it was conceivable to study turbulence in three dimensions, in the late 1960s, the problem arose of how to simulate the dissipation of turbulent kinetic energy cascaded to scales too small to represent explicitly. Doug was very well acquainted with J. Smagorinsky's work on this subject, and was very helpful in advising me on how to apply Smagorinsky's method to small-scale turbulence. Doug had already done his own research on this problem, and so could recommend a coefficient of proportionality between the magnitude of the subgrid-scale eddy coefficient and the resolvable strain rate.

By the 1970s, my interests had turned towards the atmospheric boundary layer and its turbulence. In so doing, it soon became clear that results from numerical integrations would be of greater value if the physical quantities involved were made non-dimensional. Here Doug's input was quite helpful to me for pointing out the use of the boundary-layer height to scale all lengths, even though that height could increase with time within the calculations. His interest in, and knowledge of, Ekman instability alerted me on what to look for in the roll-like structures that emerged from the numerical calculations for neutral and slightly unstable planetary boundary layers.

In the mid 1970s my interests turned towards the stratocumulus-topped planetary boundary layer, partly as a result of Doug's earlier study on this topic (Lilly, 1968). It had taken quite a while for his work on this to sink into my consciousness, but with Doug on hand to explain, from time to time, how radiative cooling from the tops of stratocumulus clouds helps drive the turbulence, I was able to make my own contribution to this topic in 1976 (Deardorff, 1976).

Throughout the time period that Doug and I were both at NCAR, our group frequently benefited from the visiting scientists that he invited there to give a talk and discuss mutual interests. Needless to say, Doug's influence within various fields of atmospheric science has continued to be contributory in all respects.

(*James Deardorff*, Oregon State University)

Doug's observational work of that time was focused on mountain waves and downslope windstorms. His paper with Joe Klemp on wave-induced downslope winds (Klemp and Lilly, 1975) was awarded an "Outstanding Publication Award" from NCAR. Don Lenschow recalls:

I first met Doug when I applied for a position at NCAR as a student nearing graduation, in 1965. He had heard from Jim Telford, who was then at Commonwealth Industrial and Research Organisation (CSIRO) in Australia, but making plans to move to the US, that it now seemed possible for the meteorological community to take advantage of new technology that would allow direct measurements of mesoscale air motions from aircraft. He saw in my 'résumé' that I had worked on the development of an air-motion sensing system on an aircraft, and hired me to help bring this technology to NCAR. He supported the development of INS-based (inertial navigation systems) air-motion sensing platforms on aircraft at NCAR, well before others in the community recognized their capabilities and applications (e.g. Lilly and Lenschow, 1974).

He also recognized limitations in other observational capabilities at that time that needed to be addressed in order to make definitive tests of hypotheses for explaining the behavior of the atmosphere. He encouraged the development of new instruments, for example, to measure temperature in clouds, and high-rate temperature and humidity for eddy fluxes in the boundary layer. He worked with NCAR and NOAA scientists on the development and deployment of Doppler radars for studying motions in convective clouds. He experimented with constant-level balloons for following air motions over the Rocky Mountains. He worked on deploying chaff from aircraft for use as radar targets to study air motions. He had a keen appreciation of the importance of technological developments in providing the tools needed for model verification and improvement, and helped to implement them at NCAR. His support led to systems that were at the cutting edge of the field and to many important research results from observational programs.

The first field program that I recall in which I participated with Doug was a marine stratocumulus study. In the late 1960s, after completing his seminal cloud-topped mixed-layer paper, he carried out a small-scale Queen-Air based field program flying out of Coronado, California to take measurements in the marine stratocumulus. Although I do not recall any published results from this study on stratocumulus, it did provide the basis for later observational studies, in terms of gaining an understanding of what this regime looked like, and how to deploy an airplane instrumented for turbulence measurements (and also ozone). Later, he worked with Wayne Schubert in deploying the NCAR Electra in the first large-aircraft deployment in the marine stratus regime. That program led to several publications, for example by Wakefield and Schubert (1976), and Brost *et al.* (1982a, b), that helped to understand this regime.

Doug had a long-term interest in studying flow over mountains – in particular the weather regimes over the Front Range that led to the wintertime Boulder wind storms. He carried out a series of airborne observational studies, including the 1970 Colorado Lee Wave Observational Study (Lilly *et al.*, 1971), and played a central role in organizing and implementing the Wave Momentum Flux Experiment (WAMFLEX). Some of the events were quite remarkable and intense (e.g. Lilly and Zipser, 1972). Over the years, he employed the NCAR Sabreliner, Buffalo, and Queen Air aircraft, as well as a WB57-F, a high-altitude military aircraft. These studies took advantage of the research platforms that he played a role in developing at NCAR and provided the observational basis for many modeling studies of mountain waves and downslope winds.

(*Don Lenschow*, NCAR, Boulder)

In the early 1980s, Doug became interested in severe supercell convective storms. In this area, his main collaborators were Joseph Klemp, Rich Rotunno, and Tzvi Gal-Chen. Along with Tzvi Gal-Chen, Doug served as editor for a book on mesoscale meteorology (Lilly and Gal-Chen, 1983).

About this same time, Doug started thinking it might be time for a change in career. In 1982, Doug and Judy visited the School of Meteorology of the University of Oklahoma (OU) and decided to move to Norman to experience university life. They did so along with Doug's long-time friend and collaborator, Tzvi Gal-Chen, and his family. I recollect Doug's mentoring during my years as a student at OU:

In 1984 I was an undergraduate and had the privilege of being in Doug's first Physical Meteorology class. I always enjoyed and appreciated his comments on my papers. He had an ability to look beyond standard measures to find the unique strengths of each of his students.

I graduated and went to the University of Wisconsin–Madison for graduate school. However, I did return to Oklahoma for my Ph.D. work and to my great honor, Dr. Lilly accepted me as his graduate student. He was an outstanding mentor, and he was never condescending. He challenged me to reach for more from myself. Sometimes he would ask me questions quite casually about our research and I would answer (smiling to myself) knowing that he knew the answer and was just testing me. Doug never made an ordeal of evaluating me.

He taught me to be a scientist by example. I watched him think and watched carefully how he approached problems. I saw that he always attended seminars and spent significant time in the library reading the latest articles. He has an encyclopedic memory, an insatiable curiosity and never ever accepts anything as true just because someone says so.

He never pushed deadlines or hurried the work. In fact, he was meticulous and deliberate in research. This was very important; the most accurate answers in science cannot be forced to reveal themselves and he recognized this. He also directed me to uphold the highest possible standards of objectivity in science. He taught me how to write and to choose my words carefully. Through his guidance, I learned patience, persistence and determination. Articles we co-authored were never submitted until they were in the finest form we could achieve. I sometimes became impatient with this, but I later understood, when the review process was smooth, why he insisted on continuing to refine the manuscript prior to submittal. He used to call those peer reviews "softballs."

One other time, when I did not initially understand his reasoning, was when he wanted me to publish my dissertation paper before completing the dissertation. I did not fully realize at the time that I had stumbled onto something new and he helped me to get it published quickly. I look back over my doctoral research mentorship and I see he almost always was two–five (or more) steps ahead of me and I never knew it. In fact, Doug did not look at things in "steps," or in a linear sequential manner at all. I believe that he viewed things in full pictures and saw the full three dimensions of everything. He was excellent at gentle steering. His approach was to lead a student half-way across the river and wait on the other side for him/her to cross the other half, perhaps with a little coaching here and there. His scientific philosophy was that no scientific topic or question is without merit. He always supported "science for science sake" and topics without direct applications, that is, fundamental basic research. Doug was not one to hand out compliments easily. Therefore, when one received one, it really meant something and he was genuinely sincere.

(*Katharine Kanak*, CIMMS, Norman)

In 1986, Doug was awarded one of OU's highest honors, the George Lynn Cross Research Professorship. From 1992 to 1995, he held the Robert Lowry Endowed Chair in Meteorology at OU, which was at the time the first endowed chair in Meteorology/Atmospheric Science in the United States.

Doug Lilly (left) together with Edward Lorenz in front of the old CIMMS building
in Norman.

In 1987, Doug became the Director of the OU Cooperative Institute for Mesoscale
Meteorological Studies (CIMMS). During that time he published a series of papers
on the numerical simulation and predictability of thunderstorms. He worked on the
novel applications of helicity concepts to modeling of severe thunderstorms and on
cirrus outflow dynamics. He also maintained his work in atmospheric turbulence
and two-dimensional turbulence as applied to atmospheric mesoscale flow motions.
Additionally, he was involved in laboratory work on simulation of atmospheric
vortices. Current Director of CIMMS, Pete Lamb recalls:

From my arrival at OU in August 1991 until Doug stepped down as CAPS Director in the
summer of 1994, I had the good fortune to occupy an office relatively adjacent to Doug's
office. The "outer office" between us housed support staff, some of whom we shared for
a couple of years. So, I was fortunate to interact with Doug on an almost daily basis for
three years. On two or three days each week, Doug would appear in my office around
5 p.m. to reflect for 10–20 minutes on the various states of CAPS, CIMMS, Meteorology
in general, OU, NOAA, the American Meteorological Society (AMS), the United States,
..., and the World. I am sure Doug's initial visits had two motivations – to make me feel
welcome and (because of his interest in other people) to see what made me "tick." Because
of Doug's encyclopedic interests, anything could be discussed during these visits. When
Doug stopped by my office, it was part of a "one–two" Gal-Chen/Lilly "combination," since

Tzvi invariably stopped by about 45 minutes earlier for essentially the same purpose. These late-afternoon visits were entirely natural because high among Doug's and Tzvi's most important pastimes was a strong interest in their colleagues' well-being and work, free-ranging thought, conversation, and debate. Doug's relinquishing of the CAPS Directorship led to him moving to another office several floors up in our building, after which I saw much less of him, unfortunately.

Doug was always very supportive of his current colleagues and keenly assessed the accomplishments and potential of possible future colleagues. Doug's mode of operation during faculty searches was a model for younger colleagues. Looking over applicants' Curriculum Vitae was just a start, to be followed by reading of some of their papers, a trip to the main University Library to consult the Science Citation Index to assess the impact of the papers, asking questions at candidates' seminars even for presentations outside of his areas of specialization, and often interviewing them informally, like when he was driving them around Norman and vicinity. Several of us learned the value (and weaknesses) of the Science Citation Index from Doug.

In 1993, Doug received the prestigious Symons Gold Medal from the Royal Meteorological Society (RMS) at their end-of-year meeting in London. Before he left for London there were ruminations about what he would say in the short presentation he had been asked to give to the "Meteorological Dining Club" at a dinner following the RMS meeting. The Club, a group of especially accomplished RMS members, has a strict "anything but religion and politics" rule for these short presentations. Doug chose to speak on the development of scientific computing during his career, and when he returned he ventured that it had been well received.

An especially revealing incident occurred when he first came into the office after that trip. When I happened to cross paths with him in our outer office very soon after his arrival that day, I inquired somewhat cavalierly "well Doug, where is the medal?" Much to my surprise, he promptly obliged by producing it from his pants' pocket. Carrying it around there was not a trivial exercise, given that the medal was neither small nor light. Clearly, Doug was appreciative of, and comfortable with, the recognition signified by the medal, and suspected that his office staff and scientific colleagues would like to see it. Indeed we did!

(*Peter Lamb*, CIMMS, University of Oklahoma, Norman)

Doug became Director of the OU Center for the Analysis and Prediction of Storms (CAPS) in 1989. Along with Kelvin Droegemeier, Doug wrote one of 11 out of 330 proposals for a Science and Technology Center that was funded by the National Science Foundation. During his time at CAPS (1989–94), he was involved in many studies of severe storms and techniques to improve their simulation, including four-dimensional data assimilation and the impacts of convective storm helicity on its predictability. Kelvin Droegemeier writes:

I recall Doug Lilly casually walking into my office during spring, 1988, carrying a four-page request for proposals issued a few days earlier by the National Science Foundation. Describing it as perhaps the best such solicitation he had ever seen, Doug hinted that we should pursue the opportunity. In Doug's classic style, he never organized any sort of formal meeting regarding the NSF solicitation, but rather began writing a vision document, that he shared with me in hard copy, on retrieving the unobserved components of the radial wind field from a single Doppler radar. I responded by beginning to pen a vision for a numerical prediction system that could make use of such information, and that would be based upon dynamically adaptive grids and new finite volume solution techniques. It was at this time I began to recognize and appreciate Doug's unique and almost artistic writing style. We submitted the proposal in 1988 and later that year were notified that we had succeeded. As Doug later commented, we "shot the bear and now had to drag it out of the woods."

Looks of "so what do we do about that" often bounced back and forth between me and Doug during the frantic start-up period. And, of course, I began to know Doug more fully as I now worked with him on a nearly daily basis. Those who also know Doug realize that he is a highly creative individual – an abstract, random thinker as Jeff Kimpel later would explain to me – and does not enjoy or even wish to be associated with management, particularly involving money! Indeed, Doug once remarked that he managed activities by passive neglect! Thus, although the first year of CAPS' budget was a whopping $900K – smaller than for all other such centers but quite large compared to anything either Doug or I had seen – our rate of spending quickly exceeded the funds available! Had Doug viewed this as a theoretical fluid dynamics problem he no doubt would have quickly appreciated our predicament!

I have to admit that Doug frustrated me during those early years, as he did members of our external advisory and site visit panels, because he was interested in doing science and not seeing the administrative picture of what CAPS could achieve as a national center. Before long, however, I began to appreciate this substantive aspect of Doug as the consummate scientist and thinker – as an individual who looks at a problem from every conceivable angle, and who must work unencumbered by the clutter and demands of bureaucracy. In that regard, I can best describe Doug as the type of scientist whom everyone wants to be around, and who draws the best out of everyone without showering praise. Perhaps this is my most enduring memory of him.

As a faculty colleague, Doug offered keen insight into challenging problems ranging from the qualifying examination to university parking! I recall how he mentored his graduate assistants in a very personal way, often providing individual lectures and tutorials on especially challenging topics. He was very tough on students, but those who understood the rigors of research appreciated it. Doug was not a conventional teacher, but rather lectured freely, often in the classic disorganized Doug Lilly style, reading from a spiral notebook and wondering how he obtained the indicated result – all the while scratching his head and laughing out loud. Yet there is no doubt that students who got 1% of what Doug Lilly had to offer reaped greater benefit than those who received 200% from most other faculty.

Words are inadequate to express the fondness I have for Doug Lilly – who without saying so indicated the extent to which he cared about others and appreciated their contributions. To say he is rare would be woefully inadequate. To say that he had a profound and enduring positive impact on the study of the atmosphere – and on untold people who were fortunate to share in his professional and personal life – gets a bit closer.

(*Kelvin Droegemeier*, CAPS Norman)

In 1995, Doug formally retired as Professor of the School of Meteorology. Nevertheless, because of his commitment and dedication to his graduate students, he stayed on to make sure his current students had completed their degrees. In his retirement, he took a part-time position as a Distinguished Senior Research Scientist, National Severe Storms Laboratory (NSSL), NOAA (from 1997 to 2002). Current Director of the NSSL, James Kimpel writes:

Doug came to the University of Oklahoma in 1982 without participating in the search process required to fill faculty positions. When I called to invite him for an interview, he said that he and Judy had already interviewed us! Months earlier they had visited Norman and the department on their own time and nickel. An understanding Provost, J. R. Morris, allowed us to hire him anyway. Doug had interviewed at five or more other universities. The University of Oklahoma is indeed fortunate that he chose us. Even before his arrival in Norman, Doug persuaded the late Tzvi Gal-Chen to join him at OU. Together their intellects underpinned the OU School of Meteorology's march, from simply a meteorology program, toward one of national prominence.

Doug said he came to a university so that he could leave behind academic progeny, i.e., little Doug Lillys. In this he succeeded having produced twelve Doctoral and six Masters graduates. During the Lilly years, the graduate student population could be divided into two camps; those who sought out Doug for advice on their research, and those who did their best to avoid him. The latter category of students prayed nightly that Doug, and especially Doug and Tzvi together, would not show up at the required departmental seminar or at the final defense of their thesis or dissertation.

Formal classroom teaching was not Doug's strong suit. He once remarked that teaching was hard work. He said it was like preparing and giving a seminar every day. It is difficult to understand why Doug was not an outstanding lecturer since his writings are so brilliantly lucid. Perhaps his capacity for abstract and random thought clashed with the concrete and sequential style of lectures preferred by most students. He responded by changing his approach to teaching toward using problem-solving exercises, student-led seminars and group projects and this was successful.

As a university faculty member, Doug took his service responsibilities seriously. Doug shouldered his share of the work and usually served with distinction and passion. He demonstrated loyalty to the School and the University. In later years Doug refused much deserved salary increases, saying that his junior colleagues needed it more. Also, he anonymously donated funds he garnered through awards and honoraria to the OU Foundation to support various student programs.

As was his style at NCAR, Doug would sometimes pepper university and national administrators with succinct, poignant memoranda on issues he felt strongly about. Usually Doug would receive a courteous response. On occasion these memoranda were misinterpreted as criticism, or worse as personal attacks. Similarly, Doug was often asked to evaluate the scientific credentials of others for tenure and promotion at universities and national laboratories. Doug would respond with brief, to the point replies, sometimes consisting of only half a page. He felt honor bound to point out a candidate's weaknesses no matter what considerable strengths he or she might possess. Fortunately, Doug often closed out these evaluation letters with a sentence ranking the candidate among others in the field. He named names. This was used to gain positive tenure and promotion decisions in cultures where recommendations were combed for damning with faint praise evidence in order to make the appropriate decision.

Shortly after retiring from the University of Oklahoma, Doug came to the National Severe Storms Laboratory as a part-time employee with CIMMS. At NSSL he interacted with the twenty plus graduate students and scientific staff on the NOAA Campus. He energized the Friday morning seminars and spent considerable time in NSSL's small library reading current journal articles. We miss you Doug. We can only hope that your passion for science and life, your good humor, and your instinct for righting all the wrongs in the world continue here in Oklahoma, *à la* Don Quixote, with us, your academic progeny.

(*James Kimpel*, NSSL, Norman)

In 1999, Doug was elected to the National Academy of Sciences (NAS), and he received his membership in an award ceremony in April 2000. He was the first scientist from Oklahoma to receive this great honor. According to statements by the Academy, it is a private, non-profit, self-perpetuating society of distinguished scholars engaged in scientific and engineering research, dedicated to the furtherance of science and technology and to their use for the general welfare. Upon the authority of the charter granted to it by the Congress in 1863, the Academy has a mandate that requires it to advise the federal government on scientific and technical matters. Election is considered one of the highest honors that can be accorded a scientist.

Doug remains fully active in meteorology (and probably always will) and has recently published two papers in the *Journal of the Atmospheric Sciences* (Lilly, 2002a, b). In addition, he participated in the Dynamics and Chemistry of Marine Stratocumulus (DYCOMS-II) field program in Summer 2002 and is an author on a paper that describes the experiment (Stevens *et al.*, 2003).

During his career, Doug traveled extensively, attending conferences all over the United States and the world. He also spent longer periods of time abroad teaching, conducting research, and collaborating with colleagues in Germany, France, United Kingdom, Australia, Russia, and China.

On October 15, 2002 Doug and Judy moved to join their daughter Carol and her family in Nebraska. He has become affiliated with the Physics Department at the University of Nebraska–Kearney. They will be greatly missed by the Oklahoma Weather Center community.

Doug's broad interest in many topics is also present in other areas of his life besides meteorology. He loves hiking, skiing, swimming, bird-watching, gardening, building kit houses, and studying any new thing, from winemaking to wild mushrooms. He loves all animals and loves raising horses. Doug is a person that breathes in all of life and seems to accept and understand the beauty and necessity of its imperfections.

Acknowledgements

Many people helped to collect information contained in this biography and the Appendices. Special thanks to Judith and Carol Lilly for providing much of the information from the early years and their comments on this biography. Thanks also to Jerry Straka, Fred Carr, Matthew Gilmore, Marcia Pallutto, James Kimpel, and William Beasley for helping to compile other information included in this biography.

References

Arakawa, A. (1963). Computational design for long-term numerical integrations of the equations of atmospheric motion. *Bull. Amer. Meteor. Soc.*, **44**, 150.

Brost, R. A., Lenschow, D. H. and Wyngaard, J. C. (1982a). Marine stratocumulus layers. Part I: Mean conditions. *J. Atmos. Sci.*, **39**, 800–817.

Brost, R. A., Wyngaard, J. C. and Lenschow, D. H. (1982b). Marine stratocumulus layers. Part II: Turbulence budgets. *J. Atmos. Sci.*, **39**, 818–836.

Deardorff, J. W., Willis, G. E. and Lilly, D. K. (1969). Laboratory investigation of non-steady penetrative convection. *J. Fluid Mech.*, **35**, 7–31.

Deardorff, J. W. (1976). On the entrainment rate of a stratocumulus-topped mixed layer. *Quart. J. Roy. Meteor. Soc.*, **102**, 563–582.

Klemp, J. B. and Lilly, D. K. (1975). The dynamics of wave-induced downslope winds. *J. Atmos. Sci.*, **32**, 320–339.

Lilly, D. K. (1960). On the theory of disturbances in a conditionally unstable atmosphere. *Mon. Wea. Rev.*, **88**, 1–17.

(1962). On the numerical simulation of buoyant convection. *Tellus*, **14**, 148–172.

(1964). Numerical solutions for the shape-preserving two-dimensional thermal convection element. *J. Atmos. Sci.*, **21**, 83–98.

(1968). Models of cloud-topped mixed layers under a strong inversion. *Quart. J. Roy. Meteor. Soc.*, **94**, 292–309.

Lilly, D. K., Pann, Y., Kennedy, P. and Tottenhoofd, W. (1971). Data catalog for the 1970 Colorado Lee Wave Observational Program. NCAR Technical Note NT/STR-72, Boulder, CO: National Center Atmospheric Research.

Lilly, D. K. and Zipser, E. J. (1972). The Front Range windstorm of 11 January 1972 – A meteorological narrative. *Weatherwise*, **25**, 56–63.

Lilly, D. K. and Lenschow, D. H. (1974). Aircraft measurements of the atmospheric mesoscales using an inertial reference system. In *Flow – Its Measurement and Control in Science and Industry*, Rodger B. Dowell, ed., Instrument Society of America, 369–377.

Lilly, D. K. and Gal-Chen, T., eds. (1983). *Mesoscale Meteorology – Theories, Observations and Models, (1982), Bonas, France.* Dordrecht: Holland; Boston: published in cooperation with NATO Scientific Affairs Division by D. Reidel Publishing Co.; Hingham, MA: sold and distributed in the USA and Canada by Kluwer Academic Publishers.

Lilly, D. K. (2002a). Entrainment into mixed layers. Part I: Sharp-edged and smoothed tops. *J. Atmos. Sci.*, **59**, 3340–3352.

(2002b). Entrainment into mixed layers. Part II: A new closure. *J. Atmos. Sci.*, **59**, 3353–3361.

Stevens, B., Lenschow, D. H., Vali, G., . . . , Lilly, D. K., *et al.* (2003). Dynamics and Chemistry of Marine Stratocumulus – DYCOMS-II. *Bull. Amer. Meteor. Soc.*, **84**, 579–593.

Wakefield, J. S. and Schubert, W. H. (1976). Design and execution of the marine stratocumulus experiment. Atmospheric Science Paper Number 256, Fort Collins: Department of Atmospheric Science, Colorado State University.

Part I

Atmospheric turbulence

1

Changing the face of small-scale meteorology

John C. Wyngaard

Department of Meteorology, Penn State University, University Park, USA

1.1 The foundations of large-eddy simulation

The first example of what is today called large-eddy simulation, or LES, is generally taken to be Deardorff's (1970) calculation of turbulent channel flow. Considering it an application of an approach used by Smagorinsky *et al.* (1965) for calculation of the general circulation of the atmosphere, Deardorff referred to it by the less elegant term "three-dimensional numerical modeling." He credited Lilly (1967) with determining its subgrid-model constants through Kolmogorov's (1941a, b) theory of the inertial subrange in three-dimensional turbulence.

Today LES is a dominant tool in turbulence research. Like our other turbulence tools it is imperfect, but it has provided a generation of researchers with insight into turbulence properties that are otherwise all but inaccessible, particularly in geophysical flows.

1.1.1 Frustration in the turbulence community

It is intriguing to view Lilly's early contributions to turbulence in the context of the community mood in the early 1960s. According to Moffatt (2002), a "sense of frustration" afflicted G. K. Batchelor and many others at that time:

These frustrations came to the surface at the now legendary meeting held in Marseille (1961) to mark the opening of the former Institut de Méchanique Statistique de la Turbulence (Favre, 1962). This meeting, for which Batchelor was a key organizer, turned out to be a most remarkable event. Kolmogorov was there, together with Obukhov, Yaglom, and Millionshchikov . . . ; von Karman and G.I. Taylor were both there – the great father figures of prewar research on turbulence – and the place was humming with all the current stars of the subject – Stan Corrsin, John Lumley, Philip Saffman, Les Kovasznay, Bob Kraichnan, Ian Proudman, and George Batchelor himself, among many others.

One of the highlights . . . was when Bob Stewart presented . . . the first convincing measurements to show several decades of a $\kappa^{-5/3}$ spectrum and to provide convincing support for Kolmogorov's (1941a, b) theory . . . . But then, Kolmogorov gave his lecture, which I recall was in the sort of French that was as incomprehensible to the French themselves as to the other participants. However, the gist was clear: He said that . . . Landau had pointed out to him a defect in the theory . . . . Kolmogorov showed that the exponent $(-5/3)$ should be changed slightly and that higher-order statistical quantities would be more strongly affected . . . .

I still see the 1961 Marseille meeting as a watershed for research in turbulence. The very foundations of the subject were shaken by Kolmogorov's presentation; and the new approaches . . . were of such mathematical complexity that it was really difficult to retain that essential link between mathematical description and physical understanding, which is so essential for real progress.

Given that Batchelor was already frustrated by the mathematical intractability of turbulence, it was perhaps the explicit revelation that all was not well with Kolmogorov's theory that finally led him to abandon turbulence in favor of other fields.

On October 21, 1961, seven weeks after the close of the Marseille meeting, the journal *Tellus* received one of Doug Lilly's first papers, "On the numerical simulation of buoyant convection" (Lilly, 1962). Its introductory section discusses "numerical experiment," a new and promising approach to turbulent-flow analysis:

The application of numerical experimentation to physical theory is generally justifiable only when more concise analytic methods have been unproductive or have reached apparent limits of usefulness, but these conditions seem to prevail in the field of turbulent fluid mechanics . . . . When the scale and energy of a system become so large that it may be considered turbulent . . . we enter a region rather poorly explained by previously available theoretical methods.

Lilly cited the advances in the understanding of turbulence that had come through theoretical analysis and experimental work, referencing Batchelor's (1953) monograph. But, he wrote:

. . . no real unifying theory exists to relate these (results) from one experimental geometry to another . . . . It should be possible to demonstrate that numerical integration of a single set of differential equations (not necessarily including the unmodified Navier–Stokes equations) . . . can yield solutions corresponding to various experimental phenomena, such as jets, puffs, wakes, and convective bubble- and plume-like thermals. Such a demonstration cannot by itself provide the desired unifying theory. The detailed statistics of the numerical solutions may, however, aid in its formulation, and in any case these statistics must provide a crucial test of such a theory, as for example, Phillips' (1956) numerical experiment aided in verifying modern theory of the atmospheric planetary circulation.

Lilly then presented a bold, three-phase "plan of attack:"

1 Develop flexible and computationally well-behaved numerical models for simulation of a large class of fluid motions;
2 test the detailed behavior of these models by means of experiments comparable with and verifiable by results of significant physical experiments; and

3 try to extend the results or generalize the models to include conditions not adequately reproducible by experiment.

This was a striking departure from conventional thinking. It was also remarkable in its timing, appearing on the heels of the frustration emerging from the Marseille turbulence conference. This was evidently coincidental. At this time Lilly's fluid-mechanical environment was meteorological; he had a position in the General Circulation Research Laboratory of the US Weather Bureau. He did not attend the Marseille meeting.

Today much, if not most, "experiment" in small-scale meteorology is numerical experiment. It is carried out to a limited extent via direct numerical simulation (DNS), the numerical simulation of all scales of motion in a turbulent flow. Most is done through (1) large-eddy simulation (LES), in which only the energy-containing range of the turbulence is resolved; and (2) mesoscale modeling, in which little or none of the turbulence is resolved.

1.1.2 Buoyant convection in two dimensions

Lilly's first results are presented in the 1962 paper. Perceiving the option of a three-dimensional numerical grid as "nearly unavailable for economic reasons," he carried out two-dimensional simulations of free convection at "low" resolution (15×30 grid squares) and "high" resolution (31×94) (Lilly, 1962). He characterized the success of these simulations as "moderate" and attributed their limitations largely to their two-dimensionality. He thought it advisable to continue as outlined in his plan of attack and mentioned several steps along that path. A key step was the "eventual development of a truly three-dimensional model."

1.1.3 A basis for three-dimensional modeling of turbulent flows

Lilly's (1967) paper "The representation of small-scale turbulence in numerical simulation experiments" provides a wide-ranging discussion of simulation issues and a remarkable set of prescriptions for their solution, some of which have yet to be fully explored. Its introduction contains his perspective on the history of the "direct numerical integration of the hydrodynamic equations:"

With some important exceptions . . . numerical simulation has been most frequently and successfully applied in the areas of large-scale meteorology and high-speed aerodynamics. The problems in these two fields . . . share the properties that they are typically two-dimensional, or nearly so, and that turbulence is either unimportant or that it can presumably be treated by fairly crude approximations. These two properties greatly simplify the numerical simulation problem, but they eliminate from consideration most other fluid dynamics problem areas . . . . In the fully three-dimensional flows, however, the interaction from the energy producing scales to those of molecular dissipation occurs in a direct and continuous process and there is no convenient scale which one may choose to separate motions of qualitatively different kinds.

He mentioned that Corrsin (1961) had stressed the futility of direct numerical solution of the equations of motion in even modest-Reynolds-number laboratory flows. Since geophysical Reynolds numbers are even larger, Lilly said, "we may dismiss such complete simulation as hopeless." He called for a more critical examination of "the mechanics of turbulent exchange" to see whether it is possible to simulate in an approximate way some of the important effects of (1) three-dimensionality and (2) molecular viscosity and diffusion, without computing them in full detail.

Previous two-dimensional simulations of turbulent convection, including his own 1962 effort, Lilly wrote, "failed to show any structures that could be identified with the irregularity and high-amplitude turbulence characteristic of the real world . . . ." Several others had also found two-dimensional numerical simulations of turbulence "partially unsatisfactory" as surrogates for three-dimensional turbulence. He discussed previous attempts to remedy this, including the use of a "sandwich" of three computational planes, rather than one plane, and adding random turbulence energy to two-dimensional computations. While neither was fully satisfactory, a "suitable justification or generalization of such quasi-three-dimensional models would represent a tremendous breakthrough."

These attempts to simulate three-dimensional turbulence in two dimensions might seem surprising today, given the strong dynamic and structural differences between two- and three-dimensional turbulence. But the computers of that time could do two-dimensional problems with no more than about $(50$–$75)^2$ grid points, and three-dimensional ones up to about $(20)^3$. Meaningful 3D calculations of real-world turbulence still awaited the development of much larger, faster computers.

Lilly (1967) then mentioned the "modern turbulence theory" based on Kolmogorov's (1941a) notion of an inertial range of the three-dimensional turbulent energy spectrum $E(\kappa)$. Using the equations from Lilly (1967) with their original numbers preceded by an "L", we have:

$$E(\kappa) = \alpha \epsilon^{2/3} \kappa^{-5/3}, \tag{L 1.5}$$

with α a universal constant. The results that Stewart presented in Marseille in 1961 had now appeared in print (Grant *et al.*, 1962), and Gibson (1963) also had found an inertial range in a much lower (but still high) Reynolds number laboratory jet. These data showed a $\kappa^{-5/3}$ inertial range separating the large-scale, energy-containing eddies from the much smaller-scale, dissipative ones, as Kolmogorov predicted. Lilly then wrote:

We now suggest that the existence and relatively simple properties of the inertial range might be used to greatly truncate the otherwise impossibly large requirements of computer resolution. Let us assume that the simulation equations are integrated for variables defined and resolvable in a scale range which includes most of the kinetic energy, and that the scale of the calculation mesh lies within the inertial range. It should then be possible to fit the explicitly calculated motion fields to the inertial range in a smooth and consistent manner. The fitting conditions would require a continuous removal of energy from the small-scale explicit motions such that (L 1.5) is maintained.

After discussing and rejecting one possible "fitting procedure," he added:

A more physically acceptable procedure is suggested by consideration of the local interactions between the explicit scale motions and those of the submesh length scales. The latter cannot be known in detail, but certain statistical probabilities can be established with the aid of the Kolmogorov spectrum function. In the following, I will describe a first and second order theory for the interactions. Most of the detailed derivations and analysis are available in an unpublished report (Lilly, 1966). For simplicity, the results are presented here for the case of an incompressible constant density fluid.

We shall sketch his approach here, again using his equations from Lilly (1967).

The dependent variables are averaged over the grid-mesh volume (considered for simplicity to be a cube of side h), the averaging denoted by an overbar:

$$\overline{F}(x_1, x_2, x_3, t) = \frac{1}{h^3} \int_{-h/2}^{h/2} \int_{-h/2}^{h/2} \int_{-h/2}^{h/2} F(x_1 + y_1, x_2 + y_2, x_3 + y_3, t)\, dy_1 dy_2 dy_3. \tag{L 2.1}$$

In mesoscale modeling one can also use an ensemble average, but the spatial average has the conceptual advantage that it allows mesoscale modeling to merge smoothly with large-eddy simulation as h decreases.

The Navier–Stokes and continuity equations for constant-density flow are

$$\frac{\partial u_i}{\partial t} + u_k \frac{\partial u_i}{\partial x_k} + \frac{\partial}{\partial x_i}\left(\frac{p}{\rho}\right) - \nu \frac{\partial^2 u_i}{\partial x_k \partial x_k} = 0, \tag{L 2.2}$$

$$\frac{\partial u_i}{\partial x_i} = 0. \tag{L 2.3}$$

Averaging these equations over the grid-mesh cube produces

$$\frac{\partial \overline{u}_i}{\partial t} + \overline{u}_k \frac{\partial \overline{u}_i}{\partial x_k} + \frac{\partial}{\partial x_i}\left(\frac{\overline{p}}{\rho} + \frac{2}{3}E\right) - \nu \frac{\partial^2 \overline{u}_i}{\partial x_k \partial x_k} = \frac{\partial \tau_{ik}}{\partial x_k}, \tag{L 2.4}$$

$$\frac{\partial \overline{u}_i}{\partial x_i} = 0, \tag{L 2.5}$$

where τ_{ik} is

$$\tau_{ik} = -(\overline{u_i u_k} - \overline{u}_i\,\overline{u}_k) + \frac{2}{3}\delta_{ik}E. \qquad \text{(L 2.6)}$$

E is the kinetic energy of the unresolved motions per unit mass,

$$E = (\overline{u_i u_i} - \overline{u}_i\,\overline{u}_i)/2. \qquad \text{(L 2.7)}$$

We shall write $\overline{u_i u_k} - \overline{u}_i\,\overline{u}_k = R_{ik}$, so Equation (L 2.6) becomes

$$\tau_{ik} = -\left(R_{ik} - \frac{2}{3}\delta_{ik}E\right),$$

the negative of the departure of R_{ij} from its isotropic form; for that reason τ_{ik} is sometimes called a deviatoric stress.

The central problem here, Lilly pointed out, is to evaluate τ_{ij} in terms of the averaged quantities. Since the grid-mesh average commutes with differentiation (away from boundaries, in the case of spatial derivatives (Ghosal and Moin, 1995)) one can write

$$\frac{\partial \overline{u_i u_j}}{\partial t} = \overline{u_i \frac{\partial u_j}{\partial t}} + \overline{u_j \frac{\partial u_i}{\partial t}},$$

which indicates how to derive the evolution equation for $\overline{u_i u_j}$. The evolution equations for $\overline{u}_i\overline{u}_j$ and E follow in similar fashion. Lilly's equation for τ_{ij} has the form

$$\frac{\partial \tau_{ij}}{\partial t} + \overline{u}_k\frac{\partial \tau_{ij}}{\partial x_k} = \frac{2}{3}E\left(\frac{\partial \overline{u}_i}{\partial x_j} + \frac{\partial \overline{u}_j}{\partial x_i}\right)$$
$$- \left[\tau_{ik}\frac{\partial \overline{u}_j}{\partial x_k} + \tau_{jk}\frac{\partial \overline{u}_i}{\partial x_k} - \frac{1}{3}\delta_{ij}\tau_{k\ell}\left(\frac{\partial \overline{u}_k}{\partial x_\ell} + \frac{\partial \overline{u}_\ell}{\partial x_k}\right)\right]$$
$$+ \text{pressure} - \text{gradient interaction} + \text{divergence of fluxes}$$
$$- \text{rate of viscous dissipation.} \qquad \text{(L 2.8)}$$

The evolution equation for E is

$$\frac{\partial E}{\partial t} + \overline{u}_i\frac{\partial E}{\partial x_i} = \text{divergence of fluxes} + \frac{1}{2}\tau_{ij}\left(\frac{\partial \overline{u}_i}{\partial x_j} + \frac{\partial \overline{u}_j}{\partial x_i}\right) - \epsilon, \quad \text{(L 2.9)}$$

with ϵ the rate of viscous dissipation per unit mass.

Lilly used the "simplest reasonable closure assumption," an eddy-viscosity model for τ_{ij}:

$$\tau_{ij} = K\left(\frac{\partial \overline{u}_i}{\partial x_j} + \frac{\partial \overline{u}_j}{\partial x_i}\right) = K D_{ij}, \qquad \text{(L 3.1)}$$

with the subgrid eddy viscosity K assumed to be positive and a function of the averaged flow variables. He adopted Smagorinsky's (1963) form

$$K = (kh)^2 D/\sqrt{2}, \tag{L 3.2}$$

with k a constant and $D = \left(D_{ij}D_{ij}\right)^{1/2}$. Lilly showed that (L 3.2) is consistent with Kolmogorov's inertial-range form (L 1.5) for scales near h provided that

$$k \simeq 0.23\alpha^{-3/4}. \tag{L 3.3}$$

In his "second-order theory" for τ_{ij} Lilly used (L 2.8) and (L 2.9) in "nontrivial but substantially simplified forms" that in steady, homogeneous turbulence yield (L 3.1)–(L 3.3).

With this derivation Lilly provided the basis for what is now known as "large-eddy simulation" (LES), a semantically precise term that originated in the engineering community, which adopted it in the 1970s. As discussed by Galperin and Orszag (1993), LES is used today in a host of geophysical research applications including boundary-layer structure, turbulent diffusion, local flows, severe storms, and oceanography. It is also widely used in engineering fluid mechanics research (Lesieur and Métais, 1996; Meneveau and Katz, 2000).

1.2 The mesoscale and LES limits

We now allow the averaging scale h to vary, subject only to the restriction that it be much larger than the scale of the dissipative eddies so that the molecular-diffusion terms in the resolvable-scale equations are negligible.

We shall call the case $h \ll \ell$, with ℓ the scale of the dominant turbulence, the "LES limit." Here the energy and flux-containing turbulence is resolved directly by the averaged equation of motion (L 2.4), as sketched in Fig. 1.1. The case $h \gg \ell$, the "mesoscale limit," is reached in mesoscale modeling. (Adding a Coriolis term to (L 2.4) presents no complications, since that term is linear in velocity, so we shall not indicate it explicitly.) In mesoscale modeling the grid-mesh element is typically much smaller in the vertical direction than in the horizontal in order to resolve some structure in the boundary layer. But since resolving three-dimensional turbulence requires a grid mesh that is smaller than ℓ in all three directions, even with fine vertical resolution essentially none of the turbulence is resolved in the mesoscale limit. The turbulence resides in the unresolved (also called subgrid-scale) fields, as also sketched in spectral terms in Fig. 1.1.

In a flow that is statistically homogeneous in two directions (such as the boundary layer over a uniform surface) spatial averaging with $h \gg \ell$ corresponds, by

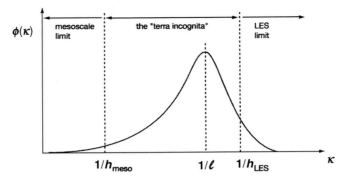

Figure 1.1. A schematic of the turbulence spectrum $\phi(\kappa)$ in the horizontal plane as a function of the horizontal wavenumber magnitude κ. Its peak is at $\kappa \sim 1/\ell$, with ℓ the length scale of the most energetic eddies. When the scale of the spatial averaging is much larger than ℓ we approach the mesoscale limit (left side) where none of the turbulence is resolved. When the scale of the spatial averaging is much smaller than ℓ we approach the LES limit (right side) where the energy-containing turbulence is resolved.

ergodicity, to ensemble averaging, and thus to time averaging in stationary conditions. That allows one to use micrometeorological observations to determine behavior in the mesoscale limit.

Equations (L 2.8) and (L 2.9) are the formal evolution equations for the deviatoric stress τ_{ij} and kinetic energy E of the unresolved motions in LES, in mesoscale modeling, and for applications over the vast range of scales in between. They are valid for all classes of filters used in LES codes. Even though in high-resolution LES the turbulent kinetic energy and fluxes are carried almost entirely by the resolved motion, the τ_{ij} term in (L 2.4) remains important. It is an essential component of the interactions that cause the cascade of kinetic energy and scalar variance from the resolved to unresolved scales (Wyngaard, 2002). Thus, models of τ_{ij} and E are required also in the LES limit.

The development of LES and mesoscale modeling has taken place largely since 1967 and, it appears, largely independently in the two fields. Until recently their spatial domains were nonoverlapping; the horizontal area of a typical boundary-layer LES domain could fit within the horizontal grid-mesh square of a typical mesoscale model. But computer power has grown to the point that very-fine-mesh mesoscale modeling is approaching the scale range of coarse-resolution LES, i.e., $h \sim \ell$. As this happens a new issue emerges: the suitability of subgrid-scale turbulence parameterization used in mesoscale models for operation in this scale range. We shall discuss this next, beginning with the simpler case of a conserved scalar.

1.3 The maintenance of subgrid-scale fluxes

1.3.1 A conserved scalar

The evolution equation for a conserved scalar c in a constant-density flow is

$$\frac{\partial c}{\partial t} + u_i \frac{\partial c}{\partial x_i} = \gamma \frac{\partial^2 c}{\partial x_i \partial x_i}. \tag{1.1}$$

The grid-mesh-averaged equation is

$$\frac{\partial \bar{c}}{\partial t} + \bar{u}_i \frac{\partial \bar{c}}{\partial x_i} + \frac{\partial f_i}{\partial x_i} = 0, \quad f_i = \overline{cu_i} - \bar{c}\,\bar{u}_i, \tag{1.2}$$

h having been assumed large enough to make the molecular destruction term negligible. f_i is the subgrid-scale flux of the scalar.

Multiplying (1.2) by $2\bar{c}$ and rearranging yields the equation for the evolution of the squared resolved scalar:

$$\frac{\partial \bar{c}^2}{\partial t} + \bar{u}_i \frac{\partial \bar{c}^2}{\partial x_i} + \frac{\partial (2 f_i \bar{c})}{\partial x_i} = 2 f_i \frac{\partial \bar{c}}{\partial x_i}. \tag{1.3}$$

The equation for the squared unresolved scalar is

$$\frac{\partial(\overline{c^2} - \bar{c}^2)}{\partial t} + \bar{u}_i \frac{\partial(\overline{c^2} - \bar{c}^2)}{\partial x_i} + \frac{\partial(\overline{u_i c^2} - \bar{u}_i \bar{c}^2 - 2 f_i \bar{c})}{\partial x_i} = -\chi - 2 f_i \frac{\partial \bar{c}}{\partial x_i}, \tag{1.4}$$

with χ the rate of destruction of $\overline{c^2}$ through molecular diffusion. Evidently the final term in (1.3) and (1.4) is the rate of transfer of squared scalar between the resolved and unresolved scales.

Wyngaard *et al.* (1971) and Deardorff (1973) presented the evolution equation for f_i. In a constant-density flow it is

$$\frac{\partial f_i}{\partial t} + \bar{u}_j \frac{\partial f_i}{\partial x_j} = -\frac{\partial \bar{u}_i}{\partial x_j} f_j - \frac{\partial \bar{c}}{\partial x_j} R_{ij}$$

$$+ \text{pressure-gradient interaction} + \text{divergence of fluxes.} \tag{1.5}$$

The first term on the right-hand side (rhs) is the rate of production of scalar flux through the amplification and rotation of existing scalar flux by the velocity gradient; it is the counterpart of the "stretching and tilting" term in the vorticity equation. The second term on the rhs is the rate of production through the interaction of Reynolds stress and the scalar gradient. The third and fourth terms are the rates of production through interaction of c and pressure gradients and through spatial rearrangement, respectively. The neglected molecular term is quite small in large-Reynolds-number turbulence (Wyngaard *et al.*, 1971). The role of sink for scalar flux then falls to the pressure-gradient term; the simplest model for it is $-f_i/T$, with T a time scale of the unresolved turbulence. A steady, homogeneous model of Equation (1.5) is therefore

$$\frac{\partial f_i}{\partial t} = 0 = -\frac{\partial \overline{u}_i}{\partial x_j} f_j - \frac{\partial \overline{c}}{\partial x_j} R_{ij} - \frac{f_i}{T}, \tag{1.6}$$

which implies the model

$$f_i = -T \left(\frac{\partial \overline{u}_i}{\partial x_j} f_j + \frac{\partial \overline{c}}{\partial x_j} R_{ij} \right). \tag{1.7}$$

In the mesoscale limit we interpret the overbar as an ensemble average, which we shall denote by brackets. Then with the classical decomposition into ensemble-mean plus fluctuating parts, $c = \langle C \rangle + c' = C + c'$, $u_i = \langle u_i \rangle + u_i' = U_i + u_i'$, we have

$$f_i = \langle cu_i \rangle - \langle c \rangle \langle u_i \rangle = \langle c'u_i' \rangle, \quad R_{ij} = \langle u_i'u_j' \rangle,$$

and (1.7) becomes

$$\langle c'u_i' \rangle = -T \left(\frac{\partial U_i}{\partial x_j} \langle c'u_j' \rangle + \frac{\partial C}{\partial x_j} \langle u_i'u_j' \rangle \right). \tag{1.8}$$

In their analysis of data from the 1968 Kansas experiment Wyngaard *et al.* (1971) found the steady, homogeneous model (1.8) to be a good representation of both the horizontal and vertical components of the potential temperature flux budget with $T \sim \ell / E^{1/2}$, a time scale of the energy-containing turbulence.

In the LES limit, when the grid-mesh average has $h \ll \ell$, the quasi-steady, homogeneous model is (1.6) with $T \sim h / E^{1/2}$, a time scale of the inertial-range turbulence.

In approximating the flux-conservation equation (1.5), Deardorff (1973) used isotropic forms for the fluxes in its production terms on its rhs, taking $f_i = 0$, $R_{ij} = \frac{2}{3} \delta_{ij} E$. Under steady, homogeneous conditions these assumptions cause the model (1.7) to reduce to the simpler downgradient-diffusion model typically used today in LES:

$$f_i \sim -\frac{2}{3} E^{1/2} h \frac{\partial \overline{c}}{\partial x_i} = -K_c \frac{\partial \overline{c}}{\partial x_i}, \tag{1.9}$$

with K_c an eddy diffusivity. With this model for f_i the rate of transfer of squared scalar in (1.3) is negative definite,

$$2 f_i \frac{\partial \overline{c}}{\partial x_i} = -2 K_c \frac{\partial \overline{c}}{\partial x_i} \frac{\partial \overline{c}}{\partial x_i}, \tag{1.10}$$

so the transfer is always from resolved to unresolved scales.

Without Deardorff's assumption of isotropy for the production terms in the f_i budget (1.5), the model (1.7) rather than (1.9) emerges. The second term on the rhs of (1.7) is the sum of three production rates, only one of which – the one that involves a positive-definite diagonal element of R_{ij} – produces a scalar flux that is necessarily directed down the scalar gradient. Thus, only one of the six production terms for scalar flux in the model (1.7) represents down-gradient diffusion.

With this fuller model of f_i the rate of transfer of squared scalar becomes

$$2 f_i \frac{\partial \overline{c}}{\partial x_i} = -2T \left[\frac{\partial \overline{u}_i}{\partial x_j} \frac{\partial \overline{c}}{\partial x_i} f_j + \frac{\partial \overline{c}}{\partial x_i} \frac{\partial \overline{c}}{\partial x_j} R_{ij} \right]. \tag{1.11}$$

It appears that this could give "backscatter," the local transfer of variance from smaller scales to larger, which is observed when $h \ll \ell$.

A steady, homogeneous model of unresolved scalar flux f_i that seems applicable across the scale range is

$$f_i = -T \left(\frac{\partial \overline{u}_i}{\partial x_j} f_j + \frac{\partial \overline{c}}{\partial x_j} R_{ij} \right), \tag{1.12}$$

with the time scale T being of order $\ell/E^{1/2}$ and $h/E^{1/2}$ in the mesoscale and LES limits, respectively. This can be written as

$$F_i = -K_{ij} \frac{\partial \overline{c}}{\partial x_j}, \tag{1.13}$$

with K_{ij} a tensor eddy diffusivity. *

1.3.2 Stress

The kinetic energy equation for the resolved motion is formed by multiplying (L 2.4) by $\overline{u}_i$ and rearranging:

$$\frac{\partial}{\partial t} \left(\frac{\overline{u}_i \, \overline{u}_i}{2} \right) + \overline{u}_j \frac{\partial}{\partial x_j} \left(\frac{\overline{u}_i \, \overline{u}_i}{2} \right) + \frac{\partial}{\partial x_j} (\overline{p}^* \, \overline{u}_j - \overline{u}_i \, \tau_{ij}) = -\frac{1}{2} \tau_{ij} D_{ij}. \tag{1.14}$$

Here $p^* = p/\rho + 2E/3$ is a modified kinematic pressure. The energy equation for the unresolved motion is (L 2.9). Each of these equations contains the term $\tau_{ij} D_{ij}/2$, which represents the rate of energy transfer between resolved and unresolved scales.

With the simplest closure approximation (L 3.1) for τ_{ij}, this energy-transfer term becomes

$$\frac{1}{2} \tau_{ij} D_{ij} = \frac{K}{2} D^2, \tag{1.15}$$

which is positive definite. Equations (L 2.9) and (1.14) then indicate that the energy transfer is always from resolved to unresolved scales. The closure (L 3.1)

---

*Wyngaard (2004) shows that data from the HATS experiment (Horst *et al.*, 2004) support this more general subgrid-scale model is commonly used in LES, where this one-way energy transfer is computationally advantageous. However, the equilibrium values of τ_{ij} components implied by (L 3.1) are clearly unphysical in the mesoscale limit. In the neutral atmospheric surface layer, for example, (L 3.1) yields $\langle u'^2 \rangle = \langle v'^2 \rangle = \langle w'^2 \rangle = 2E/3$, which agrees poorly with observations.

The next level of approximation for τ_{ij} uses the steady, homogeneous form of its conservation equation (L 2.8). Assuming local isotropy in its viscous term, this is

$$\frac{\partial \tau_{ij}}{\partial t} = 0 = \frac{2}{3} E D_{ij}$$

$$- \left[\tau_{ik} \frac{\partial \overline{u}_j}{\partial x_k} + \tau_{jk} \frac{\partial \overline{u}_i}{\partial x_k} - \frac{1}{3} \delta_{ij} \tau_{k\ell} D_{k\ell} \right] - \frac{\tau_{ij}}{T}, \qquad (1.16)$$

with T again a time scale of the unresolved turbulence. This yields the closure

$$\tau_{ij} = \frac{2}{3} E T D_{ij} - T \left[\tau_{ik} \frac{\partial \overline{u}_j}{\partial x_k} + \tau_{jk} \frac{\partial \overline{u}_i}{\partial x_k} - \frac{1}{3} \delta_{ij} \tau_{k\ell} D_{k\ell} \right]. \qquad (1.17)$$

The rate of energy transfer from the resolved to unresolved scales is now

$$\frac{1}{2} \tau_{ij} D_{ij} = \frac{1}{3} E T D^2 - \frac{T}{2} D_{ij} \left[\tau_{ik} \frac{\partial \overline{u}_j}{\partial x_k} + \tau_{jk} \frac{\partial \overline{u}_i}{\partial x_k} \right], \qquad (1.18)$$

which could give backscatter. In the mesoscale limit, (1.17) implies that the turbulent velocity variances in the neutral surface layer are

$$\langle u'^2 \rangle = T \left(-2\langle u'w' \rangle \frac{\partial U}{\partial z} - \frac{2\epsilon}{3} \right) + \frac{2E}{3},$$

$$\langle v'^2 \rangle = \langle w'^2 \rangle = T \left(-\frac{2\epsilon}{3} \right) + \frac{2E}{3}. \qquad (1.19)$$

With the proper choice of T these can be made to agree fairly well with observations, the main discrepancy being that $\langle v'^2 \rangle$ is observed to be somewhat larger than $\langle w'^2 \rangle$.

1.3.3 Implications for modeling

We have shown that the simplest steady, homogeneous closure of the conservation equations for turbulent stress and scalar flux yields subgrid-scale turbulence models that are more complex than those generally used today.

In highly resolved LES, where $h \ll \ell$, the subgrid models carry little flux; their principal role is extracting energy and scalar variance from the resolved scales. The eddy-diffusivity subgrid model commonly used in LES, which emerges from the subgrid-flux conservation equations when a number of production terms are dropped, is quite effective in this transfer role; it is not clear that the additional production terms would have any strong effects. But the $h \ll \ell$ constraint for LES has not typically been met in severe-storm modeling (Bryan *et al.*, 2003), and is never met very near the surface in boundary-layer applications of LES because the horizontal integral scale of vertical velocity varies like z there. The more complex subgrid model discussed here could impact such applications.

The spatial resolution of mesoscale models has improved continuously over the past 30 years, and grid meshes as fine as 1 km are now used in research applications. In such cases h can be in the large-scale end of the energy-containing range of boundary-layer turbulence, which means that the $h \gg \ell$ constraint typically implicit in the subgrid turbulence models used in mesoscale modeling is also violated. This raises two questions:

- How can super-high-resolution mesoscale modeling and coarse-resolution LES be carried out optimally, given that each is presently beyond the design range of its subgrid model?
- How can the subgrid models in LES and in mesoscale modeling be made to converge in the region where $h \sim \ell$?

These questions concern the numerical modeling scale range that we shall call the "terra incognita."

1.4 The "terra incognita"

1.4.1 Background

Wyngaard (1982) and Bryan *et al.* (2003) have discussed qualitatively the behavior of the averaged equation of motion (L 2.4) as the averaging scale h varies. To simplify the summary here we use the eddy-viscosity closure (L 3.1) with $K \sim E^{1/2}\ell_s$, where ℓ_s is the length scale of the unresolved turbulence, and we assume the averaged flow is horizontally homogeneous and has velocity and length scales U and L. The Reynolds number Re of the averaged flow is of order

$$Re = \frac{UL}{K} \sim \frac{UL}{E^{1/2}\ell_s}. \tag{1.20}$$

In the mesoscale limit the velocity and length scales $E^{1/2}$ and ℓ_s of the unresolved turbulence are u and ℓ, the scales of the turbulence, which in turn are of the order of U and L, the scales of the averaged flow. The Reynolds number of the averaged flow is then $O(1)$ and below the value required for transition to turbulence. Thus, coarse-resolution mesoscale model output fields are nonturbulent.

In the LES limit ℓ_s is the averaging scale h, which lies in the inertial subrange of large Reynolds number turbulence. We can write

$$E \simeq \int_{1/h}^{\infty} E(\kappa)\,d\kappa \sim \epsilon^{2/3} h^{2/3}, \tag{1.21}$$

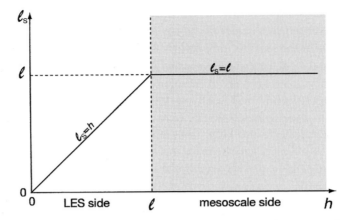

Figure 1.2. The suggested behavior of the length scale ℓ_s of the unresolved turbulence as a function of the averaging scale h. On the LES side, $0 \le h \le \ell$, $\ell_s = h$. On the mesoscale side, $\ell \le h \le \infty$, $\ell_s = \ell$.

which, with $\epsilon \sim u^3/\ell$, yields

$$E^{1/2} \sim u \left(\frac{h}{\ell}\right)^{1/3}, \quad K \sim E^{1/2}h \sim uh \left(\frac{h}{\ell}\right)^{1/3}, \quad Re \sim \frac{u\ell}{K} \sim \left(\frac{\ell}{h}\right)^{4/3}. \quad (1.22)$$

Since $\ell/h \gg 1$, Re is large and LES output fields are turbulent.

Presumably the transition to turbulence occurs when the averaging scale h is of the order of ℓ. Here h is probably too large to lie in the inertial range, for which LES subgrid models are designed, and probably smaller than averaging scales for which mesoscale subgrid models are designed.

1.4.2 A unified closure concept

The discussion in Section 1.3 suggests using a single closure with its length scale chosen to transition between ℓ on the mesoscale side and h on the LES side (Fig. 1.2):

$$\ell_s = \ell, \; h \ge \ell; \quad \ell_s = h, \; h \le \ell.$$

We can then sketch how the simplest eddy-diffusivity closure (L 3.1) with $K \sim E^{1/2}\ell_s$ and ℓ_s so tailored could perform between the mesoscale and LES limits. On the mesoscale side (the far right of the h-axis in Fig. 1.2) the averaging scale h exceeds the turbulence scale ℓ, so $\ell_s = \ell$. Equation (1.17) for unresolved energy E, using $\epsilon \sim E^{3/2}/\ell$ and ignoring constants of $O(1)$, reduces to

$$E = \ell^2 D^2. \quad (1.23)$$

As $h \to \ell$ we approach the energy-containing range; we enter it when h becomes less than ℓ. At that point the curve becomes $\ell_s = h$ (Fig. 1.2). The energy equation (1.23) becomes $E = h^2 D^2$, and further decreases in h cause Re, the Reynolds number of the averaged flow, Equation (1.20), to increase as h^{-2}. When Re exceeds its critical value the averaged flow becomes turbulent.

In the LES limit (near the origin in Fig. 1.2) Re grows as $(\ell/h)^{4/3}$ (Equation (1.22)) and when Re is large enough the turbulent LES fields approach Re-independence (Fig. 1.1).

We can now consider the implications of using the fuller closure (1.17) rather than the simplest eddy-diffusivity closure (L 3.1), and using (1.12) rather than (1.9) for scalars. In the mesoscale limit the significant turbulent fluxes and mean-field gradients are those in the vertical, so the two closures are essentially equivalent. In the LES limit the two closures are quite different, with the fuller closure having a tensor eddy diffusivity in contrast to the scalar eddy diffusivity of the simpler closure. But here the subgrid model carries essentially no flux; it simply transfers energy and scalar variance at a rate that is made correct through the choice of the constant in the eddy diffusivity. The fuller model would need to have its time scale set in that way as well. Once that is done the two should perform similarly, although there could be subtle differences – e.g., backscatter in the fuller model.

In the "terra incognita" where $h \sim \ell$ the two closures are quite different, suggesting that the model performance could be quite different as well.

1.4.3 The roles of buoyancy and turbulent transport

We have shown that a subgrid turbulence model more complex than the usual K-closure emerges naturally from the conservation equations described in Section 3. However, those equations could need buoyancy and turbulent-transport terms to be useful in geophysical applications with $h \sim \ell$. Including buoyancy is straightforward: through the Boussinesq approximation, for example, the equation of motion (L 2.2) gains a buoyant acceleration term $g\Theta/\Theta_0$, with g the acceleration of gravity, Θ_0 a background potential temperature profile, and Θ a deviation from this profile. This generates buoyant-production terms in (L 2.8) for stress τ_{ij}, (L 2.9) for energy E, and (1.5) for scalar flux f_i.

We can assess the importance of these buoyant-production terms as follows. We define $\theta(h)$, the intensity scale of temperature fluctuations of spatial scale h, as the rms value of $\Theta - \overline{\Theta}$ for averaging scale h. Its counterpart for the velocity field is $u(h)$. Their Kolmogorov inertial-range scaling is (Tennekes and Lumley, 1972)

$$u(h) = f(\epsilon, h) = (\epsilon h)^{1/3}, \quad \theta(h) = g(\chi, \epsilon, h) = \chi^{1/2}\epsilon^{-1/6}h^{1/3}. \quad (1.24)$$

These results hold also for h in the energy-containing range, where the intensity scales are θ and u. That is, using $\epsilon \sim u^3/\ell$, $\chi \sim \theta^2 u/\ell$, Equation (1.24) yields for $h = \ell$

$$u(\ell) = u, \quad \theta(\ell) = \theta. \tag{1.25}$$

It follows that

$$\theta(h) = \theta \left(\frac{h}{\ell}\right)^{1/3}, \quad u(h) = u \left(\frac{h}{\ell}\right)^{1/3}. \tag{1.26}$$

When the buoyant- and gradient-production terms in the stress and energy conservation equations are scaled with h, $u(h)$, and $\theta(h)$, their ratio becomes a scale-dependent turbulent Richardson number:

$$Ri(h) = \frac{g\, \theta(h) h}{\Theta_0 \, [u(h)]^2} = \frac{g\, \theta \ell}{\Theta_0 \, u^2} \left(\frac{h}{\ell}\right)^{2/3} = Ri_e \left(\frac{h}{\ell}\right)^{2/3}. \tag{1.27}$$

Thus, when $h \to \ell$ this turbulent Richardson number becomes Ri_e, that for the energy-containing range, which in atmospheric turbulence can be $O(1)$ in both stable and unstable stratification (Wyngaard, 1992). In the inertial subrange, where $h/\ell \ll 1$, (1.27) says the direct effects of buoyancy on the unresolved turbulence budgets are small.

An equation for a second moment II, say, has the general form

$$\frac{\partial II}{\partial t} + \overline{u}_i \frac{\partial II}{\partial x_i} = \frac{\partial III_i}{\partial x_i} + \cdots.$$

The first term on the right is a "turbulent transport" (flux divergence) term that can be important, particularly in the mesoscale limit in unstably stratified conditions. There are two extremes in modeling the turbulent-transport term in such a second-moment equation (Zeman, 1982). The simplest is to model it through gradient diffusion,

$$III_i = -K \frac{\partial II}{\partial x_i}, \quad K \sim E^{1/2}\mathcal{L},$$

with $\mathcal{L} \sim \ell$ and h in the mesoscale and LES limits, respectively. The most complex approach is to write the rate equation for III,

$$\frac{\partial III}{\partial t} + \overline{u}_i \frac{\partial III}{\partial x_i} = \text{rhs},$$

and calculate directly some of the terms on the right-hand side (including the buoyancy term) while approximating others. This approach can be quite effective in the unstable boundary layer (Zeman, 1982). Again, the prescribed behavior of the length scale could allow a smooth transition of the turbulent-transport model in the LES limit.

Acknowlededgements

I am grateful to Doug Lilly for being so thoroughly and delightfully himself over the 25 years I have been privileged to know him, and to Evgeni Fedorovich and Bjorn Stevens for inviting me to prepare this paper. This work was supported in part by the National Science Foundation under grant ATM-0222421.

References

Batchelor, G. K. (1953). *The Theory of Homogeneous Turbulence*, Cambridge, UK: Cambridge University Press.

Bryan, G. H., Wyngaard, J. C. and Fritsch, J. M. (2003). On adequate resolution for the simulation of deep moist convection. *Mon. Wea. Rev.*, **131**, 2394–2416.

Corrsin, S. (1961). Turbulent flow. *Amer. Scientist*, **49**, 300–324.

Deardorff, J. W. (1970). A numerical study of three-dimensional turbulent channel flow at large Reynolds numbers. *J. Fluid Mech.*, **41**, 453–480.

(1973). Three-dimensional numerical modeling of the planetary boundary layer. In *Workshop on Micrometeorology*, D. A. Haugen, ed., Boston: American Meteorological Society.

Favre, A., ed. (1962). *Méchanique de la Turbulence*, No. 108. Paris, France: Centre National de la Recherche Scientifique.

Galperin, B. and Orszag, S. A. eds. (1993). *Large-Eddy Simulation of Complex Engineering and Geophysical Flows*, Cambridge: Cambridge University Press.

Ghosal, S., and Moin, P. (1995). The basic equations for the large-eddy simulation of turbulent flows in complex geometries. *J. Comp. Phys.*, **118**, 24–37.

Gibson, M. M. (1963). Spectra of turbulence in a round jet. *J. Fluid Mech.*, **15**, 161–173.

Grant, H. L., Stewart, R. W. and Moilliet, A. (1962). Turbulence spectra from a tidal channel. *J. Fluid Mech.*, **12**, 241–263.

Horst, T. W., Kleissl, J., Lenschow, D. *et al.* (2004). HATS: Field observations to obtain spatially filtered turbulence fields from crosswind arrays of sonic anemometers in the atmospheric surface layer. *J. Atmos. Sci.*, **61**, in press.

Kolmogorov, A. N. (1941a). The local structure of turbulence in an incompressible fluid with very large Reynolds number. *CR Acad. Sci. URSS*, **30**, 301–305.

(1941b). Dissipation of energy in locally isotropic turbulence. *CR Acad. Sci. URSS*, **32**, 16–18.

Lesieur, M. and Métais, O. (1996). New trends in large-eddy simulations of turbulence. *Ann. Rev. Fluid Mech.*, **28**, 45–82.

Lilly, D. K. (1962). On the numerical simulation of buoyant convection. *Tellus*, **14**, 148–172.

(1966). On the application of the eddy viscosity concept in the inertial sub-range of turbulence. NCAR Manuscript No. 123, Boulder, Co: National Center for Atmospheric Research.

(1967). The representation of small-scale turbulence in numerical simulation experiments. In *Proc. IBM Scientific Computing Symposium on Environmental Sciences*, IBM Form No. 320-1951, 195–210.

Meneveau, C. and Katz, J. (2000). Scale-invariance and turbulence models for large-eddy simulation. *Ann. Rev. Fluid Mech.*, **32**, 1–32.

Moffatt, H. K. (2002). G. K. Batchelor and the homogenization of turbulence. *Ann. Rev. Fluid Mech.*, **34**, 19–35.

Phillips, N. A. (1956). The general circulation of the atmosphere: a numerical experiment. *Quart. J. Roy. Meteor. Soc.*, **82**, 123–164.

Smagorinsky, J. (1963). General circulation experiments with the primitive equations: Part I, The basic experiment. *Mon. Wea. Rev.*, **91**, 99–164.

Smagorinsky, J., Manabe, S. and Holloway, J. L. (1965). Numerical results from a nine-level general circulation model of the atmosphere. *Mon. Wea. Rev.*, **93**, 727–768.

Tennekes, H., and Lumley J. L. (1972). *A First Course in Turbulence*, Cambridge, MA: Massachusetts Institute of Technology.

Wyngaard, J. C., Coté O. R. and Izumi, Y. (1971). Local free convection, similarity, and the budgets of shear stress and heat flux. *J. Atmos. Sci.*, **28**, 1171–1182.

Wyngaard, J. C. (1982). Planetary boundary layer modeling. In *Atmospheric Turbulence and Air Pollution Modelling*, F. T. M. Nieuwstadt and H. Van Dop, eds., Netherlands: Reidel, 69–106.

(1992). Atmospheric turbulence. *Ann. Rev. Fluid Mech.*, **24**, 205–233.

(2002). On the mean rate of energy transfer in turbulence. *Phys. Fluids*, **14**, 2426–2431.

(2004). Toward numerical modeling in the Terra Incognita. *J. Atmos. Sci.*, **61**, in press.

Zeman, O. (1982). Progress in the modeling of planetary boundary layers. *Ann. Rev. Fluid Mech.*, **13**, 253–272.

2

Phenomenological hunts in two-dimensional and stably stratified turbulence

James C. McWilliams

Department of Atmospheric and Oceanic Sciences and Institute of Geophysics and Planetary Physics,
University of California, Los Angeles, USA

2.1 Introduction

There are many distinctive turbulent regimes in nature that arise due to the various physical influences of velocity shear, density gradient and gravity, boundary configuration, (planetary) rotation, ionization, etc. At high Reynolds number (i.e., $Re = VL/\nu$, where V and L are characteristic velocity and length scales and ν is the kinematic viscosity), the generic turbulent behaviors are to evince cascades of velocity and scalar variance that act to (1) broaden their wavenumber spectra and effect dissipation of variance; (2) spatially transport momentum and scalars; and (3) develop coherent structures. The particular manifestations of these behaviors, however, are highly regime dependent. From this perspective the classical (Kolmogorov's) regime of isotropic, homogeneous, uniform-density, three-dimensional (3D) turbulence seems no more than typically distinctive, except insofar as it might emerge as universal behavior at sufficiently small scales beneath the control of the physical influences listed above. Even this hypothesis of universality, however, is contradicted in some regimes including two-dimensional (2D) and, perhaps, stably stratified turbulence, the subjects of this essay.

Turbulence is a tough scientific problem because of its mathematical intractability at the fundamental level of the Navier–Stokes equations and its experimental inaccessibility due to the complexity of flow patterns and difficulty in mimicking nature in the laboratory (e.g., achieving a high enough value of Re). So the rise of modern computers and their application to fluid dynamics have complemented theory and measurement and thereby greatly expanded our understanding of turbulence, even though it remains only a partially solved problem.

Douglas Lilly is both a pioneer and homesteading practitioner of computational studies of turbulence for the more than three decades that this approach has been feasible. He worked in several regimes, some of which are addressed in other chapters in this book. The focus here is on the phenomena that occur in a 2D fluid and in a 3D, stably stratified fluid without any accompanying large-scale shear, asymptotically as $Re \rightarrow \infty$. These regimes have in common a high degree of anisotropy compared to the Kolmogorov regime, and even approximately share the same governing equation (as explained in Section 2.4); however, their behaviors are quite different, as we will see. This essay is a perspective on the historical paths by which at least the qualitative behaviors in these regimes have come to be fairly well understood. It is also an appreciation of Lilly's contributions.

2.2 Computational turbulence

The natural phenomena of turbulence are firmly believed to be equivalent to the solutions of the Navier–Stokes partial differential equation (PDE) system at large Re. With this premise the two necessary ingredients for computational simulation of turbulence are the mathematics of discretization for the continuous PDE and the technology of computers with sufficient power to encompass the many degrees of freedom that arise in turbulence. These ingredients are not unrelated since discrete approximations converge only as the degrees of freedom become infinite, even at a fixed large value of Re. Both ingredients steadily evolved over the twentieth century. In particular, computational speed has followed Moore's law of exponential growth with time with a growth rate that itself has increased (Moore, 2003). This has created an ever expanding capability for computational fluid dynamics (CFD) – may it continue!

The term CFD is most widely used in the engineering community where the primary goal is computational design of devices with high precision and reliability. In the G(eophysical)FD community the primary goal continues to be phenomenological discovery and interpretation of measurements, with the precision of natural simulations a more futuristic goal. Due to the difficulty of natural turbulence problems, computational studies have provided many important discoveries about the cascades, transport, and coherent structures in different regimes; no doubt many more will come.

As one of the pioneers of GFD turbulence simulation, Lilly had to work through various practical methodological issues. These included spatial operator discretization, time-integration stability, and subgrid-scale parameterization – i.e., representing by physical approximation the effects of unresolved small-scale motions on the larger-scale simulated flow evolution when the phenomena of interest occur at larger Re than can be encompassed in a feasible computation – see Lilly (1961, 1965, 1966, 1967, 1975, 1992, 1997), Fox and Lilly (1972), Scotti *et al.* (1993), and Wong and Lilly (1994).

2.3 Two-dimensional turbulence

The governing PDE system for 2D, incompressible, viscous flow is

$$\frac{\partial \zeta}{\partial t} + (\mathbf{u} \cdot \boldsymbol{\nabla}) \zeta = \nu \nabla^2 \zeta + F$$

$$\mathbf{u} = (u, v) = \left(-\frac{\partial \psi}{\partial y}, \frac{\partial \psi}{\partial x} \right)$$

$$\zeta = \frac{\partial v}{\partial x} - \frac{\partial u}{\partial y} = \nabla^2 \psi. \tag{2.1}$$

Here $\mathbf{u}(x, y, t)$ is the 2D velocity; $\boldsymbol{\nabla}$ is the 2D gradient operator; F is a specified random torque; ψ is the streamfunction; and ζ is the vorticity. Simple boundary conditions are periodicity in x and y.

Of course, a 2D flow cannot be expected to occur widely in 3D nature. Nevertheless, for various reasons associated with the thinness of Earth's atmosphere and ocean, its typically stable density stratification, and its rotation (i.e., the Taylor–Proudman effect), 2D turbulence can be argued to be more relevant to large-scale flows than classical 3D turbulence (McWilliams, 1983). At the least, it embodies an extreme form of the observed anisotropy in nature: $H \ll L$ and $W \ll V$, where H and W are characteristic vertical (i.e., parallel to gravity) length and velocity scales, and L and V are their horizontal counterparts. An additional, historical attraction of 2D fluid dynamics is that it presents a smaller computational problem than does 3D. A collection of early CFD papers is in *The Physics of Fluids*, vol. 12, *Supplement II*, 1969: most of the papers are in fact 2D simulations even though the target phenomena are 3D, but an exception is Lilly's (1969) first simulation of 2D turbulence (along with Batchelor's, 1969).

When $\nu = F = 0$, (2.1) requires the conservation of two quadratic integrals, the kinetic energy and enstrophy:

$$E = \frac{1}{2} \int \int \mathbf{u}^2 \, dx \, dy, \quad \mathcal{E} = \frac{1}{2} \int \int \zeta^2 \, dx \, dy. \tag{2.2}$$

This implies that any turbulent evolutionary tendency to broaden the wavenumber spectrum for $\mathbf{u}$ leads to an inverse cascade of energy density toward larger scales together with a forward cascade of enstrophy density toward smaller scales and eventual viscous dissipation for any $\nu \neq 0$ (Batchelor, 1953). (In contrast, classical 3D turbulent evolution has a primarily forward cascade of kinetic energy toward small-scale dissipation.) Thus, the energy dissipation rate will become vanishingly small as $\nu \to 0$ ($Re \to \infty$), while the enstrophy dissipation rate will develop over time to a finite value. In the presence of a sustained forcing ($F \neq 0$ within a limited range of wavenumbers around k_F), two different equilibrium, inviscid, inertial-range shapes are predicted for the isotropic kinetic-energy spectrum in 2D turbulence (Kraichnan, 1967):

$$E(k) \sim k^{-5/3}, \quad k \ll k_F \quad \text{inverse energy cascade;}$$
$$E(k) \sim k^{-3}, \quad k \gg k_F \quad \text{forward enstrophy cascade.} \tag{2.3}$$

Note that this leads to a non-equilibrium divergence of the energy with time as it is continually generated by F and inverse-cascades to ever larger scales (or accumulates at the finite domain scale) without dissipation as $Re \to \infty$.

This is the conceptual framework within which Lilly (1969, 1971, 1972) made his computational simulations of 2D turbulence. Their purposes were to test these ideas and to assess the implications of sensitive dependence in 2D turbulence (i.e., small perturbations between two realizations amplify in time at an exponential rate) for predictability limits in initial-value problems, with intended implications for weather forecasting. He sensibly controlled the energy divergence in forced solutions by adding a linear drag term, $-C\zeta, C > 0$, to the rhs of the vorticity equation in (2.1), justifying it as an Ekman drag against the missing vertical boundaries. The conclusions from his simulations, broadly speaking, confirmed the extant theories about the inverse and forward cascades, the predicted inertial-range spectrum shapes, and the relative weakness of the energy dissipation rate. A new discovery was the growth of intermittency (non-Gaussianity) in the ζ and $\nabla\zeta$ fields in the freely decaying simulations ($F = 0$), although the associated magnitudes (e.g., of kurtosis) were modest. In these papers several remarks indicate Lilly's lack of confidence in the adequacy of the numerical resolution, related to the modest values of Re that were feasible. Nevertheless, they comprised a considerable achievement that allowed the conclusion, "numerical simulation is now apparently capable of adding to our fundamental understanding of turbulent processes at an affordable cost" (Lilly, 1971, p. 414). These simulation results provided the standard for 2D turbulence up through the comprehensive review article by Kraichnan and Montgomery (1980).

How has our understanding of 2D turbulence changed since then? There have been, of course, many quantitative refinements allowed by more powerful computers. But perhaps the most important change is the realization that the generic behavior of turbulence is the development of coherent structures at high-Re values. The case for this was first and best made in 2D turbulence because of its computational affordability (McWilliams, 1984; Borue, 1993, 1994); however, this phenomenon has since been confirmed in many turbulent regimes, albeit with different typical flow patterns in different regimes. In hindsight the characteristic long-lived, isolated, axisymmetric, monopole patterns in $\zeta(x, y)$ – coherent vortices – are not evident in Lilly's published figures. (He has subsequently remarked that he did see indications of them in evidence not published.) A statistical measure of coherent

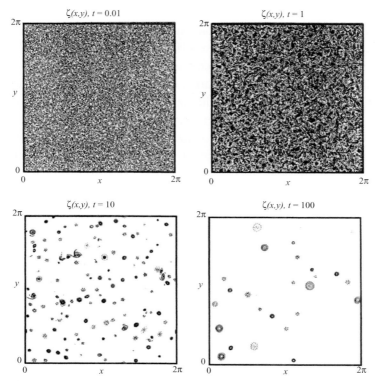

Figure 2.1. Vortex emergence and evolution in computational 2D turbulence, as seen in $\zeta(x, y)$ at sequential times, with random, spatially smooth initial conditions. Solid contours are for positive ζ, and dashed ones are for negative ζ. The contour interval is twice as large in the first panel as in the others. The times are non-dimensionalized based on an advective scaling, L/V. (Adapted from McWilliams, 1984.)

vortex presence is intermittency in ζ, first noted in Lilly (1971), and very large values of its kurtosis have since been shown in various circumstances (especially in free decay).

The cascades and dissipation in 2D turbulence co-exist with vortex emergence, movement, and mergers (Fig. 2.1). From smooth initial conditions or random forcing, coherent vortices emerge by axisymmetrization; move around under each other's far-field circulation (similarly to point vortices); occasionally form opposite-parity couples for brief intervals; and merge when two vortices of the same vorticity parity move close enough together. With time and in the absence of new generation by forcing, the vortices become fewer, larger, and sparser in space (McWilliams, 1990a), and they undergo close encounters less frequently. Since during close encounters the vortices change in ways other than simple movement and reversible deformation, the overall evolutionary rate for the spectrum shape and vortex population becomes ever slower with time, even though the kinetic energy does not diminish appreciably. Enstrophy dissipation occurs primarily during emergence and merger events: filaments of vorticity are stripped off of the vortices and continue to elongate irreversibly until their transverse scale comes under the control of viscous diffusion, and the enstrophy they contain is thereby dissipated. The filamentation is induced by the differential velocity field (i.e., shear, strain rate) of one vortex acting on another, which increases in magnitude as the vortices come closer together.

In 2D turbulence the aggregate evolutionary behavior of the flow (e.g., scalar transport, cascade, and dissipation rates) is governed by the coherent vortices, unless overcome by forcing and damping or other competing dynamic influences. This is shown by artificial suppression of the vortices that alters the cascade and dissipation rates (McWilliams, 1990b). It has also been shown by construction of a vortex-based dynamic system that reproduces the aggregate evolutionary behavior of 2D turbulence rather well (Carnevale *et al.*, 1991; Weiss and McWilliams, 1993). These results inspire the hypothesis of system evolutionary control by the coherent structures in all turbulent regimes, although this idea has not yet been fully tested.

Further computational exploration at higher values of *Re* (e.g., Bracco *et al.*, 2000) and at least some testing by the passage of time have not qualitatively altered the perceived phenomenology. This summary of 2D turbulence studies in the past two decades is, of course, highly abbreviated (and personal); it certainly does not comprise a review of the subject. Many scientists have found that 2D turbulence continues to be an important and efficient context for testing a variety of ideas about turbulence, even if it is not wholly defensible as a natural regime.

2.4 Stably stratified turbulence

The governing PDE system for 3D, stably stratified, incompressible, viscous, diffusive flow in the Boussinesq approximation is

$$\frac{\partial \mathbf{u}}{\partial t} + (\mathbf{u} \cdot \nabla)\mathbf{u} = -\nabla\phi + \hat{\mathbf{z}}b + \nu\nabla^2\mathbf{u} + \mathbf{f}$$

$$\nabla \cdot \mathbf{u} = 0$$

$$\frac{\partial b}{\partial t} + (\mathbf{u} \cdot \nabla)b + N^2 w = \kappa\nabla^2 b. \tag{2.4}$$

Here $\hat{\mathbf{z}}$ is the opposite direction to gravity; $\mathbf{u}(x, y, z, t) = (u, v, w)$ is the 3D velocity; ∇ is the 3D gradient operator; $\overline{\rho}(z)$ and $\overline{p}(z)$ are the background density stratification and pressure in hydrostatic balance; $\phi = (p - \overline{p})/\rho_0$ is the normalized pressure anomaly; $\mathbf{f}$ is a specified random force; $N^2 = -(g/\rho_0)\,d\overline{\rho}/dz$ is the squared buoyancy frequency of the background stratification, assumed here to be a constant; $b = -(g/\rho_0)(\rho - \overline{\rho})$ is the buoyancy anomaly; and κ is the buoyancy diffusivity. Simple boundary conditions are periodicity for $\mathbf{u}$, ϕ, and b in x, y, and z.

The regime of stratified turbulence has a small Froude number, $Fr = V/NH \ll 1$, with N the buoyancy frequency; $Re \gg 1$; moderate Prandtl number, $Pr = \nu/\kappa = \mathcal{O}(1)$; and some conditioning of the initial and forcing data so that the flow is primarily both horizontally oriented and horizontally non-divergent, hence anisotropic. The latter condition is necessary to distinguish stratified turbulence from the more isotropic, weakly nonlinear, internal gravity waves that are another class of solutions to (2.4) when $Fr \ll 1$, $Re \gg 1$, and $Pr \sim 1$. The presumption, supported by both observational and computational experience, is that these two solution classes are usually weakly interacting. A mathematical representation of this distinction is by a Helmholtz decomposition of the horizontal velocity, $\mathbf{u}_\perp = (u, v, 0)$, into its vertically rotational and horizontally divergent components, with the vertical velocity related to the latter by incompressibility:

$$\mathbf{u}_\perp = \hat{\mathbf{z}} \times \nabla_\perp \psi + \nabla_\perp \chi, \qquad w = -\int \nabla_\perp^2 \chi \, dz, \qquad (2.5)$$

where $\nabla_\perp$ is the horizontal gradient operator. Thus, stratified turbulence has $|\chi| \ll |\psi|$ and $|w| \ll |\mathbf{u}_\perp|$, while gravity waves have $|\psi| \ll |\chi|$.

When $\nu = \kappa = \mathbf{f} = 0$, again there are conserved integrals of motion, analogous to (2.2), the total energy and potential enstrophy:

$$E = \frac{1}{2} \int \int \int (\mathbf{u}^2 + b^2/N^2) \, dx \, dy \, dz, \qquad \mathcal{E} = \frac{1}{2} \int \int \int q^2 \, dx \, dy \, dz, \quad (2.6)$$

where $q = (\nabla \times \mathbf{u}) \cdot (N^2 \hat{\mathbf{z}} + \nabla b)$ is the potential vorticity. However, unlike for 2D flow, the second integral is not a quadratic functional of the primary dependent variables, so no strong inferences can be made about the direction of the turbulent cascades. Similarly, no useful equilibrium inertial-range prediction can be made. Finally, internal gravity waves have no manifestation in q (to leading order in Fr); so their evolution is not well constrained by enstrophy conservation. For all these reasons simple theoretical arguments about the evolution of stratified turbulence have been illusive.

In nature, this stably stratified regime is a common one. It is often identified by a local measure, the gradient Richardson number,

$$Ri(\mathbf{x}) = \left(N^2 + \frac{\partial b}{\partial z} \right) \Big/ \left(\frac{\partial \mathbf{u}_\perp}{\partial z} \right)^2, \qquad (2.7)$$

being large, which is understood to preclude vertically overturning motions. It may also be characterized by the Ozmidov scale, $L_O = (\epsilon/N^3)^{1/2}$ (ϵ is the kinetic-energy dissipation rate), being larger than the Kolmogorov viscous scale, $L_\nu = (\nu^3/\epsilon)^{1/4}$, with the implication that motions with $L > L_O$ are stratified turbulence while those with $L_O > L > L_\nu$ are classical, isotropic 3D turbulence; however, note that this transition to universal turbulence at small scales depends on the dissipation rate being sufficiently large in stratified turbulence, $\epsilon > \nu N^2$, which is not known to be true a priori. Finally, this regime is, by definition, characterized by weak influence by Earth's rotation with large Rossby number, $Ro = V/fL \gg 1$ (with f the Coriolis frequency), hence it has $L \ll NH/f$, the Rossby deformation radius. (When $L \sim NH/f$, the regime is called geostrophic turbulence.)

An early posing of the stratified turbulence problem was as the collapse of isotropic turbulence (e.g., in the wake of a towed object in a stratified fluid) once the turbulent energy dissipates to a stage when it cannot overcome the potential energy barrier of the stratification to induce overturning motions. The ensuing phenomena are quasi-linear gravity waves radiating away from the wake and a field of long-lived, thin $(\lambda = H/L \ll 1)$, nonlinearly evolving, horizontally recirculating "pancake vortices," or "vortical modes," that remain behind (Lin and Pao, 1979). Gage (1979) noted the anisotropic character of measured atmospheric mesoscale wind and its approximate kinetic-energy spectrum shape, $E \sim k_\perp^{-5/3}$, and proposed the interpretation that this is a consequence of an inverse energy cascade from a forcing due to cumulus convection by a turbulent dynamics somehow made 2D by the stable stratification. Lilly (1983) performed an asymptotic scaling analysis of (2.4) as $Fr \to 0$ and derived the 2D vorticity equation (2.1) as the leading-order dynamic balance, with the understanding that it holds independently in each vertical layer. This result largely framed the central issues in subsequent stratified turbulence research (see, e.g., the review by Riley and Lelong, 2000):

- Is layerwise 2D turbulence a uniformly valid approximation in stratified turbulence?
- Is the dynamic coupling to gravity waves and overturning motions weak?
- Is there an inverse energy cascade?
- How large is the energy dissipation rate?
- How much are natural motions with $NH/f > L > L_0$ like pancake vortices?

Lilly and Petersen (1983) added to the empirical evidence for a $\sim k_\perp^{-5/3}$ mesoscale kinetic-energy spectrum, as well the $\sim k_\perp^{-3}$ spectrum at larger scales (attributed, simplistically, to a forward enstrophy cascade of 2D-like turbulence from a source at synoptic scales induced by Earth's rotation or, more soundly, to geostrophic turbulence, sometimes called $2\frac{1}{2}$D turbulence). Lilly (1989) showed that 2D turbulence with F concentrated at two well-separated scales can indeed exhibit this type of contiguous dual inertial-range structure.

Lilly (1983, p. 757) also sowed seeds of doubt in his asymptotic approximation by the argument that the sensitive dependence of 2D turbulence would cause divergence in time between flow patterns in adjacent layers, hence a shrinking vertical scale, hence a growth in Fr, hence eventually a small local Ri value and overturning motions by Kelvin–Helmholtz instability. McWilliams (1985) extended the $Fr \to 0$ asymptotic analysis to derive consistent leading-order balance or "slaving" relations to ζ and ψ among all the other dependent variables, e.g., cyclostrophic and hydrostatic balance for the pressure and buoyancy anomalies:

$$\nabla_{\perp}^2 \phi = 2\left[\frac{\partial^2 \psi}{\partial x^2}\frac{\partial^2 \psi}{\partial y^2} - \left(\frac{\partial^2 \psi}{\partial x \partial y}\right)^2\right], \quad \frac{\partial \phi}{\partial z} = b. \qquad (2.8)$$

This suggests the possibility that a limiting vertical scale would emerge under evolution through conservative dynamic coupling between adjacent layers (a type of vortex stretching), thereby preserving the uniform validity of the $Fr \ll 1$ regime.

Laboratory measurements (e.g., Browand *et al.*, 1987; Spedding, 2002) and computational simulations (Riley *et al.*, 1981; Herring and Metais, 1989; Metais and Herring, 1989; Kimura and Herring, 1996) confirmed some important aspects of the layerwise 2D hypothesis for stratified turbulence: in free decay an initially small Fr remains small, and pancake vortices routinely emerge but with less intermittency than the vortices in 2D turbulence. But they contradicted some others: the energy dissipation rate is large, and inverse energy cascade is not seen. The structure of the energy dissipation was associated with the large vertical shear, $|\partial \mathbf{u}_{\perp}/\partial z|$, that develops at the vertical edges of the pancake vortices. This can be viewed as a forward energy cascade in vertical wavenumber. It implies a viscous dynamic selection of a limiting vertical scale, and it inspired a simple extension of the 2D vorticity equation in (2.1) to include vertical eddy diffusion, $\nu(\partial^2 \zeta/\partial z^2)$, and to downplay the importance of the horizontal diffusion that plays an essential role in enstrophy dissipation in 2D turbulence (Majda and Grote, 1997). The absence of demonstrable inverse cascade in stratified turbulence undermined Gage's interpretation of the atmospheric mesoscale wind spectrum; this led Lilly *et al.* (1998) to invoke a more complex interaction with cumulus convection to partially salvage the interpretation.

Until recently experiments and simulations have only been made for modest values of Re; furthermore, most studies have been for freely decaying stratified turbulence where $Re(t)$ steadily decreases due to energy dissipation, making it difficult to assess the inertial equilibrium state. Several recent studies (deBruynKops *et al.*, 2003; Laval *et al.*, 2003; Reasor *et al.*, 2004; Riley and deBruynKops, 2003; K. B. Winters, personal communication) have made computational simulations at larger Re values, and each concludes that, for a given small value of Fr, the flow will evolve to have some locally small values of Ri and associated overturning motions if Re is large enough. The term "hot spots" has been suggested for these events. A scaling argument balancing production and dissipation in the kinetic-energy balance suggests that the transition is for $Re > \mathcal{O}(Fr^{-2}\lambda^{-2})$ with apsect ratio, $\lambda = H/L \ll 1$, and $Re = VL/\nu$ defined in terms of the most energetic horizontal length scale, L (Riley and deBruynKops, 2003). This phenomenon resolves the issue of whether the $Fr \to 0$ asymptotic approximation is uniformly valid at large Re: it is not. On the other hand, each study also shows that the overturning regions are highly intermittent and that the pancake vortices remain the dominant flow structures even after local overturning arises. This suggests that some kind of annealing occurs after a local rupture of the otherwise balanced evolution of stratified turbulence (cf. (2.8)). Evidently the $Fr \rightarrow 0$ approximation continues to be at least highly germane even if not precisely correct.

This local breakdown of pancake vortex dynamics is not the only new phenomenon that arises in stratified turbulence at large Re. This is demonstrated by the computational simulation in Laval *et al.* (2003) for randomly forced, quasi-equilibrium turbulence. Its forcing **f** is chosen to project only onto accelerations of the rotational component of **u**, i.e., $\mathbf{f} = \hat{\mathbf{z}} \times \nabla \Phi$ for a random large-scale potential, Φ. A simulation control path is designed to scan the behavior as a function of Re. The forcing magnitude, N^2, and ν (with $Pr = 1$) are chosen so that the bulk Froude number associated with the kinetic-energy spectrum peak is small and stays constant at $Fr \approx 0.08$ while the Taylor Reynolds number, R_λ (based on the rms velocity and the length scale that is the square root of the ratio of kinetic energy and kinetic enstrophy), is held fixed for time intervals of $\Delta \tau = 100$ eddy turnover times, L/V, during which the flow comes into statistical equilibrium between upward steps of $\Delta R_\lambda \approx 100$ (Fig. 2.2). Experience has shown that the Taylor Reynolds number is the more germane measure of the turbulent development, compared to the energy-containing Reynolds number, Re (though they approximately scale as $Re = Re_\lambda^2$ as $Re \rightarrow \infty$).

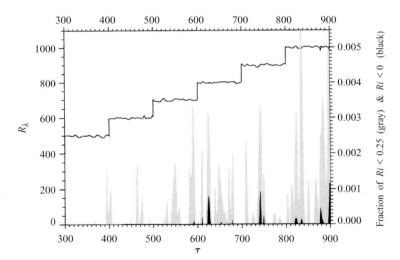

Figure 2.2. Experimental path in $R_\lambda(\tau)$ for forced stratified turbulence with a fixed Fr and a step-wise decreasing viscosity. Also shown are time series of the volume fraction of the domain with local $Ri < 0.25$ (filled gray area) and with local $Ri < 0$ (filled black area). τ is a non-dimensional time, normalized by the eddy turnover time. There is occurrence of $Ri < 0.25$ for $0 < \tau < 300$. (Adapted from Laval *et al.*, 2003.)

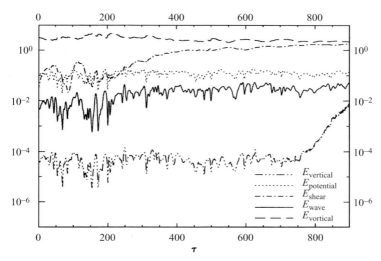

Figure 2.3. Non-dimensional time evolution of components of the energy (defined in the text with quotation marks) for the same simulation as Fig. 2.2. (Adapted from Laval *et al.*, 2003.)

Time histories of various energy components (Fig. 2.3) show evidence of three successive transitions with Re. At small Re (i.e., for $\tau \leq 200$ and $R_\lambda < 300$), the pancake vortices wholly dominate the solution as indicated by the "vortical" kinetic energy (associated with ψ) dominating the divergent (or "wave") kinetic energy (χ) and the "potential" energy (b/N). The "vertical" velocity (w) contribution to kinetic energy is especially small. Furthermore, the spontaneously generated mean "shear" velocity, $\langle \mathbf{u}_\perp \rangle (z, t)$ (where angle brackets denote horizontal average), has a kinetic energy no larger than the "potential energy." No overturning occurs, and the local Ri values are large everywhere (Fig. 2.2). The **first transition** from this archetypal pancake regime occurs around $R_\lambda = 400$ starting at $\tau \approx 200$: the shear energy starts to grow and continues to do so as R_λ further increases, although it approximately equilibrates when R_λ is held constant for longer intervals. This phenomenon has also been simulated by Smith and Waleffe (2002) for Fr values below an $\mathcal{O}(1)$ critical value. The **second transition** occurs around $R_\lambda = 600$ starting at $\tau \approx 400$. It is not particularly evident in the energy components (Fig. 2.3) but is evident in the distribution function for local Ri values (Fig. 2.2): a tail of rare small values develops – initially with $Ri < 0.25$, the classical onset value for Kelvin–Helmholtz instability (Miles, 1961), and subsequently with $Ri < 0$, indicating a gravitationally unstable density profile caused by overturning motions. An illustration of a local overturning event in the shear layer vertically between two pancake flows in opposite horizontal directions is shown in Fig. 2.4. It has a nearly periodic overturning structure along the surface of maximum shear, reminiscent of

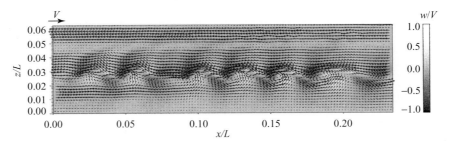

Figure 2.4. A local overturning event in forced stratified turbulence (as in Fig. 2.2) with $R_\lambda = 1000$. Plotted are instantaneous vertical velocity (gray scale) and $(0, v, w)$ velocity (vectors) in a (y, z) plane; both are normalized by the rms velocity, V, indicated by the reference arrow in the upper left. L is the horizontal extent of the domain. (Courtesy of J.-P. Laval.)

a Kelvin–Helmholtz event (Cortesi *et al.*, 1998; Caulfield and Peltier, 2000). The **third transition** occurs around $R_\lambda = 900$ starting at $\tau > 700$: the "vertical" energy begins to grow (Fig. 2.3). This occurs without any evident accompanying growth in "wave" and "potential" energies, so it is something other than a gravity-wave outbreak. In fact it occurs in w at the largest vertical and horizontal scales in the domain. We have provisionally interpreted it as an example of a negative eddy-viscosity instability, analogous to the result of Dubrulle and Frisch (1991) for a uniform-density flow, although this interpretation requires confirmation.

Given the newness of the results about high-Re transitions in stratified turbulence, it is premature to attempt a definitive resolution of the issues framed by Lilly (1983) (see the bullet list on p. 42). However, it does seem justifiable to forecast that the issues, as phrased, will prove to have ambiguous and somewhat subtle answers. Pancake vortices do comprise the dominant behavior of stratified turbulence even at large Re values, but they do generate other types of motion that are seemingly distinct from internal gravity waves (i.e., small-scale overturning and large-scale horizontal and vertical flows). The overturning motions allow the possibility of a transition to a classical Kolmogorov regime on scales smaller than the vortical motions if their occurrence proves to be not too intermittent. The large-scales flows represent a kind of inverse energy cascade whose spatial scale content is not yet well determined (because of domain-size limitations in the present simulations). But this is certainly a different kind of inverse cascade than the scale-by-scale progression in 2D turbulence.

2.5 Final remarks

The problems of turbulence are addressable by computational simulation both qualitatively – in terms of the manifested phenomena and dominant flow patterns – and quantitatively. In geophysical fluid dynamics the phenomenological hunts have been and will continue to be important scouting activities. For 2D turbulence the most surprising parts of the hunt may be over, although such a forecast should only be made humbly. For stably stratified turbulence the hunt still continues, and the outcome may be much more complex than we now envision. Nevertheless, great progress has been made on both problems during the last 30^+ years, and Douglas Lilly will always be known as a mighty hunter.

References

Batchelor, G. K. (1953). *The Theory of Homogeneous Turbulence*, Cambridge: Cambridge University Press.
  (1969). Computation of the energy spectrum in homogeneous two-dimensional turbulence. *The Physics of Fluids*, vol. 12, Suppl. II, 233–239.
Borue, V. (1993). Spectral exponents of enstrophy cascade in stationary two-dimensional homogeneous turbulence. *Phys. Rev. Lett.*, **71**, 3967–3970.
  (1994). Inverse energy cascade in stationary two-dimensional homogeneous turbulence. *Phys. Rev. Lett.*, **72**, 1475–1478.
Bracco, A., McWilliams, J. C., Murante, G., Provenzale, A. and Weiss, J. B. (2000). Revisiting two-dimensional turbulence at modern resolution. *Phys. Fluids*, **12**, 2931–2941.
Browand, F. K., Guyomar, D. and Yoon, S. C. (1987). The behavior of a turbulent front in a stratified fluid: experiments with an oscillating grid. *J. Geophys. Res.*, **92**, 5329–5341.
Carnevale, G. F., McWilliams, J. C., Pomeau, Y., Weiss, J. B. and Young, W. R. (1991). Evolution of vortex statistics in two-dimensional turbulence. *Phys. Rev. Lett.*, **66**, 2735–2737.
Caulfield, C. P. and Peltier, W. R. (2000). The anatomy of the mixing transition in homogeneous and stratified free shear layers. *J. Fluid Mech.*, **413**, 1–47.
Cortesi, A. B., Yadigaroglu, G. and Banerjee, S. (1998). Numerical investigation of the formation of three-dimensional structures in stably-stratified mixing layers. *Phys. Fluids.*, **10**, 1449–1473.
deBruynKops, S. M., Riley, J. J. and Winters, K. B. (2003). Reynolds and Froude number scaling in stably-stratified flows. In: *IUTAM Symp. on Reynolds Number Scaling in Turbulent Flow*, A. J. Smits, ed., Kluwer Academic Publishers.
Dubrulle, B. and Frisch, U. (1991). Eddy viscosity of parity-invariant flow. *Phys. Rev. A*, **59**, 1187–1226.
Fox, D. G. and Lilly, D. K. (1972). Numerical simulation of turbulent flows. *Rev. Geophys. Space Phys.*, **10**, 51–72.
Gage, K. S. (1979). Evidence for a $k^{-5/3}$ law inertial range in mesoscale two-dimensional turbulence. *J. Atmos. Sci.*, **36**, 1950–1954.
Herring, J. R. and Metais, O. (1989). Numerical experiments in forced stably stratified turbulence. *J. Fluid Mech.*, **202**, 97–115.
Kimura, Y. and Herring, J. R. (1996). Diffusion in stably stratified turbulence. *J. Fluid Mech.*, **328**, 253–269.
Kraichnan, R. (1967). Inertial ranges in two-dimensional turbulence. *The Physics of Fluids*, vol. 10, 1417–1423.
Kraichnan, R. and Montgomery, D. (1980). Two-dimensional turbulence. *Rep. Prog. Physics*, **43**, 547–619.

Laval, J. -P., McWilliams, J. C. and Dubrulle, B. (2003). Forced stratified turbulence: Successive transitions with Reynolds number. *Phys. Rev. E*, **68**, 036308/1– 036308/8.

Lilly, D. K. (1961). A proposed staggered-grid system for numerical integration of dynamic equations. *Mon. Wea. Rev.*, **89**, 59–66.

(1965). On the computational stability of numerical solutions of time-dependent nonlinear geophysical fluid dynamics problems. *Mon. Wea. Rev.*, **93**, 11–26.

(1966). On the application of the eddy viscosity concept in the inertial sub-range of turbulence. NCAR Manuscript No. 123, Boulder, CO: National Center for Atmospheric Research.

(1967). The representation of small-scale turbulence in numerical simulation experiments. In *Proc. IBM Scientific Computing Symposium on Environmental Sciences*, H. H. Goldstein, ed., IBM Form No. 320–1951, 195–210.

(1969). Numerical simulation of two-dimensional turbulence. *The Physics of Fluids*, vol. 12, *Suppl. II*, 240–249.

(1971). Numerical simulation of developing and decaying two-dimensional turbulence. *J. Fluid Mech.*, **45**, 395–415.

(1972). Numerical simulation of two-dimensional turbulence. Part I: Models of statistically steady turbulence and Part II: Stability and predictability studies. *Geophys. Fluid Dyn.*, **3**, 289–319, and **4**, 1–28.

(1975). Some comments on fine-mesh modeling. In *Proc. Severe Environmental Storms and Mesoscale Experiment Project*, Project Sesame, NOAA-ERL, 277–280.

(1983). Stratified turbulence and the mesoscale variability of the atmosphere. *J. Atmos. Sci.*, **40**, 749–761.

Lilly, D. K. and Petersen, E. L. (1983). Aircraft measurements of atmospheric kinetic energy spectra. *Tellus*, **35A**, 379–382.

Lilly, D. K. (1989). Two-dimensional turbulence generated by energy sources at two scales. *J. Atmos. Sci.*, **46**, 2026–2030.

(1992). A proposed modification of the Germano subgrid-scale closure method. *Phys. Fluids A*, **4**, 633–635.

(1997). Introduction to "Computational design for long-term numerical integration of the equations of fluid motion: Two-dimensional incompressible flow. Part I." *J. Comp. Phys.*, **135**, 101–102.

Lilly, D. K., Bassett, G., Droegemeier, K. and Bartello, P. (1998). Stratified turbulence in the atmospheric mesoscales. *Theor. Comp. Fluid Dyn.*, **11**, 139–153.

Lin, J. T. and Pao, Y. H. (1979). Wakes in stratified fluids. *Ann. Rev. Fluid Mech.*, **11**, 317–338.

Majda, A. J. and Grote, M. J. (1997). Model dynamics and vertical collapse in decaying strongly stratified flows. *Phys. Fluids*, **9**, 2932–2940.

McWilliams, J. C. (1983). On the relevance of two-dimensional turbulence to geophysical fluid motions. *J. de Mechanique*, **Numero Special**, 83–97.

(1984). The emergence of isolated, coherent vortices in turbulent flow. *J. Fluid Mech.*, **146**, 21–43.

(1985). A note on a uniformly valid model spanning the regimes of geostrophic and isotropic, stratified turbulence: balanced turbulence. *J. Atmos. Sci.*, **42**, 1773–1774.

(1990a). The vortices of two-dimensional turbulence. *J. Fluid Mech.*, **219**, 361–385.

(1990b). A demonstration of the suppression of turbulent cascades by coherent vortices in two-dimensional turbulence. *Phys. Fluids A*, **2**, 547–552.

Metais, O. and Herring, J. R. (1989). Numerical simulations of freely evolving turbulence in stably stratified fluids. *J. Fluid Mech.*, **239**, 157–194.

Moore, G. E. (2003). No exponential is forever . . . but we can delay "forever". In *IEEE International Solid State Circuit Conference, Feb. 10, 2003*.
[Also: http://www.intel.com/research/silicon/mooreslaw.htm]

Miles, J. W. (1961). On the stability of heterogeneous shear flows. *J. Fluid Mech.*, **10**, 496–508.

Reasor, P. D., Montgomery, M. T. and Grasso, L. D. (2003). A new look at the problem of tropical cyclones in vertical shear flow: Vortex resiliency. *J. Atmos. Sci.*, **61**, 3–22.

Riley, J. J., Metcalf, R. W. and Weissman, M. A. (1981). Direct numerical simulations of homogeneous turbulence in density-stratified fluids. In *Nonlinear Properties of Internal Waves*, B. J. West, ed., American Institute of Physics, 79–112.

Riley, J. J. and Lelong, M. -P. (2000). Fluid motions in the presence of strong stable stratification. *Ann. Rev. Fluid Mech.*, **32**, 613–657.

Riley, J. J. and deBruynKops, S. M. (2003). Dynamics of turbulence strongly influenced by buoyancy. *Phys. Fluids*, **15**, 1–13.

Scotti, A., Meneveau, C. and Lilly, D. K. (1993). Generalized Smagorinsky model for anisotropic grids. *Phys. Fluids A*, **5**, 2306–2308.

Smith, L.M. and Waleffe, F. (2002). Generation of slow, large scales in forced rotating, stratified turbulence. *J. Fluid Mech.*, **451**, 145–168.

Spedding, G. R. (2002). Vertical structure in stratified wakes with high initial Froude number. *J. Fluid Mech.*, **454**, 71–112.

Weiss, J. B. and McWilliams, J. C. (1993). Temporal scaling behavior of decaying two-dimensional turbulence. *Phys. Fluids A*, **5**, 608–621.

Wong, V. C. and Lilly, D. K. (1994). A comparison of two dynamic subgrid closure methods for turbulent thermal convection. *Phys. Fluids A*, **6**, 1016–1023.

3

Energy dissipation in large-eddy simulation: dependence on flow structure and effects of eigenvector alignments

Chad Higgins, Marc B. Parlange

Center for Environmental and Applied Fluid Mechanics, and Department of Geography and Environmental Engineering, Johns Hopkins University, Baltimore, USA

and

Charles Meneveau

Center for Environmental and Applied Fluid Mechanics, and Department of Mechanical Engineering, Johns Hopkins University, Baltimore, USA

3.1 Introduction

Most numerical simulations of turbulent flow use grid spacings that far exceed the viscous scale at which turbulent kinetic energy is dissipated into heat. Large-eddy simulation (LES), requires closure models to account for the turbulent motions occurring at small (the so-called subgrid) scales. These motions are responsible for mixing and they interact with the large-scale motions in a way that tends, typically, to transfer kinetic energy to smaller scales in the turbulent energy cascade. This transfer must be reproduced accurately by subgrid-scale (SGS) closures in order to prevent overdamping of resolved scales, or insufficient damping which can lead to spurious instabilities. Lilly (1967) was the first to combine this insight with concepts from the phenomenological theory of 3D turbulence to provide quantitative answers to several important parameterization issues in LES. Our goals in this article are to review briefly Lilly's pioneering contribution, and to reinterpret certain variables using geometric tools.

Forty years ago, Smagorinsky (1963) proposed a simple eddy-viscosity model based on local variables characterizing the motions at the length scale of the computational grid. In this model, the deviatoric part of the SGS stress tensor, τ_{ij}, where

$$\tau_{ij} = \widetilde{u_i u_j} - \tilde{u}_i \tilde{u}_j, \tag{3.1}$$

is set proportional to the strain-rate tensor, $\tilde{S}_{ij} = \frac{1}{2}(\partial_i \tilde{u}_j + \partial_j \tilde{u}_i)$, characterizing the rate of local deformation of the resolved velocity field. In these expressions a tilde denotes spatial filtering at a length scale Δ. The model is written as:

$$\tau_{ij} - \frac{1}{3}\tau_{kk}\delta_{ij} = -2\nu_\mathrm{T}\tilde{S}_{ij}. \tag{3.2}$$

The constant of proportionality is the eddy viscosity ν_T which is written as $\nu_T = \lambda^2 |\tilde{\mathbf{S}}|$, where $|\tilde{\mathbf{S}}| = (2\tilde{S}_{ij}\tilde{S}_{ij})^{1/2}$. Here λ is a mixing length scale, while $\lambda|\tilde{\mathbf{S}}|$ is a characteristic velocity scale estimated from the shear scale $|\tilde{\mathbf{S}}|$ and the mixing length λ. The mixing length must be chosen judiciously. For locations far from boundaries and in the absence of buoyancy and rotation effects, the only length scale available to characterize the local turbulence structure of the simulated flow is the filter scale, Δ. Dimensionally it follows that $\lambda = c_s\Delta$, where c_s is a dimensionless model parameter. This parameter must be specified in LES, and has been the subject of much attention in the literature (Deardorff, 1971; Mason, 1994; Piomelli, 1999; Meneveau and Katz, 2000).

In Section 3.2, we review Lilly's classic argument linking c_s to the universal Kolmogorov constant c_K. In Section 3.3 we discuss some dependencies between the local SGS dissipation and parameters characterizing the structure of the resolved-scale motions. In particular, we review field experimental data showing that the SGS dissipation is correlated with axisymmetric expanding motions at the resolved scales. In Section 3.4, we present a geometric view of the tensor contraction between SGS stress and strain-rate tensors in terms of the alignment angles among their respective eigenvectors. In Section 3.5, we combine observational evidence about most likely alignment angles among eigenvectors with the expressions for SGS dissipation and, using these empirical inputs, present a prediction of the preferred SGS dissipation as function of the structure parameter of the resolved scales. A discussion is presented in Section 3.6.

3.2 The Smagorinsky–Lilly model parameter

In a ground-breaking paper, Lilly (1967) showed how c_s could be evaluated from basic knowledge of turbulence, and thus c_s is often referred to as the "Smagorinsky–Lilly" constant in the literature. Central to Lilly's development was the realization that the most important effect of the SGS model upon the dynamics of the large-scale structures is the amount of kinetic energy the model extracts. Hence, the energetics of the flow computed in an LES takes on a special role. Lilly (1967) derives the transport equation for the subgrid kinetic energy $E = \frac{1}{2}\tau_{kk}$ and obtains:

$$\frac{\partial E}{\partial t} + \tilde{u}_k \frac{\partial E}{\partial x_k} - \nu \left[\frac{\partial^2 E}{\partial x_k^2} - \left(\widetilde{\frac{\partial u_i}{\partial x_k}} \right)^2 + \left(\frac{\partial \tilde{u}_i}{\partial x_k} \right)^2 \right]$$

$$= -\tau_{ij}\tilde{S}_{ij} - \frac{\partial}{\partial x_k} \left(\frac{\widetilde{u_k u_i^2}}{2} - \frac{\tilde{u}_k \tilde{u}_i^2}{2} - \tilde{u}_i \widetilde{u_k u_i} + \tilde{u}_i^2 \tilde{u}_k + \frac{\widetilde{u_k p}}{\rho} - \frac{\tilde{u}_k \tilde{p}}{\rho} \right) \quad (3.3)$$

where $\tilde{S}_{ij} = \frac{1}{2}(\partial_j \tilde{u}_i + \partial_i \tilde{u}_j)$. Taking the ensemble average of this equation (denoted below by angled brackets), and assuming steady-state conditions, one obtains the equality of molecular dissipation of SGS kinetic energy and its rate of production:

$$\nu \left[\left\langle \widetilde{\left(\frac{\partial u_i}{\partial x_k} \right)^2} \right\rangle - \left\langle \left(\frac{\partial \tilde{u}_i}{\partial x_k} \right)^2 \right\rangle \right] = -\langle \tau_{ij} \tilde{S}_{ij} \rangle. \tag{3.4}$$

The quantity $-\langle \tau_{ij} \tilde{S}_{ij} \rangle$ is interpreted as the mean flux of kinetic energy from the range of resolved scales into the SGS range, and also appears as a sink in the equation for resolved kinetic energy, $\frac{1}{2} \tilde{u}_k \tilde{u}_k$. When Δ is in the inertial range, the first term in the lhs of (3.4) dominates and equals ε, the overall rate of dissipation by viscosity. Hence, we can write $\varepsilon = -\langle \tau_{ij} \tilde{S}_{ij} \rangle$.

Lilly then makes the next important step in his derivation by replacing τ_{ij} with the Smagorinsky closure. One obtains the expression

$$\varepsilon = 2^{3/2} (c_s \Delta)^2 \langle (\tilde{S}_{ij} \tilde{S}_{ij})^{3/2} \rangle \tag{3.5}$$

as a condition for the Smagorinsky model to extract kinetic energy from the resolved scales at the correct rate. Two more assumptions are required to complete Lilly's original argument: (1) that at the grid scale Δ the turbulence exhibits a universal Kolmogorov spectrum $E(k) = c_K \varepsilon^{2/3} k^{-5/3}$ with turbulence statistics that are isotropic; this assumption is justified when Δ pertains to the inertial range of turbulence; and (2) the third-order statistics of the strain-rate magnitude may be approximated with its second-order moment as

$$\langle (\tilde{S}_{ij} \tilde{S}_{ij})^{3/2} \rangle \approx \langle \tilde{S}_{ij} \tilde{S}_{ij} \rangle^{3/2}. \tag{3.6}$$

The latter assumption is not explicitly stated in Lilly's paper since he did not elaborate explicitly on the nature of statistical averaging underlying the argument. The accuracy of this assumption was recently tested with Direct Numerical Simulation (DNS) data by Cerutti *et al.* (2000) and deviations on the order of 20% were observed in the inertial range (the correction factor β of that paper). Also Novikov (1990) has speculated that small-scale intermittency could introduce a further dependence of c_s upon Δ/l, where l is the integral length scale. Equation (3.6) still leaves the task of evaluating the second-order moment and strain-rate tensor contraction $\langle \tilde{S}_{ij} \tilde{S}_{ij} \rangle$. Using standard techniques from isotropic turbulence analysis, it is straightforward to show that

$$\langle \tilde{S}_{ij} \tilde{S}_{ij} \rangle = \int\limits_0^{\pi/\Delta} k^2 E(k) \, dk \tag{3.7}$$

where, as in Lilly (1967), a spherical spectral sharp filter is used to cut off the integration in wavenumber space at a wavenumber π/Δ. Substituting this into (3.5) and using the Kolmogorov spectrum $E(k) = c_K \varepsilon^{2/3} k^{-5/3}$ and solving for the coefficient c_s one obtains Lilly's result:

$$c_s = \frac{1}{\pi} \left(\frac{2}{3c_K} \right)^{3/4} \sim 0.165 \text{ (for } c_K = 1.6). \tag{3.8}$$

The analysis has been reproduced in tutorial detail in Pope (2001).

When using computational meshes that are unequal in each Cartesian direction (e.g., $\Delta_1 < \Delta_2 < \Delta_3$), the above derivation can be repeated, but now using an anisotropic 3D filter (Scotti *et al.*, 1993) such as a parallelepiped of sides $2\pi/\Delta_1$, $2\pi/\Delta_2$, and $2\pi/\Delta_3$ in Fourier space. The integrals in Fourier space are more complicated but can be evaluated numerically for exact evaluations of the coefficient. Lilly's central contribution to the Scotti *et al.* (1993) paper was to recognize that the integrations are greatly simplified if an ellipsoidal domain is used in Fourier space instead of a rectangular one. To zeroth order in $\log(a_i)$ (where $a_1 = \Delta_1/\Delta_3$ and $a_2 = \Delta_2/\Delta_3$ are the two grid aspect ratios), one can then show analytically that Δ in the definition of ν_T must be replaced with a length scale based on the cell volume,

$$\Delta_{eq} = (\Delta_1 \Delta_2 \Delta_3)^{1/3}. \tag{3.9}$$

This expression was already proposed on heuristic grounds by Deardorff (1970) and is often used in LES. Lilly's argument published in Scotti *et al.* (1993) thus serves as a formal justification to the often-used cube-root length scale and clearly demonstrates that it is to be preferred over other heuristic proposals that have occasionally been made over the years. For large filter anisotropies, Scotti *et al.* (1993) show that in addition to the use of Δ_{eq}, c_s should be replaced with $c_s f(a_1, a_2)$, where

$$f(a_1, a_2) \approx \cosh\left\{ \frac{4}{27}[(\ln a_1)^2 - \ln a_1 \ln a_2 + (\ln a_2)^2] \right\}^{1/2}.$$

The developments above relied upon statistical averaging to define the mean SGS dissipation, $-\langle \tau_{ij} \tilde{S}_{ij} \rangle$. The fact that the average is non-zero is related to subtle relationships among turbulent motions which lead to non-zero correlation among the tensors τ_{ij} and $\tilde{S}_{ij}$. This correlation is a third-order moment (since τ_{ij} is quadratic with velocity and $\tilde{S}_{ij}$ is linear) similar to the third-order velocity structure function that has important dynamic significance. Hence, it is of interest to explore more precisely the nature of these correlations, to which the remaining parts of this chapter are dedicated.

3.3 Field experimental studies of SGS dissipation

Before describing the field experiments, we introduce several variables to be measured from the data. The local SGS dissipation of turbulent kinetic energy is written as $\Pi = -\tau_{ij}\tilde{S}_{ij}$. We decompose both the filtered strain rate $\tilde{S}_{ij} = 0.5(\partial_i\tilde{u}_j + \partial_j\tilde{u}_i)$, and the deviatoric SGS stress into their respective eigenvectors and eigenvalues by the transform $A = Q_A\Lambda_A Q_A^{\mathrm{T}}$ where A is an arbitrary symmetric tensor, Q_A is a matrix containing the eigenvectors of A, and Λ_A is a diagonal matrix containing the corresponding eigenvalues of A on its diagonal.

The eigenvalues are named according to their magnitudes as $\alpha \geq \beta \geq \gamma$, and satisfy the condition $\alpha + \beta + \gamma = 0$. This requires $\alpha \geq 0$, $\gamma \leq 0$, and β is either positive or negative. Eigenvectors are named by their corresponding eigenvalues: $\boldsymbol{\alpha}$ is the extensive eigenvector, $\boldsymbol{\gamma}$ is the contractive eigenvector, and $\boldsymbol{\beta}$ is the intermediate eigenvector. To focus attention on the geometric alignment it is of interest to scale out the magnitudes of stress and strain rate and define a dimensionless dissipation, according to

$$\Pi^* = \frac{\Pi}{|\tilde{\mathbf{S}}||\boldsymbol{\tau}|} = \frac{-\tilde{S}_{ij}\tau_{ij}}{|\tilde{\mathbf{S}}||\boldsymbol{\tau}|}, \tag{3.10}$$

where $|\tilde{\mathbf{S}}| = \sqrt{\tilde{S}_{ij}\tilde{S}_{ij}} = \sqrt{\alpha_{\tilde{S}}^2 + \beta_{\tilde{S}}^2 + \gamma_{\tilde{S}}^2}$ and $|\boldsymbol{\tau}| = \sqrt{\tau_{ij}\tau_{ij}} = \sqrt{\alpha_{-\tau}^2 + \beta_{-\tau}^2 + \gamma_{-\tau}^2}$ (note that henceforth $|\tilde{\mathbf{S}}|$ does not include the factor $\sqrt{2}$ that is usually included in the definition of $|\tilde{\mathbf{S}}|$ for the Smagorinsky model). One can show that Π^* is now bounded between -1 and 1, and only characterizes the geometric nature of the stress–strain relationship. We now wish to study possible dependencies of this quantity with the geometric structure of the resolved strains. Following Lund and Rogers (1994) we characterize the geometric structure of the resolved strains using the so-called strain state parameter

$$s^* = \frac{-3\sqrt{6}\,\alpha_{\tilde{S}}\beta_{\tilde{S}}\gamma_{\tilde{S}}}{\left(\alpha_{\tilde{S}}^2 + \beta_{\tilde{S}}^2 + \gamma_{\tilde{S}}^2\right)^{3/2}}. \tag{3.11}$$

The strain state parameter is useful since it indicates the type of motions occurring at the location of the measured filtered strain rate. For example, $s^* = 1$ corresponds to axisymmetric extension (i.e., $\alpha_{\tilde{S}} = \beta_{\tilde{S}} > 0$, $\gamma_{\tilde{S}} < 0$), $s^* = 0$ corresponds to plane strain (i.e., $\beta_{\tilde{S}} = 0$), and $s^* = -1$ corresponds to axisymmetric contraction (i.e., $\alpha_{\tilde{S}} > 0$, $\beta_{\tilde{S}} = \gamma_{\tilde{S}} < 0$). Similar expressions exist to relate the non-dimensional eigenvalues of the SGS stress to the stress state parameter $s^*_{-\tau} = -3\sqrt{6}\alpha_{-\tau}\beta_{-\tau}\gamma_{-\tau}(\alpha_{-\tau}^2 + \beta_{-\tau}^2 + \gamma_{-\tau}^2)^{-3/2}$. The strain state parameter, s^*, is bounded between -1 and 1 for incompressible flow, and the stress state parameter, $s^*_{-\tau}$, is bounded between -1 and 1 when the deviatoric part of the SGS stress $\tau_{ij}^{\mathrm{d}} = \tau_{ij} - \frac{1}{3}\tau_{kk}\delta_{ij}$ is used instead of τ_{ij} in the analysis. Inverse relations also exist

(Lund and Rogers, 1994) that express the non-dimensional eigenvalues in terms of the structure parameter:

$$\beta_{\tilde{S}}^* = \frac{\sqrt{6}\beta_{\tilde{S}}}{\sqrt{\alpha_{\tilde{S}}^2 + \beta_{\tilde{S}}^2 + \gamma_{\tilde{S}}^2}} = 2\cos\left(\frac{5}{3}\pi + \frac{1}{3}\cos^{-1}(s^*)\right),$$

$$\alpha_{\tilde{S}}^* = \frac{\sqrt{6}\alpha_{\tilde{S}}}{\sqrt{\alpha_{\tilde{S}}^2 + \beta_{\tilde{S}}^2 + \gamma_{\tilde{S}}^2}} = -\cos\left(\frac{5}{3}\pi + \frac{1}{3}\cos^{-1}(s^*)\right)$$

$$+ \sqrt{3}\left|\sin\left(\frac{5}{3}\pi + \frac{1}{3}\cos^{-1}(s^*)\right)\right|,$$

$$\gamma_{\tilde{S}}^* = \frac{\sqrt{6}\gamma_{\tilde{S}}}{\sqrt{\alpha_{\tilde{S}}^2 + \beta_{\tilde{S}}^2 + \gamma_{\tilde{S}}^2}} = -\cos\left(\frac{5}{3}\pi + \frac{1}{3}\cos^{-1}(s^*)\right)$$

$$- \sqrt{3}\left|\sin\left(\frac{5}{3}\pi + \frac{1}{3}\cos^{-1}(s^*)\right)\right|. \tag{3.12}$$

Several previous studies of the full three-dimensional structure of SGS dissipation and alignment between the filtered strain-rate tensor and the SGS stress tensor eigendirections have been performed. Tao *et al.* (2002) studied alignments in the turbulent flow in a square duct using holographic particle image velocimetry (HPIV). Higgins *et al.* (2003) studied the flow in the unstable atmospheric boundary layer with arrays of sonic anemometers. Horiuti (2001) used DNS to study alignments. Despite the large disparity in length scales and flow conditions between the studies, they showed strikingly similar qualitative results with preferred orientations of the eigenvectors of $\tilde{S}_{ij}$ and τ_{ij} (see Section 3.4 for further discussion of their results).

In this chapter we present data from the same experimental setup as used in Higgins *et al.* (2003) and discussed in detail in Porté-Agel *et al.* (2001). Two vertically separated horizontal arrays of sonic anemometers were deployed in Davis, California, to obtain spatial measurements of the temperature, T, and the full three-component velocity vector. The upper array contained five sonic anemometers while the lower array contained seven sonic anemometers. Horizontal separations between sonics was 0.4 m and the vertical spacing between the two arrays was 0.51 m. Data were acquired at a temporal resolution of 20 Hz. The friction velocity $u_* = (\langle u'w' \rangle^2 + \langle v'w' \rangle^2)^{1/4}$ and the Monin–Obukhov length $L = \frac{-\langle T \rangle u_*^3}{\kappa g \langle T'w' \rangle}$ were used to classify the data into subsets according to the values of z/L, where z is the average height of the sensors above the ground ($z = 3.9$ m). Primes denote fluctuating quantities, $\langle \ldots \rangle$ represents averaging over time, κ is von Kármán's constant ($\kappa = 0.4$) and g is the acceleration of gravity. Atmospheric conditions are classified as having near neutral stability when $|z/L| \leq 0.02$. The friction velocity for the data used was $u_* = 0.27$ m s^{-1}. The present segment represents about 30 minutes of data and $\sim 40,000$ time realizations, and is distinctly different from the data used in Higgins *et al.* (2003), who used data collected from the convectively unstable atmosphere.

For the direct calculation of the SGS stress, the field measurements are filtered in the spanwise horizontal direction with a box filter, and in the streamwise direction with a Gaussian filter. Taylor's hypothesis is invoked to convert the temporal data record into a streamwise spatial record for streamwise filtering. The filter size, Δ, used throughout the present work corresponds to five times the instrument spacing, i.e., $\Delta \cong 2$ m. For the purposes of this analysis, this scale is considered to fall below the turbulence integral scale (since $\Delta/z < 1$). Spectra shown in Higgins *et al.* (2003) confirm that $\Delta = 2$ m falls broadly within the $k^{-5/3}$ region. No filtering is performed in the vertical direction. For consistency, gradients are calculated with finite differences over a distance of approximately $\Delta/5$ in all three directions. Then $\tilde{S}_{ij}$ and τ_{ij} are computed according to their respective definitions: $\tilde{S}_{ij} = \frac{1}{2}(\partial_j \tilde{u}_i + \partial_i \tilde{u}_j)$, and $\tau_{ij} = \widetilde{u_i u}_j - \tilde{u}_i \tilde{u}_j$. For a complete description of this approximate filtering technique and applications to atmospheric datasets, see Porté-Agel *et al.* (2001), Tong *et al.* (1999), and Horst *et al.* (2004). For applications to wind-tunnel laboratory data from arrays of hot-wire anemometers, see Cerutti and Meneveau (2000), and Kang and Meneveau (2002).

A probability distribution function (PDF) of non-dimensional dissipation for atmospheric sonic anemometer data under near neutral stability is presented in Fig. 3.1. Figure 3.2 shows the conditional PDF of Π^* as function of the parameter

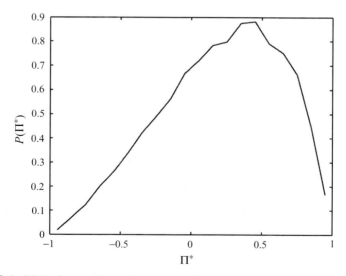

Figure 3.1. PDF of normalized dissipation, Π^*, from the near-neutral atmospheric surface layer. The mean normalized dissipation is positive ($\langle \Pi^* \rangle = 0.2$) and the most likely normalized dissipation is 0.4.

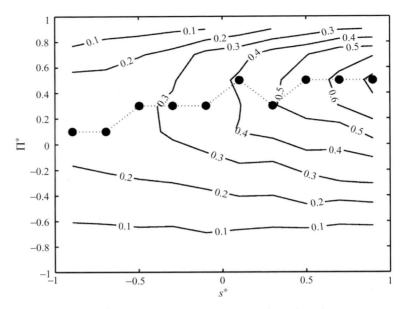

Figure 3.2. Joint PDF of normalized dissipation, Π^*, and strain state parameter. To calculate the joint PDF, the s^* axis was divided into 10 bins and the Π^* axis was divided into 10 bins. The mode normalized dissipation, conditioned on strain state parameter (represented by the symbols and dotted line), tends to increase as s^* increases. Similar trends were shown in Tao *et al.* (2002), but for a different normalization of SGS dissipation.

s^*. Also shown as symbols and dotted line is the mode value of Π^* at given s^*. It shows that the SGS dissipation tends to increase in regions of large s^*, where resolved motions are of the axisymmetric extension type. In the following section we seek to understand this trend in terms of preferred orientations among the two tensors.

Figure 3.3 shows the PDFs of the two structure parameters s^* and $s^*_{-\tau}$ obtained from the present data. Both PDFs peak at $s^* = s^*_{-\tau} = 1$, indicating preferential occurrence of axisymmetric extensional motions, and a preferential axisymmetric contractive stress field. Probability density functions of s^* and $s^*_{-\tau}$ were also presented in both Tao *et al.* (2002) and Higgins *et al.* (2003). Both studies showed that the most likely strain-rate state correspond to $s^* = 1$. A most likely value of $s^* = 1$ was also obtained from DNS of unfiltered turbulence at lower Reynolds numbers and smaller scales by Lund and Rogers (1994) and from multi-component hot-wire data by Tsinober *et al.* (1992). Also in agreement with present results, Tao *et al.* (2002) and Higgins *et al.* (2003) found that the most likely state of the negative SGS stress is $s^*_{-\tau} = 1$. Note that the peak is particularly pronounced for the SGS stress structure, where more than half the data correspond to $s^*_{-\tau} > 0.64$.

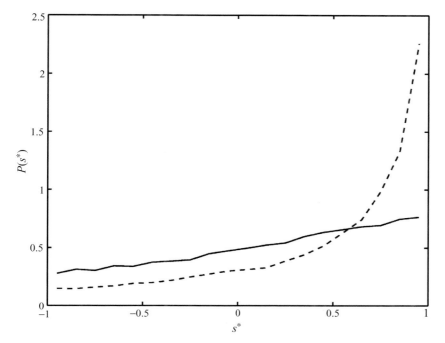

Figure 3.3. PDFs of s^* (solid line) and $s^*_{-\tau}$ (dashed line) showing that both have a most likely value of 1. This indicates that the eigenvectors of both the filtered strain rate, $\tilde{S}_{ij}$, and the negative SGS stress, $-\tau_{ij}$, are in a state of axisymmetric extension. This behavior was already noted in DNS data (Lund and Rogers, 1994), hot-wire anemometer data (Tsinober *et al.*, 1992), HPIV data in a square duct (Tao *et al.*, 2002) and in the atmospheric surface layer under unstable conditions (Higgins *et al.*, 2003).

3.4 Geometric view of stress–strain rate correlation

The contraction of (3.10) can be expanded in terms of the eigenvalues and eigenvectors of the two tensors as follows:

$$
\begin{aligned}
\Pi = \; & \alpha_{\tilde{S}}\alpha_{-\tau}(\boldsymbol{\alpha}_{\tilde{S}}, \boldsymbol{\alpha}_{-\tau})^2 + \alpha_{\tilde{S}}\beta_{-\tau}(\boldsymbol{\alpha}_{\tilde{S}}, \boldsymbol{\beta}_{-\tau})^2 + \alpha_{\tilde{S}}\gamma_{-\tau}(\boldsymbol{\alpha}_{\tilde{S}}, \boldsymbol{\gamma}_{-\tau})^2 \\
& + \beta_{\tilde{S}}\alpha_{-\tau}(\boldsymbol{\beta}_{\tilde{S}}, \boldsymbol{\alpha}_{-\tau})^2 + \beta_{\tilde{S}}\beta_{-\tau}(\boldsymbol{\beta}_{\tilde{S}}, \boldsymbol{\beta}_{-\tau})^2 + \beta_{\tilde{S}}\gamma_{-\tau}(\boldsymbol{\beta}_{\tilde{S}}, \boldsymbol{\gamma}_{-\tau})^2 \\
& + \gamma_{\tilde{S}}\alpha_{-\tau}(\boldsymbol{\gamma}_{\tilde{S}}, \boldsymbol{\alpha}_{-\tau})^2 + \gamma_{\tilde{S}}\beta_{-\tau}(\boldsymbol{\gamma}_{\tilde{S}}, \boldsymbol{\beta}_{-\tau})^2 + \gamma_{\tilde{S}}\gamma_{-\tau}(\boldsymbol{\gamma}_{\tilde{S}}, \boldsymbol{\gamma}_{-\tau})^2, \quad (3.13)
\end{aligned}
$$

where $(\boldsymbol{\alpha}, \boldsymbol{\beta})$ denotes the cosine of the angle between two vectors $\boldsymbol{\alpha}$ and $\boldsymbol{\beta}$. Non-dimensionalizing with the SGS stress and strain-rate magnitudes yields:

$$
\begin{aligned}
\Pi^* = \frac{1}{6} \big[& \alpha_{\tilde{S}}^*\alpha_{-\tau}^*(\boldsymbol{\alpha}_{\tilde{S}}, \boldsymbol{\alpha}_{-\tau})^2 + \alpha_{\tilde{S}}^*\beta_{-\tau}^*(\boldsymbol{\alpha}_{\tilde{S}}, \boldsymbol{\beta}_{-\tau})^2 + \alpha_{\tilde{S}}^*\gamma_{-\tau}^*(\boldsymbol{\alpha}_{\tilde{S}}, \boldsymbol{\gamma}_{-\tau})^2 \\
& + \beta_{\tilde{S}}^*\alpha_{-\tau}^*(\boldsymbol{\beta}_{\tilde{S}}, \boldsymbol{\alpha}_{-\tau})^2 + \beta_{\tilde{S}}^*\beta_{-\tau}^*(\boldsymbol{\beta}_{\tilde{S}}, \boldsymbol{\beta}_{-\tau})^2 + \beta_{\tilde{S}}^*\gamma_{-\tau}^*(\boldsymbol{\beta}_{\tilde{S}}, \boldsymbol{\gamma}_{-\tau})^2 \\
& + \gamma_{\tilde{S}}^*\alpha_{-\tau}^*(\boldsymbol{\gamma}_{\tilde{S}}, \boldsymbol{\alpha}_{-\tau})^2 + \gamma_{\tilde{S}}^*\beta_{-\tau}^*(\boldsymbol{\gamma}_{\tilde{S}}, \boldsymbol{\beta}_{-\tau})^2 + \gamma_{\tilde{S}}^*\gamma_{-\tau}^*(\boldsymbol{\gamma}_{\tilde{S}}, \boldsymbol{\gamma}_{-\tau})^2 \big]. \quad (3.14)
\end{aligned}
$$

Equation (3.14) contains nine distinct angles (inner products) and six eigenvalues. Yet, the alignment between the eigenvectors of two symmetric tensors is fixed with only three Euler angles. Instead of Euler angles, which do not give uniform probability densities when computing their joint probability distributions for random data, we follow the approach of Tao *et al.* (2002) who introduced three specific angles that have uniform measure for random data. To reduce the degrees of freedom in the dissipation equation, we must express each of the nine individual dot products in (3.14) as a function of these three distinct angles. Briefly, the analysis performed by Tao *et al.* (2002) and Higgins *et al.* (2003) fixed the relative orientation between two tensors with a triplet of angles:

$$\theta = \cos^{-1}|(\boldsymbol{\alpha}_{-\tau}, \boldsymbol{\alpha}_{\tilde{s}})|, \phi = \cos^{-1}\left(|(\boldsymbol{\alpha}_{-\tau}^{\mathrm{p}}, \boldsymbol{\beta}_{\tilde{s}})||\boldsymbol{\alpha}_{-\tau}^{\mathrm{p}}|^{-1}\right),$$

$$\zeta = \cos^{-1}\left(|(\boldsymbol{\gamma}_{\tilde{s}}^{\mathrm{p}}, \boldsymbol{\gamma}_{-\tau})||\boldsymbol{\gamma}_{\tilde{s}}^{\mathrm{p}}|^{-1}\right).$$

Here $\boldsymbol{\alpha}_{-\tau}^{\mathrm{p}}$ is the projection of $\boldsymbol{\alpha}_{-\tau}$ onto the $\boldsymbol{\gamma}_{\tilde{s}} - \boldsymbol{\beta}_{\tilde{s}}$ plane and $\boldsymbol{\gamma}_{\tilde{s}}^{\mathrm{p}}$ is the projection of $\boldsymbol{\gamma}_{\tilde{s}}$ onto the $\boldsymbol{\gamma}_{-\tau} - \boldsymbol{\beta}_{-\tau}$ plane. The angle triplets were calculated for each point in the dataset, and then a 3D joint probability density function of the three angles was computed. By interpreting the modes in the joint PDF, Tao *et al.* (2002), and Higgins *et al.* (2003) were able to deduce the most likely relative orientation of the SGS stress with the filtered strain-rate eigendirections.

To simplify the trigonometry required to express the nine inner products in (3.14) in terms of the three above angles, we circumscribe the set of eigendirections given by the filtered strain rate, and the SGS stress with the unit sphere. Each eigenvector is a unit vector; therefore, each eigenvector can be represented as a point on the unit sphere. The intersection of the sphere and a plane defined by any two eigenvectors forms a great circle that connects the two respective points on the sphere. The arc-length between two points (defined by a great circle) on the unit sphere is identical to the angle between the corresponding vectors. With spherical geometry, the problem is no longer one of finding angles in Cartesian coordinates, but is instead finding distances on the unit sphere. Once this transformation is made, we can use the standard tools of spherical trigonometry to find distances on the sphere. The Law of Cosines for spherical triangles is given by:

$$\cos a = \cos b \cos c + \sin b \sin c \cos A. \tag{3.15}$$

Lower-case letters represent the sides of the spherical triangle and upper-case letters represent the angles opposite of their respective side. The Law of Cosines for spherical triangles will be used to express all of the dot products in (3.14) as functions of the known angle triplet. To complete the final formulation, and to make the geometry as general as possible, it is necessary to redefine the angles used by Tao *et al.* (2002) and Higgins *et al.* (2003) so that a point can be located anywhere on the sphere relative to an eigenvector coordinate system. We will use the following definitions for the angles:

$$\theta = \cos^{-1}(\alpha_{\tilde{s}}, \alpha_{-\tau}), \tag{3.16}$$

$$\phi = (\gamma_{\tilde{s}} \times \beta_{\tilde{s}}) \cdot (\alpha^{\text{p}}_{-\tau} \times \alpha_{\tilde{s}}) \cos^{-1} \frac{(\alpha^{\text{p}}_{-\tau}, \beta_{\tilde{s}})}{|\alpha^{\text{p}}_{-\tau}|}, \tag{3.17}$$

$$\zeta = (\gamma_{-\tau} \times \beta_{-\tau}) \cdot (\gamma^{\text{p}}_{\tilde{s}} \times \beta_{-\tau}) \cos^{-1} \frac{(\gamma^{\text{p}}_{\tilde{s}}, \gamma_{-\tau})}{|\gamma^{\text{p}}_{\tilde{s}}|}. \tag{3.18}$$

The above definitions ensure that the angles are defined relative to a consistent coordinate system, and can vary from $-\pi$ to π. The nine dot products in (3.14) are now given by the following set of equations:

$$(\alpha_{-\tau}, \alpha_{\tilde{s}})^2 = \cos^2\theta, \tag{3.19}$$

$$(\gamma_{\tilde{s}}, \alpha_{-\tau})^2 = \sin^2\theta \sin^2\phi, \tag{3.20}$$

$$(\gamma_{-\tau}, \gamma_{\tilde{s}})^2 = \cos^2\zeta(1 - \sin^2\theta \sin^2\phi), \tag{3.21}$$

$$(\gamma_{-\tau}, \alpha_{\tilde{s}})^2 = \frac{(\cos\theta \sin\phi \cos\zeta + \cos\phi \sin\zeta)^2 \sin^2\theta}{1 - \sin^2\theta \sin^2\phi}, \tag{3.22}$$

$$(\beta_{-\tau}, \gamma_{\tilde{s}})^2 = 1 - (\alpha_{-\tau}, \gamma_{\tilde{s}})^2 - (\gamma_{-\tau}, \gamma_{\tilde{s}})^2, \tag{3.23}$$

$$(\beta_{\tilde{s}}, \alpha_{-\tau})^2 = 1 - (\alpha_{\tilde{s}}, \alpha_{-\tau})^2 - (\gamma_{\tilde{s}}, \alpha_{-\tau})^2, \tag{3.24}$$

$$(\beta_{\tilde{s}}, \gamma_{-\tau})^2 = 1 - (\gamma_{-\tau}, \alpha_{\tilde{s}})^2 - (\gamma_{-\tau}, \gamma_{\tilde{s}})^2, \tag{3.25}$$

$$(\beta_{-\tau}, \alpha_{\tilde{s}})^2 = 1 - (\alpha_{\tilde{s}}, \alpha_{-\tau})^2 - (\gamma_{-\tau}, \alpha_{-\tau})^2, \tag{3.26}$$

$$(\beta_{-\tau}, \beta_{\tilde{s}})^2 = 1 - (\beta_{\tilde{s}}, \alpha_{-\tau})^2 - (\beta_{\tilde{s}}, \gamma_{-\tau})^2. \tag{3.27}$$

Equation (3.14) is first simplified by using the angle relationships in (3.23)–(3.27) (those relationships do not require any predefined angles or geometry) and we are left with

$$\Pi^* = \frac{1}{6}[(\alpha_{\tilde{s}}, \alpha_{-\tau})^2(\alpha^*_{\tilde{s}} - \beta^*_{\tilde{s}})(\alpha^*_{-\tau} - \beta^*_{-\tau}) + (\alpha_{\tilde{s}}, \gamma_{-\tau})^2(\alpha^*_{\tilde{s}} - \beta^*_{\tilde{s}})(\gamma^*_{-\tau} - \beta^*_{-\tau})$$
$$+ (\gamma_{\tilde{s}}, \alpha_{-\tau})^2(\gamma^*_{\tilde{s}} - \beta^*_{\tilde{s}})(\alpha^*_{-\tau} - \beta^*_{-\tau}) + (\gamma_{\tilde{s}}, \gamma_{-\tau})^2(\gamma^*_{\tilde{s}} - \beta^*_{\tilde{s}})(\gamma^*_{-\tau} - \beta^*_{-\tau})$$
$$- 3\beta^*_{\tilde{s}}\beta^*_{-\tau}], \tag{3.28}$$

which will be the starting point of our analysis. To give a complete picture of the final equation form we express all non-dimensional eigenvalues in terms of s^* and

$s^*_{-\tau}$ using the relationships in (3.12), and the angle relations in (3.19)–(3.22):

$$
\Pi^* = \frac{1}{6}[-3\cos(\Gamma) + \sqrt{3}|\sin(\Gamma)|][-3\cos(\Gamma_\tau) + \sqrt{3}|\sin(\Gamma_\tau)|]\cos^2\theta
$$

$$
+ \frac{1}{6}[-3\cos(\Gamma) + \sqrt{3}|\sin(\Gamma)|][-3\cos(\Gamma_\tau) - \sqrt{3}|\sin(\Gamma_\tau)|]
$$

$$
\times \frac{(\cos\theta\sin\phi\cos\zeta + \cos\phi\sin\zeta)^2\sin^2\theta}{1 - \sin^2\theta\sin^2\phi}
$$

$$
+ \frac{1}{6}[-3\cos(\Gamma) - \sqrt{3}|\sin(\Gamma)|][-3\cos(\Gamma_\tau) + \sqrt{3}|\sin(\Gamma_\tau)|]\sin^2\theta\sin^2\phi
$$

$$
+ \frac{1}{6}[-3\cos(\Gamma) - \sqrt{3}|\sin(\Gamma)|][-3\cos(\Gamma_\tau) - \sqrt{3}|\sin(\Gamma_\tau)|]
$$

$$
\times \cos^2\zeta(1 - \sin^2\theta\sin^2\phi) - 12\cos\Gamma\cos\Gamma_\tau, \tag{3.29}
$$

where $\Gamma = \frac{5}{3}\pi - \frac{1}{3}\cos^{-1}s^*$ and $\Gamma_\tau = \frac{5}{3}\pi - \frac{1}{3}\cos^{-1}s^*_{-\tau}$. Equation (3.29) can be used to investigate the effect of alignment and stress/strain state on dissipation; however, for simplicity, (3.28) is a more natural starting point.

Recall that by definition $\alpha \geq \beta \geq \gamma$. This set of inequalities allows us to determine the signs of the terms containing angles in (3.28):

$$
(\boldsymbol{\alpha}_{\tilde{S}}, \boldsymbol{\alpha}_{-\tau})^2(\alpha^*_{\tilde{S}} - \beta^*_{\tilde{S}})(\alpha^*_{-\tau} - \beta^*_{-\tau}) \geq 0
$$

$$
(\boldsymbol{\alpha}_{\tilde{S}}, \boldsymbol{\gamma}_{-\tau})^2(\alpha^*_{\tilde{S}} - \beta^*_{\tilde{S}})(\gamma^*_{-\tau} - \beta^*_{-\tau}) \leq 0
$$

$$
(\boldsymbol{\gamma}_{\tilde{S}}, \boldsymbol{\alpha}_{-\tau})^2(\gamma^*_{\tilde{S}} - \beta^*_{\tilde{S}})(\alpha^*_{-\tau} - \beta^*_{-\tau}) \leq 0
$$

$$
(\boldsymbol{\gamma}_{\tilde{S}}, \boldsymbol{\gamma}_{-\tau})^2(\gamma^*_{\tilde{S}} - \beta^*_{\tilde{S}})(\gamma^*_{-\tau} - \beta^*_{-\tau}) \geq 0. \tag{3.30}
$$

With these constraints, we can deduce alignments of filtered strain-rate and SGS stress eigendirections that maximize or minimize energy dissipation for all possible stress/strain states (s^* and $s^*_{-\tau}$). Eliminating negative terms $(\boldsymbol{\alpha}_{\tilde{S}}, \boldsymbol{\gamma}_{-\tau}) = (\boldsymbol{\gamma}_{\tilde{S}}, \boldsymbol{\alpha}_{-\tau}) = 0$ and maximizing positive terms $(\boldsymbol{\alpha}_{\tilde{S}}, \boldsymbol{\alpha}_{-\tau}) = (\boldsymbol{\gamma}_{\tilde{S}}, \boldsymbol{\gamma}_{-\tau}) = 1$ will yield a maximum dissipation for all possible states of the stress or strain. This maximum is of course attained by the alignment corresponding to the eddy-viscosity model (see Fig. 3.5(a)). The resulting normalized dissipation is given by:

$$
\Pi^* = \frac{1}{6}(\alpha^*_{\tilde{S}}\alpha^*_{-\tau} + \beta^*_{\tilde{S}}\beta^*_{-\tau} + \gamma^*_{\tilde{S}}\gamma^*_{-\tau}) = \cos\Gamma\cos\Gamma_\tau + |\sin\Gamma\sin\Gamma_\tau|. \tag{3.31}
$$

Equation (3.31) represents an eddy-viscosity behavior, $\Pi^* = 1$, only when $\Gamma_\tau = \Gamma$ (i.e., the stress state and strain state parameters are equal). A plot of the maximum dissipation for all stress–strain state combinations is shown as the upper surface in Fig. 3.4. The short thick line on the upper surface in Fig. 3.4 represents the eddy-viscosity model.

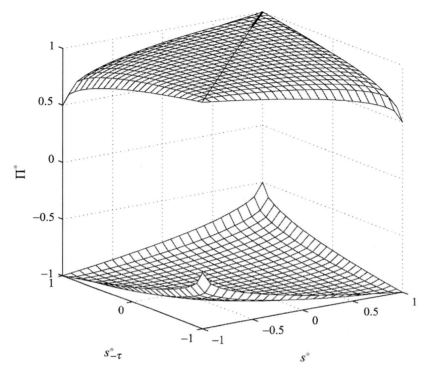

Figure 3.4. Upper and lower bounds on normalized dissipation for each strain state and SGS stress state combination as deduced from (3.28). The diagonal line at the upper surface corresponds to the eddy-viscosity closure. This line is also the global maximum of dissipation.

Note that the alignment that produces a minimum bound on the normalized dissipation for all SGS stress states and strain states can also be deduced from (3.28). If an alignment is chosen so that the positive terms are eliminated, i.e., $(\alpha_{\tilde{S}}, \alpha_{-\tau}) = (\gamma_{\tilde{S}}, \gamma_{-\tau}) = 0$, and the negatives are maximized, i.e., $(\alpha_{\tilde{S}}, \gamma_{-\tau}) = (\gamma_{\tilde{S}}, \alpha_{-\tau}) = 1$, we will have set the lower bound on dissipation for all possible SGS stress and strain state combinations. The alignment that yields this minimum is when the contractive direction of the filtered strain-rate tensor, $\gamma_{\tilde{S}}$, is aligned with the extensive direction of the (negative) SGS stress tensor, $\alpha_{-\tau}$, and the two intermediate eigendirections are aligned. An interpretive sketch of this alignment is presented in Fig. 3.5(b). Such an alignment yields a normalized dissipation given by:

$$\Pi^* = \frac{1}{6}(\alpha_{\tilde{S}}^* \gamma_{-\tau}^* + \beta_{\tilde{S}}^* \beta_{-\tau}^* + \gamma_{\tilde{S}}^* \alpha_{-\tau}^*) = \cos \Gamma \cos \Gamma_\tau - |\sin \Gamma \sin \Gamma_\tau| \quad (3.32)$$

which is the minimum for all possible stress–strain state combinations. A plot of this minimum is presented as the lower surface in Fig. 3.4.

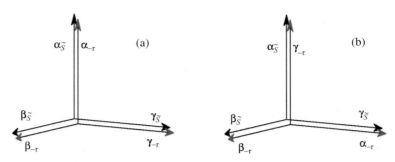

Figure 3.5. Interpretive sketches that give rise to (a) maximum and (b) minimum dissipation. The maximum dissipation is achieved when the vector alignment implied by the eddy-viscosity model is realized (a). The minimum dissipation is achieved by keeping the intermediate eigendirections aligned and pairing extensive and contracting directions together (b). Global minimization of the dissipation function also requires that a functional relationship between strain state and SGS stress state be specified. The global maximum dissipation occurs when the eigenvectors are aligned as in (a) and $s^* = s^*_{-\tau}$. This is achieved when $\tau_{ij} = -\lambda \tilde{S}_{ij}$. The global minimum occurs when the eigenvectors are aligned as in (b) and $s^* = -s^*_{-\tau}$. This is achieved when $\tau_{ij} = \lambda \tilde{S}_{ij}$.

The form of (3.28) can be further simplified if either the filtered strain rate or the SGS stress exhibits axisymmetric contraction or extension. For example, in the most likely case of axisymmetric extension in the filtered strain rate, $s^* = 1$, the non-dimensional eigenvalues have the property $\alpha^*_{\tilde{S}} = \beta^*_{\tilde{S}}$. Two of the required angles then drop from (3.28). When the tensor's eigenvector composition is axisymmetric, all of the directional information is described by the axis of symmetry, including the dissipation.

3.5 Dissipation from observed alignments

Tao *et al.* (2002) and Higgins *et al.* (2003) found two relative orientations of the filtered strain rate and the SGS stress that are highly likely. The two alignments that these studies reported are shown in Fig. 3.6. The atmospheric data used in our study did not contain sufficient points to allow us to obtain statistically converged joint PDFs of the three angles and so we rely on these earlier results. The alignment configuration of Fig. 3.6(a) represents the primary configuration, while the alignment in Fig. 3.6(b) represents the secondary configuration. Each corresponds to a unique alignment of the eigenvectors, but in both alignment configurations, the angle between the two contracting directions (the angle between $\gamma_{\tilde{S}}$ and $\gamma_{-\tau}$) is approximately the same (about 30°), and the contracting direction $\gamma_{-\tau}$ is perpendicular to the intermediate direction of the filtered strain rate, $\beta_{\tilde{S}}$.

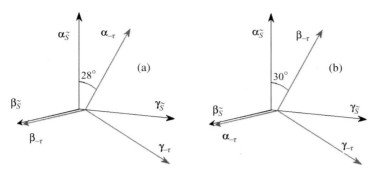

Figure 3.6. (a) Primary and (b) secondary alignment configurations between the filtered strain rate and the SGS stress reported by Tao *et al.* (2002) and Higgins *et al.* (2003). The alignments reflect a bimodal behavior with a characteristic angle between the contracting directions of approximately 30°.

In (3.29), the only variable composed of purely filtered scale quantities is the strain state parameter s^*. Therefore, to explore the dependence of dissipation on filtered scale quantities, we must choose an eigenvector alignment and a value for the SGS stress state parameter.

We have seen that $s^*_{-\tau} = 1$ is the most likely value of this parameter. Using $s^*_{-\tau} = 1$, only the alignment of the axis of symmetry, $\boldsymbol{\gamma}_{-\tau}$, with the filtered strain rate is needed to completely specify the alignment of the filtered strain rate and the SGS stress tensor. This will require only two distinct angles. As mentioned before, it was found in Tao *et al.* (2002) and Higgins *et al.* (2003) that $\boldsymbol{\gamma}_{-\tau}$ is perpendicular to $\boldsymbol{\beta}_{\tilde{S}}$ in both of the likely alignment configurations. Substituting these two conditions into (3.28) ($s^*_{-\tau} = 1$ and $\boldsymbol{\gamma}_{-\tau} \perp \boldsymbol{\beta}_{\tilde{S}}$), and using (3.25), the normalized dissipation becomes

$$\Pi^* = \frac{1}{2}[(\boldsymbol{\gamma}_{\tilde{S}}, \boldsymbol{\gamma}_{-\tau})^2(\alpha^*_{\tilde{S}} - \gamma^*_{\tilde{S}}) - \alpha^*_{\tilde{S}}], \tag{3.33}$$

which is a function of a single angle, namely the angle $(\boldsymbol{\gamma}_{\tilde{S}}, \boldsymbol{\gamma}_{-\tau})$ that is approximately the same in both peaks of the alignment PDF. We can then use the most likely value of this angle as observed from data $(\boldsymbol{\gamma}_{\tilde{S}}, \boldsymbol{\gamma}_{-\tau})^2 \approx \cos^2 30° = 0.75$ (the value reported by Higgins *et al.*, 2003). Equation (3.33) then reduces to:

$$\Pi^* = 0.25 \left\{ 2.0 \cos \left[\frac{5}{3}\pi - \frac{1}{3}\cos^{-1}(s^*) \right] + \sqrt{3} \left| \sin \left[\frac{5}{3}\pi - \frac{1}{3}\cos^{-1}(s^*) \right] \right| \right\}. \tag{3.34}$$

The value of Π^* therefore varies in a range between about 0.125 when $s^* = -1$ to about 0.625 when $s^* = 1$.

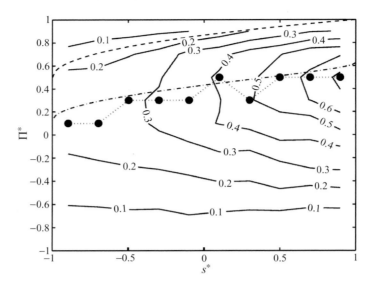

Figure 3.7. Contour plot of the joint PDF between normalized dissipation, Π^*, and the strain state parameter (solid black lines). Symbols and dotted line: the measured mode dissipation as a function of s^* (same as in Fig. 3.2). Dashed line: prediction using $s^*_{-\tau} = 1$ and the eddy-viscosity alignment. Dash–dot line: present prediction using $s^*_{-\tau} = 1$, $\gamma_{-\tau} \perp \beta_{\bar{s}}$, and $\zeta = 30°$, showing good agreement with the data.

The dash–dot line in Fig. 3.7 denotes the prediction based on the three above assumptions [Equation (3.34)], and is compared to the mode values measured from the data (symbols and dotted line, same as Fig. 3.2).

Picking larger values for $(\gamma_{\bar{s}}, \gamma_{-\tau})^2$ moves the alignment closer to the one presumed by the eddy-viscosity model. Specifically, for $s^*_{-\tau} = 1$, the prediction for perfect alignment (Fig. 3.5(a)) is shown as a dashed line (this is equal to the upper surface shown in Fig. 3.4 along the line $s^*_{-\tau} = 1$). In addition, the Smagorinsky model gives a constant prediction, $\Pi^* = 1$, for all values of s^* which represents an even greater over-prediction of dissipation. Implications of these observations in terms of improved subgrid models will be discussed in the next section.

3.6 Discussion and conclusions

In this chapter we have reviewed Lilly's ground-breaking development in LES, which recognized the central importance of the SGS dissipation, or contraction between the SGS stress and resolved strain-rate tensors. The original work focused upon the ensemble average value of the SGS dissipation as a means of deriving the model parameter for the Smagorinsky model. We remark in passing that Lilly's reasoning was extended to more complicated systems, for instance to account for buoyancy (Deardorff, 1971), or viscous effects (Voke, 1996). Another extension was proposed in Cerutti *et al.* (2000) in which the dissipation of enstrophy rather than kinetic energy was shown to be useful to quantify model parameters for hyper-viscosity models.

We have analyzed field experiment data collected under near-neutral conditions, and found that the local SGS dissipation tends to increase in regions of axisymmetric extension. In examining more closely the stress–strain alignments, we have written down an expression for the non-dimensional form of the SGS dissipation. It depends on only five independent parameters (three angles and two non-dimensional structure parameters). The expression allows the investigation of dissipation caused by a particular alignment and stress/strain state configurations (i.e., the local state of the flow). Alignments of eigendirections that give nontrivial limits on dissipation were deduced for every possible stress–strain state.

Using three observations (obtained by inspection of the data and knowledge of the alignment structure presented in Fig. 3.6) the behavior of the normalized dissipation with respect to the strain state is well reproduced. Specifically, the results from Section 3.5 imply that any attempt to reduce the dissipation estimated by the Smagorinsky model with an adjustment to the eddy-viscosity coefficient will have no effect. Recall from Section 3.4 that the eddy-viscosity model gives $\prod^* = 1$ by definition. This is a result of the local non-dimensionalization that scales out tensor magnitudes. The difference in dissipation seen here (Fig. 3.7) from that given by the Smagorinsky model is a result of structural differences in the SGS stress and the filtered strain rate only. If we wished to modify the Smagorinsky model so that it better reproduced the measurements, we would have to modify model structure through either: (1) the eigenvector alignments with a rotation matrix; or (2) by modifying the strain state, s^*, within the strain-rate eigensystem. The former is quite complicated but has the potential to produce the desired result. The latter is simpler, but the reduction in dissipation is limited to the upper surface in Fig. 3.4. Thus, it seems that manipulating the local state of strain alone cannot achieve a great enough reduction in modeled dissipation to match the measured dissipation behavior as a function of s^*. The results show the potential of interpreting turbulent parameters within a geometric framework, and make a clear and immediate connection between the local flow structure and the resulting dissipation.

Acknowledgements

It is with great pleasure that we dedicate this chapter to Professor Douglas K. Lilly. He has been a true pioneer in many areas of geophysical fluid mechanics, and in the area of LES of turbulent flows in particular. We thank F. Porté-Agel, W.E. Eichinger and M. Pahlow for their efforts in data collection and acknowledge the financial support of NSF-ATM 01300766 and EPA STAR Fellowship.

References

Cerutti, S., Meneveau, C. and Knio, O. M. (2000). Spectral and hyper eddy viscosity in high-Reynolds-number turbulence. *J. Fluid Mech.*, **421**, 307–338.

Cerutti, S. and Meneveau, C. (2000). Statistics of filtered velocity in grid and wake turbulence. *Phys. Fluids*, **12**, 1142–1165.

Deardorff, J. W. (1970). A numerical simulation of 3-dimensional turbulent channel flow at large Reynolds numbers. *J. Fluid Mech.*, **41**, 453–480.

  (1971). On the magnitude of the subgrid-scale eddy coefficient. *J. Comp. Phys.*, **7**(1), 120–133.

Higgins, C. W., Parlange, M. B. and Meneveau, C. (2003). Alignment trends of velocity gradients and subgrid-scale fluxes in the turbulent atmospheric boundary layer. *Boundary-Layer Meteorol.*, **109**(1), 59–83.

Horiuti, K. (2001). Alignment of eigenvectors for strain rate and subgrid-scale stress tensors. In *Direct and Large Eddy Simulation IV*, B. J. Geurts, R. Friedrich and O. Métais, eds., Kluwer Academic Publishers, 67–72.

Horst, T. W., Kleissl, J., Lenschow, D. H. *et al.* (2004). HATS: Field observations to obtain spatially-filtered turbulence fields from crosswind arrays of sonic anemometers in the atmospheric surface layer. Submitted to *J. Atmos. Sci.*, **61**, in press.

Kang, H. S. and Meneveau, C. (2002). Universality of large eddy simulation model parameters across a turbulent wake behind a heated cylinder. *J. Turbulence*, **3** 032 (http://jot.iop.org/).

Lilly, D. K. (1967). The representation of small-scale turbulence in numerical simulation experiments. In *Proc. IBM Scientific Computing Symposium on Environmental Sciences, November 14–16, 1966*, Thomas J. Watson Research Center, Yorktown Heights, NY, H. H. Goldstein, ed., IBM Form No. 320–1951, 195–210.

Lund, T. S. and Rogers, M. M. (1994). An improved measure of strain state probability in turbulent flows. *Phys. Fluids*, **6**(5), 1838–1847.

Mason, P. J. (1994). Large eddy simulation: A critical review of the technique. *Quart. J. Roy. Meteor. Soc.*, **120**, 1–26.

Meneveau, C. and Katz, J. (2000). Scale invariance and turbulence models for large-eddy simulation. *Ann. Rev. Fluid Mech.*, **319**, 353–385.

Novikov, E. A. (1990). The effects of intermittency on statistical characteristics of turbulence and scale similarity of breakdown coefficients. *Phys. Fluids A*, **2**(5), 814–820.

Piomelli, U. (1999). Large eddy simulation: achievements and challenges. *Prog. Aerosp. Sci.*, **35**(4), 335–362.

Pope, S. B. (2001). *Turbulent Flows*, Cambridge, UK: Cambridge University Press.

Porté-Agel, F., Parlange, M. B., Meneveau, C., Eichinger, W. E. and Pahlow, M. (2000). Subgrid-scale dissipation in the atmospheric surface layer: Effects of stability and filter dimension. *J. Hydrometeorol.*, **1**, 75–87.

Porté-Agel, F., Parlange, M. B., Meneveau, C. and Eichinger, W. E. (2001). A priori field study of the subgrid-scale heat fluxes and dissipation in the atmospheric surface layer. *J. Atmos. Sci.*, **58**, 2673–2697.

Scotti, A., Meneveau, C. and Lilly, D. K. (1993). Generalized Smagorinsky model for anisotropic turbulence. *Phys. Fluids A*, **5**(9), 2306–2308.

Smagorinsky, J. (1963). General circulation experiments with the primitive equations. I. The basic experiment. *Mon. Wea. Rev.*, **91**, 99–164.

Tao, B., Katz, J. and Meneveau, C. (2002). Statistical geometry of subgrid-scale stresses determined from holographic particle image velocimetry measurements. *J. Fluid Mech.*, **457**, 35–78.

Tong, C., Wyngaard, J. C. and Brasseur, J. G. (1999). Experimental study of the subgrid-scale stresses in the atmospheric surface layer. *J. Atmos. Sci.*, **56**, 2277–2292.

Tsinober, A., Kit, E. and Dracos, T. (1992). Experimental investigation of the field of velocity-gradients in turbulent flows. *J. Fluid Mech.*, **242**, 169–192.

Voke, P. R. (1996). Subgrid-scale modeling at low mesh Reynolds number. *Theor. Comp. Fluid Dyn.*, **8**(2), 131–143.

4

Dreams of a stratocumulus sleeper

David A. Randall and Wayne H. Schubert

Department of Atmospheric Science Colorado State University Fort Collins, USA

4.1 Foggy recollections

When we were students at UCLA in the early 1970s, the California stratus deck frequently floated over our heads. There on the beautiful campus under the clouds, we studied Doug Lilly's (1968; hereafter L68) paper about cloud-topped mixed layers under strong inversions. The paper was recommended to us by our mentor, Professor Akio Arakawa, who recognized the relevance of Lilly's insights to climate dynamics. Whereas spectacular supercells leap from the boundary layer to the tropopause in a single bound, L68 analyzed "wimpy" stratus and stratocumulus clouds that are only a few hundred meters thick and barely manage to precipitate. L68 was a "sleeper." It received little attention at first, but over the decades since then it has picked up many citations (417 as of June 2003), and it forms the groundwork for several currently thriving lines of research. L68's emergence as a classic research paper stems in part from the climatic importance of the cloud regimes it dealt with, but more importantly from the amazing prescience of Lilly's ideas and the clarity with which he expressed them.

Several ingredients, acquired over a number of years, came together in a two-week period during the summer of 1965 to produce the remarkable L68 paper. The first was personal experience and interest. Lilly's high-school physics teacher, at Sequoia Union High School in Redwood City, California, ran a weather club, which Lilly joined with enthusiasm. After becoming the club's student leader, Lilly began to evolve, by his own description, into a

weather junkie, keeping daily weather records, making forecasts, and testing their accuracy. When I learned to drive, I did something akin to tornado chasing, within the limits of California weather. Several times I took the car and drove up to the top of the Coast Range to the west of us, with a thermometer, and observed the big temperature increase when you go above the stratus layer.

The second ingredient was a simple, accurate conceptual model based on observational work, especially Petterssen's (1938) account of the radiative driving of California stratocumulus clouds. In his insightful paper, Petterssen argued that stratocumulus clouds are not caused by the direct cooling from the upwelling water, but rather that the boundary layer air is unstable and stratocumulus forms because of convection under the temperature inversion. Petterssen realized that outgoing radiation from the top of the moist layer is effective in maintaining the temperature inversion and the instability of the boundary-layer air. The third ingredient was the mixed-layer modeling approach, one of the topics of study at the second (1960) summer study program in geophysical fluid dynamics at the Woods Hole Oceanographic Institution, a program to which Lilly had been sent by his boss Joseph Smagorinsky. This remarkable program, now approaching its 45th consecutive year, afforded Lilly the chance to interact with oceanographers and fluid dynamicists, including Stewart Turner, who was applying mixed-layer modeling to the problem of the ocean thermocline. The fourth and final ingredient was moist thermodynamics, no doubt mastered by Lilly during the research leading to his dissertation "On the Theory of Disturbances in a Conditionally Unstable Atmosphere." A clear understanding of moist thermodynamics is crucial for generalizing dry mixed-layer theory so that it can apply to cloud-topped mixed layers.

In the summer of 1965, nine months after joining the National Center for Atmospheric Research (NCAR), Lilly was invited by Stanley Rosenthal to visit the National Hurricane Research Laboratory, then collocated on the University of Miami campus with the National Hurricane Center and the Rosenstiel School of Marine and Atmospheric Science. It was during this two-week visit that Lilly, spending most of his time in the university library, generated the important parts of the 1968 paper, a nearly final version of which became available in June 1967 as NCAR Manuscript No. 386. The NCAR manuscript contains more detail concerning the radiation calculations than does L68. It became evident that Lilly's approach to mixed-layer modeling was also relevant to the experimental work of James Deardorff and Glen Willis, and the paper by Deardorff *et al.* (1969) soon followed.

It is not our purpose here to comprehensively review the research edifice that the atmospheric science community has built on L68. Instead we begin, in Sections 4.2 and 4.3, by giving our perspective on why L68 is still being actively discussed 35 years after its publication. As part of this discussion, we compare modern observations of the large-scale atmospheric circulations associated with marine subtropical stratocumulus regimes with the data that was available when L68 was published. In Sections 4.4 and 4.5, we offer a new analysis of issues that were raised for the first time by L68: the effects of radiative and evaporative cooling on entrainment. Our approach is formulated in terms of what we call the "effective inversion strength."

We conclude with some comments about the relevance of L68's conceptual framework to the design of global circulation models.

4.2 Why are stratocumulus clouds so prevalent over the eastern subtropical oceans?

On the basis of climatological data from millions of ship reports during the period 1885–1933 (McDonald, 1938) and radiosonde data collected during research cruises in 1949–1952, Neiburger *et al.* (1961) described the peculiar structure of the lower troposphere off the coast of California. A strong subsidence inversion hangs a few hundred meters above the sea. Below the inversion, moist-conservative variables are vertically homogenized within a "marine layer." Especially during the summer months, a thin but remarkably persistent sheet of stratus or stratocumulus clouds hugs the inversion base. As the marine-layer air circulates towards the southwest, around the subtropical high, it encounters gradually warming sea surface temperatures (SSTs) and gradually weakening subsidence. A couple of thousand kilometers downstream from the California coast, the stratus deck breaks up into shallow cumuli.

In light of the widespread use of the Neiburger *et al.* analysis for theoretical and modeling work, it is of interest to compare their surface wind climatology with recent SeaWinds observations. SeaWinds is a microwave scatterometer on the QuikSCAT satellite, launched in July 1999. QuikSCAT streamlines and isotachs of the surface wind (calibrated to the 10 m level), averaged for July of the years 1999–2002, are shown in Plate I(a). For comparison, Fig. 35 of Neiburger *et al.* is reproduced in Plate I(b). The Neiburger *et al.* figure is a redrafted version of the northeast Pacific region in Charts 9 and 21 of McDonald (1938). A glaring omission in the Neiburger *et al.* report is that the isotach units are not given. Many users of the figure have no doubt assumed that the units are m s^{-1} or knots, but reference to the original Chart 21 of McDonald confirms that the isotachs are in Beaufort units. In Plate I(b) we have converted the Beaufort units to m s^{-1} with the arrow notation, e.g., $4 \rightarrow 7.2$ meaning 4 Beaufort units is equivalent to 7.2 m s^{-1}. The QuikSCAT data reveal that the strongest surface winds along the coast average more than 10 m s^{-1}, just off Cape Mendocino, California. A secondary maximum occurs just northwest of San Nicolas Island, site of the 1987 FIRE Marine Stratocumulus Experiment (Albrecht *et al.*, 1988). Somewhat surprisingly, these isotach maxima near Cape Mendocino and San Nicolas Island do not appear in the isotach analysis of Neiburger *et al.*[†] According to QuikSCAT data, an isotach maximum of 9 m s^{-1}

[†] In contrast to Chart 21, Charts 9 and 29 of the McDonald atlas do show isotach maxima just off the coast of northern California. Apparently, the spatial smoothing used in Chart 21 (5 degrees in latitude and longitude) is enough to eliminate the feature.

is also found over a large region of the diffluent northeasterly trades east and south of Hawaii. Another interesting feature is the shadow of light winds in the lee of the Hawaiian Islands. The Neiburger *et al.* analysis does show an isotach maximum east and south of Hawaii, but its magnitude is weaker. Because the QuikSCAT winds are the product of a remote sensing algorithm, it is of interest to investigate their accuracy. Recently, Chelton *et al.* (2001) compared QuikSCAT winds to surface winds from the TAO buoy array for three months during 1999. Their conclusion was that the QuikSCAT winds are approximately 0.74 m s^{-1} weaker than TAO-observed winds. Even more recently, Bourassa *et al.* (2003) compared QuikSCAT winds with surface wind measurements from six different research ships. Their conclusion was that the QuikSCAT winds were about 0.14 m s^{-1} weaker than the ship-observed winds. Based on the these results, we conclude that the wind speeds given in the Neiburger *et al.* climatology are probably slightly too weak and too smooth spatially.

The divergence field associated with the QuikSCAT winds of Plate I(a) is shown by the color analysis in Plate II. For comparison, the July mean divergence estimates of Neiburger *et al.* (their Fig. 37) are shown by the solid black isolines, labeled in units of 10^{-6} s^{-1}. Although the QuikSCAT surface wind data in Plate I(a) is presented at 0.5 degree latitude/longitude resolution, for clarity the QuikSCAT divergence field presented in Plate II has been subjected to a 7-element boxcar average, which effectively eliminates features smaller than approximately 150 km and also reduces the peak magnitudes. Plate II reveals that the surface winds are divergent over nearly the whole area; the exceptions are the region influenced by midlatitude disturbances in the northern part of the figure, a small region in the lee of the Hawaiian Islands, and the northern edge of the ITCZ in the southeastern part of the figure. The largest divergence is found along the North American coast, with peak values of $6.5 \times 10^{-6} \text{ s}^{-1}$. Two lobes of enhanced divergence extend westward, with a region of weak divergence between them. The northern lobe of enhanced divergence lies between $40°$ N and $45°$ N, in the surface flow accelerating eastward on the north side of the subtropical high. The southern lobe of enhanced divergence lies in the diffluent, accelerating northeasterly trades midway between Hawaii and southern California. The agreement between the QuikSCAT divergence field and the earlier estimate of Neiburger *et al.* (1961) is generally good, although the QuikSCAT data reveal a larger north–south extent of strong divergence along the coast and a more extensive northern lobe of strong divergence.

A few years after the study of Neiburger *et al.* was published, satellite imagery revealed similar cloud systems west of South America, west of Namibia in southern Africa, and to some extent west of Europe and Australia (e.g., Hubert, 1966). We began to speak of "marine subtropical stratocumulus cloud" *regimes*.

L68 explained why such regimes exist. Surface evaporation is promoted by the persistent winds of the subtropical highs. Coastal and equatorial upwelling bring cold water to the surface, chilling the air. Boundary-layer turbulence mixes the moist cool air upward, but subsidence and the strong inversion cap the mixing. If the marine layer's top climbs high enough so that the relative humidity there reaches 100%, marine subtropical stratocumulus clouds (MSCs) form. Through mechanisms explained by L68 and discussed below, cloud formation invigorates the turbulence, and so favors a deepening of the marine layer and the cloud itself.

At the top of the marine layer, the temperature increases sharply upward by 10 K or more, and the mixing ratio of water vapor decreases upward by as much as 90%, all within an interfacial zone that is much less than 100 m thick. The inversion is a "battle-front" that marks the collision of dry, high-potential-temperature air that has smoothly subsided from the upper troposphere, with humid, cool air that is being turbulently mixed upward from near the sea surface. "Hot–dry" is coming down, "cool–wet" is going up, and where they collide they agree to disagree.

It is amazing that the subsiding air of the subtropical high actually crosses the spectacular boundary at the marine-layer top. Turbulent *entrainment* prevents the interface itself from being advected downward by the subsidence. Entrainment is *not* mixing. Mixed particles move in both directions across an interface, but entrained particles move in only one direction: from the quiet air into the turbulent air. Entrainment is the active "annexation" of non-turbulent air by a growing turbulence. An observational demonstration that the marine-layer top is a region of turbulent entrainment was first provided for the northeast trades of the Pacific Ocean in the classic paper by Riehl *et al.* (1951).

As the air crosses the top of the marine layer, it is very rapidly transformed from hot–dry to cool–wet. The moistening is caused by a strong convergence of the turbulent moisture flux. The cooling is due to a strong divergence of the radiative energy flux, combined with the evaporative chilling associated with a strong convergence of the turbulent flux of liquid water; without such cooling the buoyancy force would prevent the air from descending into the marine layer.

At this point it is useful to introduce the moist static energy:

$$h \equiv c_p T + gz + Lq, \tag{4.1}$$

where c_p is the heat capacity of air at constant pressure, T is temperature, g is the acceleration of gravity, z is height, L is the latent heat of condensation, and q is the mixing ratio of water vapor. We also define the mixing ratio of total (vapor plus liquid) water,

$$r \equiv q + \ell, \tag{4.2}$$

where ℓ is the mixing ratio of liquid water. The definitions (4.1) and (4.2) are useful because h is materially conserved under both dry adiabatic and moist adiabatic processes (with or without precipitation) in the absence of radiative heating, while r is also materially conserved under both dry adiabatic and moist adiabatic processes in the absence of precipitation.

L68 showed that the sharp turbulent and radiative flux divergences and the interfacial entrainment rate satisfy two simple relations:

$$(F_h)_{\mathrm{B}} = -E\,\Delta\bar{h} + \Delta\bar{R}, \tag{4.3}$$

$$(F_r)_{\mathrm{B}} = -E\,\Delta\bar{r}. \tag{4.4}$$

Here F_η is the turbulent flux of a quantity η, the subscript B denotes a level at the top of the turbulent boundary layer, E is the entrainment mass flux, $\Delta\eta$ is the upward increase of a quantity η across the top of the marine layer, an overbar represents a horizontal average, and R is the net upward flux of energy due to radiation. A positive value of $\Delta\bar{R}$ represents intense radiative cooling within the thin interfacial layer. L68 credits Petterssen (1938) with the key insight that the longwave contribution to $\Delta\bar{R}$ (typically on the order of 50 to 100 W m^{-2}) plays a key role in the physics of MSCs. L68 emphasized that by cooling the top of the marine layer, $\Delta\bar{R}$ promotes convection below. The convection effectively distributes the effects of the radiative cooling over the entire depth of the marine layer, and in so doing it also drives the turbulence of the marine layer.

The subsidence and cold-water characteristic of MSC regimes are compatible with either a very shallow cloud-free marine layer or a somewhat deeper cloud-topped marine layer (Randall and Suarez, 1984). If a thin cloud forms, cloud processes including cloud-top radiative cooling promote faster entrainment that leads to a further deepening of the layer and a thicker cloud. Conversely, if a transient increase in subsidence pushes the top of the marine layer below the condensation level so that the cloud disappears, the rate of entrainment decreases, favoring a further decrease in the depth of the marine layer and tending to prevent the cloud layer from re-establishing itself.

L68 deals with interactions among turbulence, moist thermodynamics, and radiative transfer. This broad scope was necessary to encompass the processes at work in MSCs. It is remarkable that even today the coupling among these processes is completely ignored in many large-scale models.

4.3 How do stratocumulus clouds interact with the global circulations of the atmosphere and ocean?

With albedoes that can reach 50%, MSCs scatter back to space a lot of solar radiation that would otherwise be absorbed by the oceans, thus tending to reduce the temperature of the water below. As discussed later, the cold water is favorable for the existence of the clouds. Coupled ocean–atmosphere models that fail to simulate MSCs suffer from large positive SST errors off the west coasts of the continents (e.g., Ma *et al.*, 1994).

The cold water on the eastern side of the Pacific basin contrasts with the warm water on the western side, especially during the La Niña phase of the ENSO cycle. This SST difference induces a surface-pressure gradient that determines the strength of the low-level trade-wind flow. As discussed by Bjerknes (1966), the trades act to reinforce the SST difference. However, the excitation and eastward propagation of equatorially trapped waves in the ocean can cause significant changes in thermocline depth and SST in the eastern Pacific. When such events occur with sufficient amplitude, an El Niño ensues, with a strong warming of the eastern-Pacific SST and a decrease in subsidence there. El Niños are associated with a decrease in MSCs, primarily over the equatorial cold tongue (Deser and Wallace, 1990). This decrease in cloud cover favors a further warming of the water in the equatorial eastern Pacific, thus providing a positive feedback on the El Niño-induced warming.

As already discussed, when MSCs are present, the top of the marine layer tends to rise, so that a deeper layer of moisture is carried into the tropics. Through this mechanism, MSCs tend to increase the supply of latent heat for the deep convection of the Intertropical Convergence Zone (ITCZ).

At the same time, however, MSCs tend to produce drizzle patterns that are quite horizontally inhomogeneous. In addition, there can be significant evaporation of the drizzle in the subcloud layer. van Zanten *et al.* (2004) and Stevens *et al.* (2004) have recently discussed airborne cloud radar measurements of drizzle rates in nocturnal marine stratocumulus west and southwest of San Diego and similar ship-based measurements of stratocumulus west of South America. From their data it is now apparent that drizzle rates are highly dependent on whether the stratocumulus convection is in the form of open or closed cells. Embedded in a large region of closed cells can be a "pocket of open cells." It is in such pockets that high drizzle rates occur, with very little drizzle observed in the region where the convection is in the closed-cell form. Averaged over the many open cells in such a pocket, it appears that the surface drizzle rate can often exceed 0.5 mm day^{-1}. Since a surface drizzle rate of 0.5 mm day^{-1} is approximately 10% of the surface evaporation rate, drizzle from MSCs can, in some situations, represent a non-negligible moisture sink for the trade-wind layer. In addition, the intensity of the drizzle is regulated to a certain extent by the availability of cloud condensation nuclei (CCN); an increase in the number density of CCN causes the available liquid water to be distributed over a larger number of smaller drops, and so decreases the drizzle rate (Albrecht, 1989). The increased number of small droplets also leads to an increase in the albedo of the clouds (Twomey *et al.*, 1984), as can be seen very clearly in ship-tracks (e.g., Liu *et al.*, 2000). These two "indirect effects" of aerosols on climate have excited a lot of interest, but they are not yet widely incorporated into atmospheric general circulation models (GCMs).

4.4 What determines the entrainment rate at the top of a cloudy turbulent layer?

Starting with L68, theories have been proposed to determine the rate of cloud-top entrainment (e.g., Lewellen and Lewellen, 1998; Lock, 1998, 2000; Lilly, 2002a, b); see the recent review by Stevens (2002). Many of these theories are variations on an approach suggested by L68, wherein a closure assumption is formulated in terms of the vertically integrated "buoyancy flux," i.e., the flux of virtual temperature. The motivation for this approach is that the convection associated with the vertically integrated buoyancy flux is the primary source of the MSC's turbulence kinetic energy (TKE). Obviously the TKE must be non-zero in order for turbulent entrainment to occur, but at the same time entrainment normally tends to reduce the buoyancy flux. All other things being equal, more TKE would imply more entrainment, but more entrainment would tend to reduce the source of TKE. This line of reasoning suggests that there is a particular entrainment rate for which the system is in balance, and L68 hypothesized that this is the entrainment rate that we observe.

Cloud-top entrainment has been a highly controversial subject, in large part because there have been very few hard facts to work with. Obviously, it is important to test the entrainment theories against observations. Unfortunately, however, the entrainment rate is very difficult to measure in the field, and almost as difficult to simulate numerically, even with LES (large-eddy simulation; see Chapter 5 by Moeng *et al.* in this book). Lacking such tests, entrainment theories have the quality of "deniable plausibility," not to be confused with the "plausible deniability" that is so useful in the political arena. Even entrainment theories that are successfully tested against data or LES are sometimes little more than curve fits, unsupported by clear physical explanations.

Despite the difficulties, there is a consensus that L68 was correct in proposing that cloud-top radiative cooling is a powerful promoter of entrainment. There is also a (perhaps somewhat weaker) consensus that L68 was correct in proposing that cloud-top evaporative cooling promotes entrainment, although the strength of this effect is still vigorously debated.

Progress can be made by:

- finding entrainment "recipes" that work increasingly well, even if the reasons for their success are incompletely understood;
- subjecting entrainment theories to increasingly challenging observational and/or numerical tests;
- deriving entrainment theories from basic assumptions that are increasingly simple and physical.

We now report an example of the third approach, in which we attempt to clarify the mechanisms by which cloud-top radiative and evaporative cooling act to increase the entrainment rate.

Our starting point is an entrainment parameterization similar to the one that has been proposed for use in the absence of clouds, following the ideas of Breidenthal and Baker (1985), Siems *et al.* (1990), and Breidenthal (1992):

$$E = \frac{b_1 \rho_B \sqrt{e_M}}{1 + b_2 \max(Ri_\Delta, 0)}. \tag{4.5}$$

Here ρ is the density of the air, e_M is the TKE vertically integrated over the planetary boundary layer (PBL), and

$$Ri_\Delta = \frac{g z_M \Delta \bar{s}_v}{c_p \bar{T}_B e_M} \tag{4.6}$$

is a Richardson number. The parameters b_1 and b_2 are usually assumed to be constants.

Observations and high-resolution numerical simulations of cloud-free boundary layers heated from below (e.g., Deardorff, 1974) suggest that $b_1 \approx 0.25$ and $b_2 \approx 0.25$. When there is no capping inversion, (4.5) simplifies to

$$E = b_1 \rho_B \sqrt{e_M}. \tag{4.7}$$

This special case is interesting, in part because it is so simple. It is relevant to our later discussion.

In (4.6), z_M is the PBL depth, and s_v is the virtual dry static energy ($\Delta \bar{s}_v$ its increment across the PBL top), which is defined by

$$s_v \equiv c_p T \left(\frac{1 + (1 + \delta)q}{1 + q + \ell} \right) + gz, \tag{4.8}$$

where $\delta = (R_v - R_a)/R_a \approx 0.608$ is a non-dimensional combination of the gas constants, R_v of water vapor and R_a of dry air. With this definition, s_v measures the density of the air, taking into account the possible presence of both water vapor and liquid water. Following the spirit of L68, we write a useful approximate formula for s, which will be used later:

$$\begin{aligned} s_v &\approx c_p T + \delta \epsilon L q - \epsilon L \ell + gz \\ &= h - (1 - \delta \epsilon) L q - \epsilon L \ell, \end{aligned} \tag{4.9}$$

where $\epsilon \equiv c_p T / L$.

In (4.5), the "velocity scale" for the entrainment rate is $\sqrt{e_M}$. The TKE appears again in the definition of Ri_Δ. Equation (4.5) is suited for use in a model that explicitly predicts e_M, as do the GCMs currently in use at Colorado State University and at the University of California at Los Angeles. The Ri_Δ term in the denominator of (4.5) represents the effects of buoyancy-induced resistance to entrainment across a statically stable interface; for a given TKE, a stronger inversion leads to slower entrainment.

The buoyancy flux at the top of a clear PBL satisfies

$$(F_{sv})_B = -E \Delta \bar{s}_v. \tag{4.10}$$

Substituting from (4.5) and (4.6), we obtain

$$(F_{sv})_B = \frac{-b_1 \rho_B \sqrt{e_M} \, \Delta \bar{s}_v}{1 + b_2 \max \left(\dfrac{g z_M \Delta \bar{s}_v}{c_p \bar{T}_B e_M}, 0 \right)}. \tag{4.11}$$

For $\Delta \bar{s}_v$ sufficiently positive, this can be approximated by

$$(F_{sv})_B \approx - \left(\frac{b_1}{b_2} \right) \left(\frac{c_p \bar{T}_B}{g z_M} \right) \rho_B e_M^{3/2}. \tag{4.12}$$

According to (4.12), the buoyancy flux at the inversion base is negative and approximately independent of $\Delta \bar{s}_v$, for a given value of e_M. This is true when the inversion is sufficiently strong. From (4.11), we see that $\left(F_{s_v} \right)_B \to 0$ in the limit as $\Delta \bar{s}_v \to 0$. For $\Delta \bar{s}_v < 0$, (4.11) gives

$$(F_{sv})_B = -b_1 \rho_B \sqrt{e_M} \, \Delta \bar{s}_v > 0. \tag{4.13}$$

This means that entrainment leads directly to convection when the entrained air is negatively buoyant.

The key hypothesis of our entrainment theory is that:

> **H1:** *Equation (4.5) holds in all cases, provided that the actual Richardson number is replaced by an "effective Richardson number" based on a suitably defined "effective inversion strength" that takes into account the effects of cloud-top cooling.*

Our physical interpretation of this assumption is that, for a given value of e_M, cloud-top cooling affects the entrainment rate by reducing the buoyancy of the air that is entering the tops of the downdrafts, making it easier for that air to sink.

We pursue this approach in a step-by-step fashion. First, consider a "smoke cloud," with radiative cooling at its top, but no moisture. This useful idealization was first studied by L68, and has recently been extensively analyzed by Bretherton *et al.* (1999; hereafter B99). A smoke cloud feels the effects of cloud-top radiative cooling, but without the additional complications associated with the phase changes of water. Radiative cooling at the top of a smoke cloud chills the air that is being entrained across the inversion, so that the newly entrained air arrives in the upper part of the mixed layer with a cooler temperature, as if it had been entrained across a weaker inversion. This effect will be incorporated by use of an effective inversion strength in (4.6), the expression for the Richardson number. Blobs of radiatively chilled air drip downward into the smoke cloud under the action of the bouyancy force, thus producing an upward buoyancy flux below the cloud top, as illustrated, for example, in Fig. 8 of B99. The upward buoyancy flux favors an increased e_M. This effect is automatically taken into account in a model that predicts e_M.

To derive an expression for the effective inversion strength, we begin with an analysis of the properties of the air that is sinking in downdrafts below the inversion; the discussion is similar to one given by Randall *et al.* (1992). Define σ as the fractional area covered by the perturbation rising motion associated with the turbulent eddies in the PBL, and write

$$\bar{\psi} = \sigma \psi_u + (1 - \sigma)\psi_d, \tag{4.14}$$

where ψ is an arbitrary intensive variable, an overbar denotes a horizontal average, and the subscripts u and d denote updrafts and downdrafts, respectively. The turbulent flux of ψ satisfies

$$F_\psi = M_c(\psi_u - \psi_d), \tag{4.15}$$

where M_c is a convective mass flux. The budget of $\bar{\psi}$ for the inversion layer can be expressed as

$$(F_\psi)_B = -E(\bar{\psi}_{B+} - \bar{\psi}_B) - \int_{z_B}^{z_{B+}} \bar{S}_\psi dz, \tag{4.16}$$

where the subscript B+ denotes a level just above the turbulent layer, and $\int_{z_B}^{z_{B+}} \bar{S}_\psi dz$ represents the effects of a concentrated source or sink of $\bar{\psi}$ inside the inversion layer. Comparing (4.15) and (4.16), we find that

$$(M_c)_B \left[(\psi_u)_B - (\psi_d)_B\right] = -E(\bar{\psi}_{B+} - \bar{\psi}_B) - \int_{z_B}^{z_{B+}} \bar{S}_\psi dz. \tag{4.17}$$

The model used here is admittedly a crude caricature of nature, in the sense that it distinguishes only two categories of air, i.e., updrafts and downdrafts, each with distinct thermodynamic properties. Obviously a real cloud layer has a much more complex structure. Nevertheless, updrafts, downdrafts, and mass fluxes do exist in nature, and our goal is to represent them simply but explicitly in our modeling framework.

We now introduce a mixing parameter, χ_E, defined by

$$(\psi_d)_B = \chi_E \bar{\psi}_{B+} + (1 - \chi_E)(\psi_u)_B + \lambda \int_{z_B}^{z_{B+}} \bar{S}_\psi dz, \tag{4.18}$$

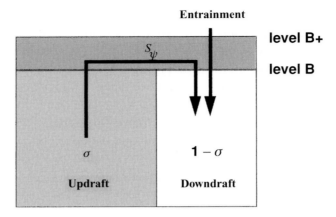

Figure 4.1. Sketch used to interpret (4.18). See text for details.

where λ is a (dimensional) coefficient to be determined below. [Note that definition (4.18) differs slightly from that used by Randall *et al.* (1992).] For $S_\psi = 0$, (4.18) reduces to

$$(\psi_{\text{d}})_{\text{B}} = \chi_E \bar{\psi}_{\text{B}+} + (1 - \chi_E)(\psi_{\text{u}})_{\text{B}} . \tag{4.19}$$

An interpretation of (4.19) is that the air found at the top of the downdraft consists of a mixture of newly entrained above-inversion air with property $\bar{\psi}_{\text{B}+}$, and air from the top of the updraft, with property $(\psi_{\text{u}})_{\text{B}}$. In other words, χ_E can be interpreted as the "mixing fraction" of newly entrained above-inversion air that contributes to $(\psi_{\text{d}})_{\text{B}}$. Obviously, for $E = 0$ we expect $\chi_E = 0$. The λ term of (4.18) represents modification of ψ by the source/sink $\bar{S}_\psi$, before the air begins to descend at the top of the downdraft. This conceptual framework is sketched in Fig. 4.1.

Using (4.16), we can rewrite (4.18) as

$$(\psi_{\text{d}})_{\text{B}} - (\psi_{\text{u}})_{\text{B}} = \frac{\chi_E(\bar{\psi}_{\text{B}+} - \bar{\psi}_{\text{B}}) + \lambda \int_{z_{\text{B}}}^{z_{\text{B}+}} \bar{S}_\psi dz}{1 - \chi_E(1 - \sigma_{\text{B}})} . \tag{4.20}$$

Comparison of (4.17) and (4.20) shows that

$$\left(\frac{(M_{\text{c}})_{\text{B}} \chi_E}{1 - \chi_E(1 - \sigma_{\text{B}})} - E\right)(\bar{\psi}_{\text{B}+} - \bar{\psi}_{\text{B}}) + \left(\frac{(M_{\text{c}})_{\text{B}} \lambda}{1 - \chi_E(1 - \sigma_{\text{B}})} - 1\right)\int_{z_{\text{B}}}^{z_{\text{B}+}} \bar{S}_\psi dz = 0. \tag{4.21}$$

Equation (4.21) holds for any ψ, including particular choices of ψ for which $S_\psi = 0$. In the case $S_\psi = 0$, (4.21) reduces to

$$\frac{(M_{\text{c}})_{\text{B}} \chi_E}{1 - \chi_E(1 - \sigma_{\text{B}})} - E = 0. \tag{4.22}$$

The quantities that appear in (4.22) are defined without reference to any particular ψ. Therefore, (4.22) must hold for any choice of ψ for which $S_\psi = 0$. We could use (4.22) to relate E to $(M_c)_B$ for given values of χ_E and σ_B. In particular, (4.22) tells us that $E = 0$ and $\chi_E = 0$ go together, as expected. For $\chi_E \ll 1$, (4.22) can be approximated by the beautifully simple relationship

$$(M_c)_B \, \chi_E \approx E. \tag{4.23}$$

In order that (4.22) be consistent with (4.21) in all cases, we must choose

$$\frac{(M_c)_B \, \lambda}{1 - \chi_E(1 - \sigma_B)} - 1 = 0, \tag{4.24}$$

or, comparing (4.22) and (4.24),

$$\lambda = \frac{\chi_E}{E} \approx \frac{1}{(M_c)_B}. \tag{4.25}$$

The second (approximate) equality in (4.25) applies for $\chi_E \ll 1$.

We use (4.16) and the first equality in (4.25) to rewrite (4.21) as

$$(\psi_d)_B - \bar{\psi}_B = \frac{\sigma_B \chi_E \left[(\bar{\psi}_{B+} - \bar{\psi}_B) + E^{-1} \int_{z_B}^{z_{B+}} \bar{S}_\psi dz \right]}{1 - \chi_E (1 - \sigma_B)}. \tag{4.26}$$

Equation (4.26) is the desired result. It is essentially an *upper boundary condition* on the properties of the air sinking in downdrafts at level B. For $\chi_E \to 0$, we get $(\psi_d)_B \to \bar{\psi}_B$ when $\bar{S}_\psi = 0$; for $\bar{S}_\psi \neq 0$, however, $(\psi_d)_B$ can differ from $\bar{\psi}_B$ even when $\chi_E \to 0$, because $\chi_E/E \to 1/(M_c)_B$ as $\chi_E \to 0$, so that

$$(\psi_d)_B - \bar{\psi}_B \to \frac{\sigma_B}{(M_c)_B} \int_{z_B}^{z_{B+}} \bar{S}_\psi dz \quad \text{as} \quad \chi_E \to 0. \tag{4.27}$$

Equation (4.27) applies in the limiting case for which the top of the turbulent layer is an impermeable "ceiling," with $E = 0$ and $\chi_E = 0$.

When ψ is the moist static energy, h, we can show that $\int_{z_B}^{z_{B+}} \bar{S}_h dz = -\Delta \bar{R}$, so that (4.26) reduces to

$$(h_d)_B - \bar{h}_B = \frac{\sigma_B \chi_E \left(\Delta \bar{h} - \dfrac{\Delta \bar{R}}{E} \right)}{1 - \chi_E (1 - \sigma_B)}. \tag{4.28}$$

We now define $(\bar{h}_{B+})_{\text{eff}}$ as the "effective" value of $\bar{h}_{B+}$ in the presence of radiative cooling, i.e., it is the value that $\bar{h}_{B+}$ would have to take, *in the absence of radiative cooling*, to make $(h_d)_B$ the same as it is when radiative cooling actually is occurring.

By analogy with (4.28), we write

$$(h_\mathrm{d})_\mathrm{B} - \bar{h}_\mathrm{B} = \frac{\sigma_\mathrm{B}\chi_E(\Delta\bar{h})_\mathrm{eff}}{1 - \chi_E(1 - \sigma_\mathrm{B})}, \tag{4.29}$$

where

$$(\Delta\bar{h})_\mathrm{eff} \equiv \left(\bar{h}_\mathrm{B+}\right)_\mathrm{eff} - \bar{h}_\mathrm{B}. \tag{4.30}$$

Comparison of (4.28) and (4.29) shows that

$$(\Delta\bar{h})_\mathrm{eff} = \Delta\bar{h} - \frac{\Delta\bar{R}}{E}, \tag{4.31}$$

so that

$$(\bar{h}_\mathrm{B+})_\mathrm{eff} = \bar{h}_\mathrm{B+} - \frac{\Delta\bar{R}}{E}. \tag{4.32}$$

It follows from (4.31) and (4.9) that the effective virtual dry static energy jump for the smoke-cloud case (with $q = \ell = 0$) is

$$(\Delta\bar{s}_\mathrm{v})_\mathrm{eff} = \Delta\bar{s}_\mathrm{v} - \frac{\Delta\bar{R}}{E}. \tag{4.33}$$

This shows explicitly that *the effect of $\Delta\bar{R} > 0$ is to make the inversion seem weaker than it really is, in terms of the effect of entrainment on the buoyancy of the air that is sinking at the tops of the downdrafts.* The amount of radiative chilling decreases as E increases, because when entrainment is more rapid the entrained air spends less time in the layer of concentrated radiative cooling. For the smoke-cloud case,

$$(F_\mathrm{sv})_\mathrm{B} = -E\,\Delta\bar{s}_\mathrm{v} + \Delta\bar{R} > 0, \tag{4.34}$$

and comparison with (4.33) shows that

$$(\Delta\bar{s}_\mathrm{v})_\mathrm{eff} < 0. \tag{4.35}$$

This means that the effective inversion strength is actually negative for the smoke cloud. As shown below, however, a stronger inversion does act to reduce the rate of entrainment across the smoke-cloud top, as expected.

In accord with **H1** (p. 82), we modify the definition of the Richardson number, (4.6), by using $(\Delta\bar{s})_\mathrm{eff}$, as defined by (4.33), in place of $\Delta\bar{s}_\mathrm{v}$:

$$(Ri_\Delta)_\mathrm{eff} = \frac{gz_\mathrm{M}}{c_p\bar{T}_\mathrm{B}e_\mathrm{M}}\left(\Delta\bar{s}_\mathrm{v} - \frac{\Delta\bar{R}}{E}\right) < 0. \tag{4.36}$$

Since $(Ri_\Delta)_\mathrm{eff} < 0$, (4.5) reduces to

$$E = b_1\rho_\mathrm{B}\sqrt{e_\mathrm{M}}. \tag{4.37}$$

To proceed, we need to determine e_M. The steady-state TKE equation can be written as a balance between buoyant production and dissipation:

$$0 = \int_{p_{B+}}^{p_S} \frac{\kappa F_{sv}}{p} dp - C\rho_M e_M^{3/2}. \tag{4.38}$$

Here p_S is the pressure at the surface of the earth, C is a non-dimensional constant, ρ_M is the vertically averaged density of the air in the PBL, and κ is the gas constant divided by c_p. According to the results of large-eddy simulations reported by Moeng and Sullivan (1994), $C \approx 1$. Following B99, we consider a smoke cloud with negligible surface heat flux, i.e.,

$$(F_{sv})_S = 0. \tag{4.39}$$

Using (4.39) and (4.34), the buoyancy flux integral for a well-mixed layer can be approximated by

$$\int_{p_{B+}}^{p_S} \frac{\kappa F_{sv}}{p} dp \approx \frac{g z_M}{2 c_p \bar{T}_B} (-E\Delta \bar{s}_v + \Delta \bar{R}). \tag{4.40}$$

Substituting (4.40) into (4.38), and rearranging, we find that the TKE satisfies

$$e_M^{3/2} = (\Delta \bar{R} - E\Delta \bar{s}_v) \left(\frac{g z_M}{2 c_p \bar{T}_B C \rho_M} \right). \tag{4.41}$$

Using (4.37) and (4.41) to eliminate e_M, we obtain a cubic equation for the entrainment rate:

$$E^3 = B(\Delta \bar{R} - E\Delta \bar{s}_v), \tag{4.42}$$

where, for convenience, we define

$$B \equiv \frac{(b_1 \rho_B)^3 g z_M}{2 c_p \bar{T}_B C \rho_M} > 0. \tag{4.43}$$

For the case studied by B99, we find that $B \approx 1.83 \times 10^{-5}$ $(\mathrm{kg\,m}^{-3})^2$. Equation (4.42) yields a single real, positive value of E. Rewriting (4.42) as

$$B = \frac{E^3}{\Delta \bar{R} - E\Delta \bar{s}_v}, \tag{4.44}$$

makes it apparent that for sufficiently large values of B, we get

$$E \approx \frac{\Delta \bar{R}}{\Delta \bar{s}_v}. \tag{4.45}$$

The large-eddy simulations of a smoke cloud presented by B99 suggest that $E \approx 0.3 \times 10^{-3}\,\mathrm{kg\,m}^{-2}\,\mathrm{s}^{-1}$, for $\Delta \bar{R} \approx 60\,\mathrm{W\,m}^{-2}$ and $\Delta \bar{s}_v \approx 7000\,\mathrm{J\,kg}^{-1}$. To make (4.42) agree with the LES results, we have to choose $b_1 \approx 0.0034$, which is much smaller that the value inferred from simulations of the cloud-free boundary layer heated from below ($b_1 \approx 0.25$). We hypothesize that the implied variation of b_1 is real, and is due to physical differences between the idealized smoke-cloud case, in which the turbulence is driven entirely by cooling from above, and a clear boundary layer that is driven entirely by heating from below. This hypothesis should be explored through additional LES-based research.

We now extend our entrainment closure to the more geophysically relevant case of a water cloud, for which both radiative and evaporative cooling occur at cloud top. Following the approach of L68, we use the fact that in a uniformly saturated layer isobaric fluctuations of water vapor and moist static energy are approximately proportional to each other, i.e.,

$$Lq' \approx \left(\frac{\gamma}{1+\gamma} \right) h', \tag{4.46}$$

where

$$\gamma \equiv \frac{L}{c_p} \left(\frac{\partial q^*}{\partial T} \right)_p, \tag{4.47}$$

and q^* denotes the saturation mixing ratio.

By combining (4.46) with (4.9), we can show that

$$(s_v)_{d,B} - (\bar{s}_v)_B = \beta[(h_d)_B - \bar{h}_B] - \epsilon L \left[(r_d)_B - \bar{r}_B \right], \tag{4.48}$$

where

$$\beta \equiv \frac{1 + (1+\delta)\gamma\epsilon}{1+\gamma} \tag{4.49}$$

is a positive non-dimensional coefficient. It is important to note that (4.48) applies only when the layer is uniformly cloudy; if there are holes in the cloud, (4.48) must be replaced by a more complicated equation, as discussed in Section 4.5 below.

By analogy with (4.28), the total water mixing ratio, r, in the downdraft satisfies

$$(r_d)_B - \bar{r}_B = \left(\frac{\sigma_B \chi_E}{1 - \chi_E(1 - \sigma_B)} \right) \Delta\bar{r}. \tag{4.50}$$

Substituting (4.28) and (4.50) into (4.48), we obtain

$$
\begin{aligned}
(s_v)_{d,B} - (\bar{s})_B &= \beta \left[\frac{\sigma_B \chi_E \left(\Delta\bar{h} - \dfrac{\Delta\bar{R}}{E} \right)}{1 - \chi_E(1 - \sigma_B)} \right] - \epsilon L \left(\frac{\sigma_B \chi_E}{1 - \chi_E(1 - \sigma_B)} \right) \Delta\bar{r} \\
&= \frac{\sigma_B \chi_E}{1 - \chi_E(1 - \sigma_B)} \left(\Delta\bar{s}_v - (\Delta\bar{s})_{\text{crit}} - \beta\frac{\Delta\bar{R}}{E} \right),
\end{aligned}
\tag{4.51}
$$

where

$$\Delta \bar{s}_v - (\Delta \bar{s})_{crit} \equiv \beta \Delta \bar{h} - \epsilon L \Delta \bar{r} \tag{4.52}$$

is a notation introduced by Randall (1980), who showed that

$$(\Delta \bar{s}_v)_{crit} = \left[\frac{1 - (1 + \delta)\epsilon}{1 + \gamma} \right] L(q_{B+}^* - q_{B+}) \tag{4.53}$$

is a positive measure of the dryness of the air above the inversion. In (4.53) the expression in [ ] is non-dimensional and positive. In a stratocumulus regime, $(\Delta \bar{s}_v)_{crit}/c_p$ is typically on the order of 5 K.

To identify the effective inversion strength and effective Richardson number, we define $(\Delta \bar{s}_v)_{eff}$ by

$$(s_v)_{d,B} - (\bar{s}_v)_B = \left(\frac{\sigma_B \chi_E}{1 - \chi_E(1 - \sigma_B)} \right) (\Delta \bar{s}_v)_{eff}, \tag{4.54}$$

where $(\Delta \bar{s}_v)_{eff} \equiv (\bar{s}_{v_{B+}})_{eff} - \bar{s}_B$. Comparison of (4.51) and (4.61) gives

$$(\Delta \bar{s}_v)_{eff} = \Delta \bar{s}_v - (\Delta \bar{s}_v)_{crit} - \beta \frac{\Delta \bar{R}}{E}. \tag{4.55}$$

Equation (4.55) shows that the effective inversion strength is reduced by evaporative cooling, by an amount $(\Delta \bar{s}_v)_{crit}$, and that it is further reduced by radiative cooling.

Replacing $\Delta \bar{s}_v$ by $(\Delta \bar{s}_v)_{eff}$ in (4.6), and substituting the result into (4.5), we obtain

$$E = \frac{b_1 \rho_B \sqrt{e_M}}{1 + b_2 \left(\frac{g z_M}{c_p \bar{T}_B e_M} \right) \max \left(\Delta \bar{s}_v - (\Delta \bar{s}_v)_{crit} - \frac{\beta \Delta \bar{R}}{E}, 0 \right)} \tag{4.56}$$

which can be rearranged to

$$E = \begin{cases} \dfrac{b_1 \rho_B \sqrt{e_M} + b_2 \left(\dfrac{g z_M}{c_p \bar{T}_B} \right) \dfrac{\beta \Delta \bar{R}}{e_M}}{1 + b_2 \left(\dfrac{g z_M}{c_p \bar{T}_B} \right) \dfrac{\Delta \bar{s}_v - (\Delta \bar{s}_v)_{crit}}{e_M}} & \text{for} \quad \Delta \bar{s}_v - (\Delta \bar{s}_v)_{crit} - \dfrac{\beta \Delta \bar{R}}{E} \geq 0 \\[4em] b_1 \rho_B \sqrt{e_M} & \text{for} \quad \Delta \bar{s}_v - (\Delta \bar{s}_v)_{crit} - \dfrac{\beta \Delta \bar{R}}{E} < 0. \end{cases} \tag{4.57}$$

We can show that

$$\Delta \bar{s}_v - (\Delta \bar{s}_v)_{crit} - \frac{\beta \Delta \bar{R}}{E} \geq 0 \implies [\Delta \bar{s}_v - (\Delta \bar{s}_v)_{crit}] b_1 \rho_B \sqrt{e_M} \geq \beta \Delta \bar{R}. \tag{4.58}$$

The effect of evaporative cooling is to reduce the denominator of (4.57). The effect of radiative cooling is to increase the numerator, just as in the case of the smoke cloud.

4.5 What are the processes that cause MSCs to break up into shallow cumuli on their western and equatorward boundaries?

If the entrainment problem has been controversial, the stratocumulus break-up problem has been even more so. Many mechanisms can lead to the destruction of a stratocumulus cloud layer, including increased subsidence and horizontal advection. One possible mechanism for the destruction of a uniform cloud layer is "cloud-top entrainment instability" (CTEI). The concept was discussed by L68, who suggested that if the inversion is not sufficiently strong, evaporatively enhanced entrainment can "run away," leading to the evaporative disruption of the uniform cloud layer. The concept of CTEI has elicited a lot of interest, and at least as much skepticism.

Randall (1980) showed that the cloud-top buoyancy flux in a uniformly cloudy layer satisfies

$$(F_{sv})_B = -E\left[\Delta \bar{s}_v - (\Delta \bar{s}_v)_{crit}\right] + \beta \Delta \bar{R}. \qquad (4.59)$$

As can be seen from (4.59), when

$$\Delta \bar{s}_v < (\Delta \bar{s}_v)_{crit}, \qquad (4.60)$$

entrainment will promote a positive buoyancy flux at the inversion base. Randall (1976, 1980) and Deardorff (1980) argued that such a positive buoyancy flux will generate additional TKE, thus promoting further entrainment, and leading to a runaway destruction of the cloud through rapid infusion of dry air, i.e., to CTEI. They hypothesized that (4.60) is the criterion for the onset of CTEI.

Observations (e.g., Kuo and Schubert, 1988) do not support this hypothesis; they indicate that the entrainment rate remains modest when (4.60) is satisfied, and that the cloud layer can survive intact or nearly so. Large-eddy simulations by Moeng (2000) agree that the entrainment rate remains small when (4.60) is satisfied but, in contrast to the observations, Moeng's results show that the fractional cloudiness decreases as $\Delta \bar{s}_v - (\Delta \bar{s}_v)_{crit}$ becomes increasingly negative. We have no explanation for the apparent discrepancy between the observations and the LES results.

Before completing our discussion of entrainment parameterization, we present some stratocumulus simulations performed with the Colorado State University General Circulation Model (CSU GCM). The model is described by Ringler *et al.* (2000). In the simulation used here, the model is running in pure climate mode; no observed initial conditions were used. The entrainment rate has been

July PBL Cloud Incidence

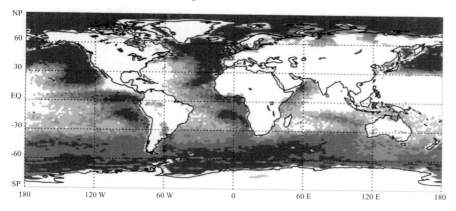

Warren Stratus Cloud Amount (JJA)

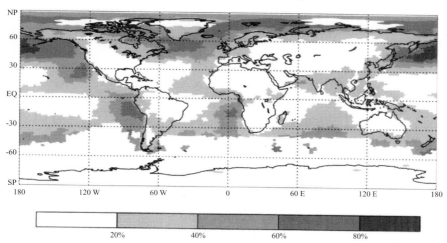

Figure 4.2. The top panel shows the July PBL stratus cloud amount as simulated with the CSU GCM. The lower panel shows the corresponding observations for June–August from Warren *et al.* (1986, 1988). The observations are not available for Antarctica and the adjacent southern ocean.

parameterized by the method described here. The GCM uses an embedded mixed-layer model, which includes the physical processes and interactions described in L68. Entrainment is parameterized using the approach that we have proposed here. When inequality (4.60) is satisfied, CTEI is assumed to occur and the stratocumulus cloudiness is set to zero. Figure 4.2 shows the model results and, for comparison, observations for June–August from Warren *et al.* (1986, 1988). The model successfully simulates the primary marine stratocumulus regimes found west of North

America, South America, and southern Africa, as well as secondary marine stratocumulus regimes in such places as the North Atlantic. Experiments have shown that if the CTEI criterion is ignored, the model produces much more extensive and less realistic boundary-layer stratus clouds.

Moeng's (2000) simulations of CTEI raise an important but neglected point: any theory of CTEI should allow for partial cloudiness. It should therefore come as no surprise that (4.59), which is based on the assumption of uniform cloudiness, cannot describe the break-up of a uniform cloud layer. In the context of the model presented in Section 4.4, a broken cloud layer has to be represented in terms of cloudy updrafts and clear downdrafts. We now derive an expression for the rate of entrainment across the top of such a partly cloudy layer. Our starting point is

$$(s_v)_{d,B} - (\bar{s}_v)_B = (h_d)_B - \bar{h}_B - L''[(r_d)_B - \bar{r}_B] + L'[(\ell_d)_B - \bar{\ell}_B], \quad (4.61)$$

where

$$L' \equiv [1 - (1+\delta)\epsilon]L \quad \text{and} \quad L'' \equiv [1 - \delta\epsilon]L,$$

which can readily be obtained from (4.9). For the case of a partly cloudy layer, with a cloudy updraft and a cloud-free downdraft,

$$(\ell_d)_B = 0 \quad \text{and} \quad \bar{\ell}_B = \sigma_B(\ell_u)_B, \quad (4.62)$$

so that (4.61) can be simplified to

$$(s_v)_{d,B} - (\bar{s}_v)_B = (h_d)_B - \bar{h}_B - L''[(r_d)_B - \bar{r}_B] - L'\sigma_B(\ell_u)_B. \quad (4.63)$$

Substituting from (4.28) and (4.50), we get

$$(s_v)_{d,B} - (\bar{s}_v)_B = \left[\frac{\sigma_B \chi_E}{1 - \chi_E(1-\sigma_B)}\right]\left(\Delta\bar{h} - \frac{\Delta\bar{R}}{E}\right)$$
$$- L''\left[\frac{\sigma_B \chi_E}{1 - \chi_E(1-\sigma_B)}\right]\Delta\bar{r} - L'\sigma_B(\ell_u)_B. \quad (4.64)$$

Comparing (4.64) with (4.54), we find that

$$\left[\frac{\sigma_B \chi_E}{1 - \chi_E(1-\sigma_B)}\right](\Delta\bar{s}_v)_{\text{eff}} = \left[\frac{\sigma_B \chi_E}{1 - \chi_E(1-\sigma_B)}\right]\left(\Delta\bar{h} - \frac{\Delta\bar{R}}{E}\right)$$
$$- L''\left[\frac{\sigma_B \chi_E}{1 - \chi_E(1-\sigma_B)}\right]\Delta\bar{r} \quad (4.65)$$
$$- L'\sigma_B(\ell_u)_B,$$

which is equivalent to

$$(\Delta\bar{s}_v)_{\text{eff}} = \Delta\bar{s}_v - \left(\frac{1-\chi_E}{\chi_E}\right)L'(\ell_u)_B - \frac{\Delta\bar{R}}{E}. \quad (4.66)$$

Equation (4.66) is analogous to (4.55).

For $\chi_E \ll 1$, we can use (4.23) to approximate (4.66) by

$$(\Delta\bar{s}_v)_{\text{eff}} \approx \Delta\bar{s}_v - \frac{1}{E}\{(M_c)_B L'(\ell_u)_B + \Delta\bar{R}\}. \tag{4.67}$$

Since $(M_c)_B(\ell_u)_B$ is the liquid water flux at level B, (4.67) states that the effective inversion strength is reduced by the cooling due to evaporation of the liquid water carried upward to the PBL top, as well as by the radiative cooling:

$$(\Delta\bar{s}_v)_{\text{eff}} \approx \Delta\bar{s}_v - \frac{1}{E}\{L'(F_\ell)_B + \Delta\bar{R}\}. \tag{4.68}$$

Using the approximation (4.67), we replace $\Delta\bar{s}_v$ by $(\Delta\bar{s}_v)_{\text{eff}}$ in (4.6), and substitute the result into (4.5), to obtain

$$E \approx \frac{b_1\rho_B\sqrt{e_M}}{1 + b_2\dfrac{gz_M}{c_p\bar{T}_B e_M}\max\left\{\Delta\bar{s}_v - \dfrac{1}{E}\left(L'(F_\ell)_B + \Delta\bar{R}\right), 0\right\}}, \tag{4.69}$$

which can be rearranged to

$$E \approx \begin{cases} \dfrac{b_1\rho_B\sqrt{e_M} + b_2\left(\dfrac{gz_M}{c_p\bar{T}_B e_M}\right)\{L'(F_\ell)_B + \Delta\bar{R}\}}{1 + b_2\left(\dfrac{gz_M}{c_p\bar{T}_B e_M}\right)\Delta\bar{s}_v} & \text{for}\quad (\Delta\bar{s}_v)_{\text{eff}} \geq 0, \\[3em] b_1\rho_B\sqrt{e_M} & \text{for}\quad (\Delta\bar{s}_v)_{\text{eff}} < 0. \end{cases} \tag{4.70}$$

By substituting (4.70) into (4.67), we can show that

$$(\Delta\bar{s}_v)_{\text{eff}} \geq 0 \quad \text{is equivalent to} \quad \Delta\bar{s}_v \geq \frac{L'(F_\ell)_B + \Delta\bar{R}}{b_1\rho_B\sqrt{e_M}}. \tag{4.71}$$

To our knowledge, (4.70) is the first equation proposed to determine the rate of entrainment at the top of a partly cloudy layer. It is suitable for use in a model that is capable of determining $(F_\ell)_B$ in a partly cloudy layer; an example is ADHOC, which is described by Lappen and Randall (2001). We envision a modified version of ADHOC in which the PBL top and the entrainment rate are introduced as explicit parameters.

A question can be raised about the liquid water flux, $(F_\ell)_B$, in a partly cloudy layer. Should it not depend on the entrainment rate, as $(F_h)_B$ and $(F_r)_B$ do, and as $(F_\ell)_B$ does in a fully cloudy layer? If $(F_\ell)_B$ did depend on E, then the entrainment rate would enter implicitly on the right-hand side of (4.70), which would therefore have to be modified in order to give a solution for E. We claim that in a partly cloudy layer $(F_\ell)_B$ is independent of E, so that (4.70) can be used as given. Our reasoning is that although the entrained air enters and alters the properties of the clear downdrafts, it has no direct effect on the cloudy updrafts. The liquid water mixing ratio in the downdrafts is zero by definition in a partly cloudy layer, and this is true regardless of the entrainment rate. The liquid water mixing ratio and other properties of the air at the tops of the updrafts are determined by processes at lower levels in the updrafts, and are not directly affected by entrainment. We can therefore determine $(F_\ell)_B$ without knowing the entrainment rate.

4.6 Conclusion: the importance of cloud-scale process-coupling for large-scale cloudiness

Thanks in large part to L68, we have a reasonably good understanding of why MSCs exist. It is therefore somewhat surprising that GCMs have had limited success in simulating them. One reason for this unfortunate situation is that MSCs are produced by closely coupled turbulent, radiative and microphysical processes. GCMs, for the most part, fail to represent this coupling. In fact, there is a trend today towards "modularization" of GCMs. In a modularized GCM, the various physical parameterizations are segregated into software "compartments" (subroutines, etc.), like animals in a zoo. The parameterizations communicate only through their mutual effects on the shared "large-scale environment," which is predicted by the GCM.

From a software engineering point of view, modularization has a certain appeal, but model development is not software engineering. One of the most revolutionary insights of L68 is that the interactions among turbulence, radiation, and phase changes are tightly coupled on small scales. For example, the turbulent entrainment rate is strongly affected by both cloud-top radiative cooling and the effects of phase changes on the buoyancy flux. The turbulence extends as high as it does only because of this coupling between the cloud layer and the turbulence. These couplings occur on small space and time scales, and *not* merely through their mutual interactions with the large-scale environment. Failure to account for such cloud-scale couplings has been a major obstacle to the simulation of MSCs with GCMs. There are many other examples of process-coupling that argue for less, not more modularity in our models (e.g., Arakawa, 2004). In its sophisticated coupling of radiative, turbulent, and microphysical processes, L68 was (at least) 35 years ahead of its time.

Acknowledgements

We are grateful to Douglas Lilly for sharing with us his memories of the background leading up to the creation of L68. We obtained his input by sneaky and underhanded means. Brian McNoldy provided Plates I and II, and Mark Branson provided Fig. 4.4. Bjorn Stevens and an anonymous reviewer made useful comments on the manuscript. Cara-Lyn Lappen, Chin-Hoh Moeng, and Richard Taft also gave helpful advice. Our work has been supported by NSF Grants ATM-9812384 and ATM-0332197 and NOAA Grant NA17RJ1228.

References

Albrecht, B. A., Randall, D. A. and Nicholls, S. (1988). Observations of marine stratocumulus clouds during FIRE. *Bull. Amer. Meteor. Soc.*, **69**, 618–626.

Albrecht, B. A. (1989). Aerosols, cloud microphysics, and fractional cloudiness. *Science*, **245**, 1227–1230.

Arakawa, A. (2004). The cumulus parameterization problem: Past, present, and future. *J. Climate*, in press.

Bjerknes, J. (1966). A possible response of the atmospheric Hadley circulation to equatorial anomalies of ocean temperature. *Tellus*, **18**, 820–829.

Bourassa, M. A., Legler, D. M., O'Brien, J. J. and Smith, S. R. (2003). SeaWinds validation with research vessels. *J. Geophys. Res.*, **108**, doi:10.1029/2001JC001028.

Breidenthal, R. E. and Baker, M. B. (1985). Convection and entrainment across stratified interfaces. *J. Geophys. Res.*, **90D**, 13055–13062.

Breidenthal, R. E. (1992). Entrainment at thin stratified interfaces: The effects of Schmidt, Richardson, and Reynolds numbers. *Phys. Fluids A*, **4**, 2141–2144.

Bretherton, C. S., MacVean, M. K., *et al.* (1999). An intercomparison of radiatively driven entrainment and turbulence in a smoke cloud, as simulated by different numerical models. *Quart. J. Roy. Meteor. Soc.*, **125**, 391–423.

Chelton, D. B., Esbensen, S. K., Schlax, M. G., *et al.* (2001). Observations of coupling between surface wind stress and sea surface temperature in the eastern tropical Pacific. *J. Climate*, **14**, 1479–1498.

Deardorff, J. W., Willis, G. E. and Lilly, D. K. (1969). Laboratory investigation of non-steady penetrative convection. *J. Fluid Mech.*, **35**, 7–31.

Deardorff, J. W. (1974). Three-dimensional numerical study of the height and mean structure of a heated planetary boundary layer. *Boundary-Layer Meteorol.*, **7**, 81–106.

(1980). Cloud top entrainment instability. *J. Atmos. Sci.*, **37**, 131–147.

Deser, C. and Wallace, J. M. (1990). Large-scale atmospheric circulation features of warm and cold episodes in the tropical Pacific. *J. Climate*, **3**, 1254–1281.

Hubert, L. F. (1966). Mesoscale cellular convection. Meteorological Satellite Laboratory Report No. 37, Washington, DC.

Kuo, H.-C. and Schubert, W. H. (1988). Stability of cloud-topped boundary layers. *Quart. J. Roy. Meteor. Soc.*, **114**, 887–916.

Lappen, C.-L. and Randall, D. A. (2001). Towards a unified parameterization of the boundary layer and moist convection. Part I. A new type of mass-flux model. *J. Atmos. Sci.*, **58**, 2021–2036.

Lewellen, D. and Lewellen, W. (1998). Large-eddy boundary layer entrainment. *J. Atmos. Sci.*, **55**, 2645–2665.

Lilly, D. K. (1968). Models of cloud-topped mixed layers under a strong inversion. *Quart. J. Roy. Meteor. Soc.*, **94**, 292–309.

(2002a). Entrainment into mixed layers. Part I: Sharp-edged and smoothed tops. *J. Atmos. Sci.*, **59**, 3340–3352.

(2002b). Entrainment into mixed layers. Part II: A new closure. *J. Atmos. Sci.*, **59**, 3353–3361.

Liu, Q., Kogan, Y. L., Lilly, D. K. *et al.* (2000). Modeling of ship effluent transport and its sensitivity to boundary layer structure. *J. Atmos. Sci.*, **57**, 2779–2791.

Lock, A. P. (1998). The parameterization of entrainment in cloudy boundary layers. *Quart. J. Roy. Meteor. Soc.*, **124**, 2729–2753.

   (2000). The numerical representation of entrainment in parameterizations of boundary layer turbulent mixing. *Mon. Wea. Rev.*, **129**, 1148–1163.

Ma, C.-C., Mechoso, C. R., Arakawa, A. and Farrara, J. D. (1994). Sensitivity of a coupled ocean–atmosphere model to physical parameterizations. *J. Climate*, **7**, 1883–1896.

McDonald, W. F. (1938). *Atlas of Climatic Charts of the Oceans*, Washington, DC: US Dept. of Agriculture, Weather Bureau.

Moeng, C.-H. and Sullivan, P. P. (1994). A comparison of shear- and buoyancy-driven planetary boundary layer flows. *J. Atmos. Sci.*, **51**, 999–1022.

Moeng, C.-H. (2000). Entrainment rate, cloud fraction, and liquid water path of PBL stratocumulus clouds. *J. Atmos. Sci.*, **57**, 3627–3643.

Neiburger, M., Johnson, D. S. and Chien, C.-W. (1961). Studies of the structure of the atmosphere over the eastern Pacific ocean in summer. I. The inversion over the eastern north Pacific ocean. *University of California Publications in Meteorology*, **1**(1), 1–94. Berkeley and Los Angeles: University of California Press.

Petterssen, S. (1938). On the causes and the forecasting of the California fog. *Bull. Amer. Meteor. Soc.*, **19**, 49–55.

Randall, D. A. (1976). *The Interaction of the Planetary Boundary Layer with Large-Scale Circulations*. Ph.D. thesis, University of California at Los Angeles.

   (1980). Conditional instability of the first kind, upside-down. *J. Atmos. Sci.*, **37**, 125–130.

Randall, D. A. and Suarez, M. J. (1984). On the dynamics of stratocumulus formation and dissipation. *J. Atmos. Sci.*, **41**, 3052–3057.

Randall, D. A., Shao, Q. and Moeng, C.-H. (1992). A second-order bulk boundary-layer model. *J. Atmos. Sci.*, **49**, 1903–1923.

Riehl, H., Yeh, T. C., Malkus, J. S. and LaSeur, N. E. (1951). The north-east trade of the Pacific Ocean. *Quart. J. Roy. Meteor. Soc.*, **77**, 598–626.

Ringler, T. D., Heikes, R. P. and Randall, D. A. (2000). Modeling the atmospheric general circulation using a spherical geodesic grid: A new class of dynamical cores. *Mon. Wea. Rev.*, **128**, 2471–2490.

Siems, S. T., Bretherton, C. S., Baker, M. B., Shy, S. and Breidenthal, R. E. (1990). Buoyancy reversal and cloudtop entrainment instability. *Quart. J. Roy. Meteor. Soc.*, **116**, 705–739.

Stevens, B. (2002). Entrainment in stratocumulus-topped mixed layers. *Quart. J. Roy. Meteor. Soc.*, **128**, 2663–2690.

Stevens, B., Vali, G., Comstock, K., *et al.* (2004). Pockets of open cells (POCs) and drizzle in marine stratocumulus. *Bull. Amer. Meteorol. Soc.*, in press.

Twomey, S. A., Piepgrass, M. and Wolfe, T. L. (1984). An assessment of the impact of pollution on global cloud albedo. *Tellus*, **36**, 356–366.

Warren, S. G., Hahn, C. J., London, J., Chervin, R. M. and Jenne, R. L. (1986). Global distribution of total cloud cover and cloud type amounts over land. NCAR/TN-273+STR, Boulder, CO: National Center for Atmospheric Research.

   (1988). Global distribution of total cloud cover and cloud type amounts over the ocean. NCAR/TN-317+STR, Boulder, CO: National Center for Atmospheric Research.

van Zanten, M. C., Stevens, B., Vali, G. and Lenschow, D. H. (2004). On drizzle rates in nocturnal marine stratocumulus. Submitted to *J. Atmos. Sci.*

5

Large-eddy simulations of cloud-topped mixed layers

Chin-Hoh Moeng, Peter P. Sullivan

National Center for Atmospheric Research, Boulder, USA

and

Bjorn Stevens

Department of Atmospheric and Oceanic Sciences, University of California, Los Angeles, USA

5.1 Introduction

With the advent of computers, scientists in the 1950s and 1960s began to explore the possibility of using numerical simulation to generate virtual laboratories for exploring specific geophysical processes in a controlled manner. Doug Lilly helped pioneer this emerging science of numerical simulation. As pointed out by Wyngaard (Chapter 1), Lilly presented a "bold, three-phase plan of attack" in which well-behaved numerical models would be developed; their fidelity would be benchmarked against known solutions; and as confidence builds they would be used to explore conditions not adequately reproducible by experiment. In the subsequent decades this strategy has become a staple of theoretical studies of turbulence. In particular, a class of numerical simulations Doug helped develop in the early 1960s has come to be known as large-eddy simulation (LES) and is now widely used in the field of planetary boundary layer (PBL) turbulence and clouds.

We begin in Section 5.2 by giving an example of the second element of Doug's plan of attack, and what we call "benchmarking." This is by no means trivial, because for turbulent flows there are no known solutions. To better appreciate this point we consider LES of the cloud-topped boundary layer which couples turbulence, radiation, and cloud processes. As cloudy boundary layers cannot be created in the laboratory, one must invariably turn to field data to construct meaningful benchmarks. Historically, field data have been collected to explore phenomenology, and thus few datasets exist to benchmark computations. The second field study of the Dynamics and Chemistry of Marine Stratocumulus (DYCOMS-II) is unique in that it was designed from the outset with the purpose of testing LES. However, even with this focused field campaign the measurements needed to design an LES to completely mimic the natural environment are difficult to make, and as we shall see this element of Doug's strategy remains in its infancy.

The third element of Doug's strategy emphasizes creative thinking. Simulations can be used to explore the parameter space that has not been (or cannot be) measured in the field; or, for a given set of parameters, simulations can be used to explore physical aspects of the solution that cannot be measured. Both tactics require theoretical guidelines. In the former, one needs a theoretical framework to help guide exploration of the parameter space, and in the latter (which we focus on here), a theoretical framework is needed to pose stimulating questions. A proper theoretical framework for the stratocumulus-topped boundary layer is Lilly's mixed-layer theory (see the review by Randall and Schubert, Chapter 4). With this theoretical framework and some early LES solutions, Lilly investigates how the structure of the cloud-top interface may affect its statistical representation (Lilly, 2002a), and how the interface property can change the entrainment rate and interface stability (Lilly, 2002b). As we show in Sections 5.3 and 5.4 both questions are impossible to attack with observations alone, but fit the third phase of Doug's "bold plan of attack" on geophysical turbulence problems.

5.2 Benchmarking

For the purpose of benchmarking LES we use data collected as part of DYCOMS-II. An overview of the experiment is given in Stevens *et al.*, (2003a); the particular case we focus on here, research flight one (RF01), is an outgrowth of an earlier study of the case described by Stevens *et al.* (2003b). The appealing aspect of DYCOMS-II is that it is predominantly nocturnal, which makes it relatively straightforward to constrain the large-scale energetics. An appealing aspect of RF01 is that the stratocumulus layer is essentially non-precipitating, and the large-scale conditions are remarkably uniform, further simplifying possible comparisons with LES. The code used throughout is the NCAR LES, which was described in Moeng (1986) and more recently in Moeng (2000).

5.2.1 Setup and initial data

The initial data and boundary forcings for the LES were derived from the RF01 measurements as reported by Stevens *et al.* (2003b). This case is characterized as a persistent, well-mixed, and slightly thickening nocturnal cloud field capped by a much warmer and drier free troposphere. The large-scale conditions were approximately constant over the measurement period (which spanned 8 hours). Input parameters for the LES are given in Table 5.1. For the vertically varying basic

Table 5.1. *Input parameters, where H_0 and L_0 denote the surface sensible and latent heat fluxes respectively, D the large-scale divergence, U_g and V_g the geostrophic wind components, z_0 the roughness height, f the Coriolis frequency, p_0 the surface pressure, and Θ_0 the reference temperature.*

H_0	15 W m^{-2}
L_0	100 W m^{-2}
D	4×10^{-6} s^{-1}
(U_g, V_g)	$(6, -4.25)$ m s^{-1}
z_0	0.035 m
f	1×10^{-4} s^{-1}
p_0	0.1 MPa
Θ_0	288 K

state we specify

$$\{q_T, \theta_l\} = \begin{cases} \{8.75 \text{ g kg}^{-1}, \quad 289.7 \text{ K}\} & z < 817 \text{ m}, \\ \{1.50 \text{ g kg}^{-1}, \quad [296.7 + (z - 817)^{1/3}] \text{ K}\} & \text{otherwise}, \end{cases} \quad (5.1)$$

where θ_l is the liquid-water potential temperature and q_T is the total water mixing ratio. In addition to the surface fluxes, radiative forcing also drives the flow; here it is calculated from a simple exponential formula shown in Equation (1) of Moeng (2000), which, using Lilly's (2002b) notation, can be approximated as follows:

$$F_R = F_i \exp\left(\frac{z - z_i}{\lambda}\right), \quad \text{for } z \leq z_i, \quad (5.2)$$

where z_i is the height of the local cloud-top interface, $F_i = 50$ W m^{-2} is the net radiative flux above cloud top derived from measurements. The decay length scale $\lambda \approx 1/(\rho_0 \kappa q_l) \approx 26$ m, given a reference air density ρ_0 of 1 kg m^{-3}, a longwave absorption coefficient $\kappa = 130$ m^2 kg^{-1}, and if the mean cloud-top liquid-water mixing ratio q_l is about 0.3g kg^{-1}. Note that these parameters and initial data differ somewhat from the observations and from our previous investigation of this case (i.e., Stevens et al., 2003b). Here changes have been made to preserve the cloud layer through the initial spin-up period of the simulation, and to compensate for the use of an assumed Boussinesq vertical structure with $\rho_0 = 1$ g kg^{-3}.

Below we explore results from three simulations, LES-(1, 2, 3), which differ only in their numerical treatment. Cases LES-(1, 2) both span a domain of

2500 m × 2500 m × 1500 m with a mesh of 96 × 96 × 400 points, and only differ in their treatment of the horizontal advection terms in the θ_l and q_T equations; LES-1 uses a pseudo-spectral method, while LES-2 uses the flux-limited upwind algorithm (Koren, 1993; Sullivan *et al.*, 1998) that is employed for the vertical of θ_l and q_T advection in all three simulations. LES-3 is identical to LES-1 except that it spans a 7500 m × 7500 m × 1500 m domain with a mesh of 200 × 200 × 400 points. [The increased size of the horizontal domain is computationally expensive, so LES-3 is integrated for only two simulation hours, as compared to four simulation hours for LES-(1,2).] Thus in the spirit of benchmarking, we ask not only if the LES can reproduce the observed structure with plausible fidelity, but also if this reproduction is sensitive (at short times) to numerical methods or the truncation of larger scales.

5.2.2 Comparison between LESs and observations

We compare the simulations only after the first hour. Before this time, the turbulence is not fully developed and the statistics are not stationary. Stationarity of the statistics is associated with invariance in the shapes of the profile statistics. Specifically, for a conserved variable, ψ, whose horizontal average, Ψ, satisfies

$$\partial_t \Psi = \partial_z \Phi, \qquad (5.3)$$

where Φ is some flux, then this condition implies that $\partial_t \partial_z \Psi$ vanishes, or equivalently Φ is linear.[†] If this condition is satisfied, the turbulent flow is near statistical equilibrium. LES is most justifiably used to study the statistical properties of turbulent flow fields; for this reason its analysis is normally confined to time periods when the turbulence is in statistical equilibrium or the so-called quasi-steady state.

In Fig. 5.1 we show how the simulations represent the cloud evolution, and the numerical effects on this evolution. The fractional cloud cover stays at about 99.5% in LES-1 and fluctuates between 85% to 92% in LES-2. The time evolution of the cloud top and base given in Fig. 5.1 are quite similar between LES-1 and LES-2 although LES-2 grows thinner compared to LES-1. Overall, the LESs compare well with the observations, although there is a tendency for cloud base to rise through the course of the simulation, in contrast to its apparent lowering in the field data. As we shall see, this is consistent with a simulated entrainment moisture flux that is larger than the observed. The simulations show that the mean cloud top rises at a rate of about 0.16 cm s^{-1}, and hence given a large-scale subsidence of ~0.32 cm s^{-1} at $z = 810$ m, this rate of PBL deepening implies an entrainment

---

[†] For instance the turbulent flux $\overline{wq_T}$ in the case when $\Psi = Q_T$, or the combination of the turbulent heat flux and the radiative flux when $\Psi = \Theta_l$. Here, w is the vertical wind velocity.

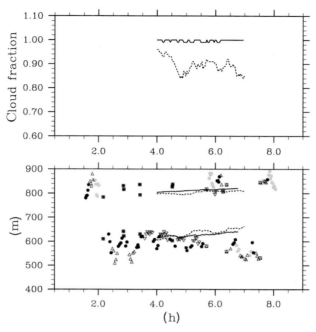

Figure 5.1. Time evolution of the fractional cloud cover (upper panel) and the mean cloud-top and cloud-base heights (lower panel) from LES-1 (solid lines), LES-2 (dotted lines), and observations adapted from Stevens *et al.* (2003b). Here the starting time of the simulation relative to the time of observation is chosen to roughly correspond to the initial data of the simulations. Note that the mean cloud-top and cloud-base heights are computed by horizontally averaging the local cloud tops and bases only over cloudy grid points.

rate of about 0.48 cm s^{-1}, which is in the range of observed values (Stevens *et al.*, 2003b).

The profile statistics from LES are computed as follows. We first compute statistics by applying horizontal $(x - y)$ averaging, then interpolate these instantaneous profile statistics to a normalized vertical coordinate $z/\overline{z_i}$ (where $\overline{z_i}$ is the horizontally averaged cloud-top height), and finally time-average these profiles between hour 1 and 4 of the simulation period. Because the vertical normalization is based on the spatially averaged cloud-top height, the transition across the cloud top is expected to be smoothed across a scale corresponding to spatial fluctuations in the simulated interfacial layer at any given time. In Section 5.3 we investigate the effect of this smoothing. Figure 5.2 shows vertical profiles of Q_T, Θ_l, and Q_l for LES-(1, 2) compared with the observed sounding taken from Stevens *et al.* (2003b). These results show that the LES provides a plausible representation of the cloud-topped mixed layer and the simulations maintain a jump structure similar to that observed. No decoupling occurs; the PBL remains well mixed throughout the simulations.

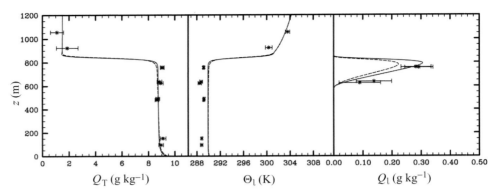

Figure 5.2. Vertical profiles of (a) total water mixing ratio, (b) liquid-water potential temperature, and (c) liquid-water mixing ratio from LES-1 (solid curves), LES-2 (dotted curves), and the observations adapted from Stevens *et al.*, 2003b.

The comparison of the mean states can be more critically evaluated by comparing the time-rate-of-change of θ_l and q_T within the mixed layer. Such comparisons (not shown here) indicate that the simulated layer is warming more rapidly (about 0.1 K h^{-1} versus the observed rate of 0.07 K h^{-1}), and not moistening as quickly (0.02 g kg^{-1} h^{-1} versus the observed rate of 0.06 g kg^{-1} h^{-1}). These differences are not large, but both contribute to the simulation tendency to raise cloud base. Such differences are consistent with simulated entrainment rates being on the upper end of the observed range.

Figures 5.3 and 5.4 show the heat fluxes, as well as fluxes of moisture and buoyancy from LES-1 and LES-2. The thick solid curves denote total (resolved plus subfilter) fluxes. The total flux of moist enthalpy ($H \equiv \rho c_p \overline{w\theta_l} + F_R$) and the total moisture flux are linear with height as required by our condition of quasi-steady state. The effect of using an upwind scheme on the horizontal advection of scalars is evident in the individual components of the resolved- and subfilter-scale (SFS) fluxes near the cloud top where the temperature and moisture gradients are large. In LES-1, the SFS heat flux is positive and the SFS moisture flux is negative, while in LES-2 these SFS fluxes reverse signs. This highlights the difficulty of treating SFS terms, which operate most effectively at the grid scale where numerical errors are most evident. Here we simply note that in LES-1 the truncation error (due to overshoots) overestimates the resolved-scale moisture flux near cloud top and is combatted by the SFS model to maintain a linear profile of the total flux.

The difference in H above the cloud layer between LES-1 and LES-2 reflects a smaller radiative flux due to a smaller fractional cloud cover in LES-2 than in LES-1. This smaller radiative forcing also results in a smaller buoyancy flux in the cloud layer of LES-2.

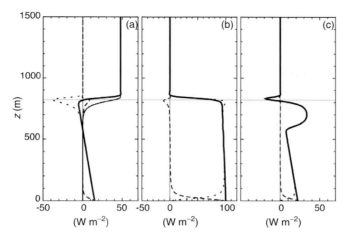

Figure 5.3. Vertical profiles, in energetic units, of (a) heat fluxes (dotted curve is the resolved-scale θ_l-flux, thin solid curve the longwave radiative flux, dashed curve the SFS θ_l-flux, thick solid curve the total heat flux H); (b) moisture fluxes (dotted curve is the resolved-scale q_T-flux, dashed curve the SFS q_T-flux, and thick solid curve the total); (c) buoyancy fluxes (dotted curve is the resolved-scale θ_v-flux, dashed curve the SFS θ_v-flux, and thick solid curve the total) from LES-1. The thin horizontal line represents the minimum θ_l-flux level. θ_v is the virtual potential temperature.

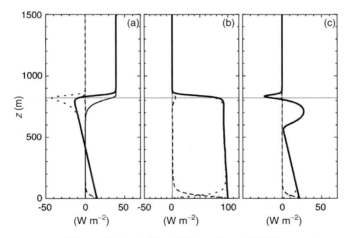

Figure 5.4. As Fig. 5.3, but from LES-2.

The mean moisture and moisture flux profiles shown above can be used to estimate w_e, the entrainment rate (velocity), because $\overline{wq_T} \equiv -w_e \Delta Q_T$ for non-precipitating cloud. From Fig. 5.2(a), we estimate a moisture jump of about -7.25 g kg^{-1}. Both Figs. 5.3(b) and 5.4(b) show that the entrainment moisture flux is about 90 W m^{-1}, which is somewhat larger than observed (see Fig. 8(a) in Stevens *et al.*,

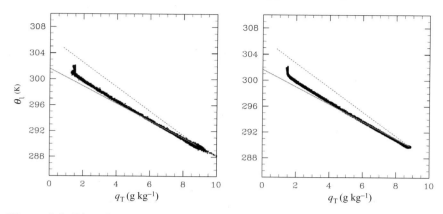

Figure 5.5. Plot of θ_l and q_T from the LES grid points between 750 m and 900 m. The thin solid line is taken from Fig. 6 of Stevens *et al.* (2003b) and the dotted line represents the Deardorff–Randall CTEI criterion. *Left panel*: LES-1; *right panel*: LES-2.

2003b). This simulated moisture flux yields an entrainment rate of about 0.50 cm s^{-1}, consistent with the estimate from the time change of the mean cloud-top height given above. Given that the entrainment rates are similar in LES-1 and LES-2, the larger radiative driving for LES-1 (implicit in the larger above-cloud values of H and a larger layer-averaged buoyancy flux) suggests that entrainment is somewhat more efficient in LES-2.

Our motivation for using the flux-limited upwind (i.e., monotone) scheme for horizontal advection of scalars in LES-2 is evident in the mixing-line plot of the LES-1 solution, shown in the left panel of Fig. 5.5. The dots are LES solutions from the layer between $z = 750$ m and 900 m (i.e., near the cloud-top regions) at hour 2. The thin solid line represents the observed mixing line (from Stevens *et al.*, 2003b) and the dotted line indicates the Deardorff–Randall cloud-top entrainment instability (CTEI) criterion (from Deardorff, 1980; Randall, 1980). The intersection between these two lines is calculated by layer-averaging the LES θ_l and q_T fields below $z = 700$ m, and is representative of the bulk mixed-layer properties, $\Theta_{l\,\mathrm{mix}} \sim 290$ K and $Q_{T\,\mathrm{mix}} \sim 8.75$ g kg^{-1}. In the LES-1 case, at the upper end of the mixing line, global minima develop (i.e., points with $q_T < 1.5$ g kg^{-1}). At the lower terminus some global maxima are evident (i.e., points with $q_T > 9$ g kg^{-1}). Neither can be justified on physical grounds. They are numerical artifacts which arise because the cloud top undulates in the presence of a mean wind so that horizontal advection also advects θ_l and q_T across sharp interfaces. The pseudo-spectral scheme used for the horizontal advection of scalars in LES-1 results in overshoot errors consistent with these extrema. We therefore implemented a flux-limited upwind scheme, which is the same as that used for vertical advection of scalars, for horizontal advection of scalars in LES-2. The extrema at both ends of the mixing line disappears in LES-2, as shown in the right panel of Fig. 5.5. Although the upwind scheme in LES-2 eliminates the spurious extrema, the bulk property of the mixing line is relatively unchanged between LES-1 and LES-2.

Because there are no spurious data points at the ends of the mixing line in LES-2, we use the right panel of Fig. 5.5 to estimate jumps in Θ_l and Q_T. The jumps of Θ_l and Q_T are computed as the differences between the end points of the mixing line (before the curve turns at the upper end), and that leads to $\Delta\Theta_l \sim 11$ K and $\Delta Q_T \sim -7.2$ g kg^{-1} at hour 2 of the simulation. Using these jumps, we obtain the Randall–Deardorff CTEI parameter $\kappa \equiv \Delta\Theta_e/(L/c_p)\Delta Q_T \sim 0.38$, where $\kappa > 0.23$ is hypothesized for break-up of cloud by Randall (1980) and Deardorff (1980). (Here $\Theta_e \sim \Theta_l + (L/c_p)Q_T$ is the equivalent potential temperature.) At the end of the simulation $\Delta\Theta_l$ grew to about 11.5 K, which results in a slightly steeper mixing line (not shown) than that shown in Fig. 5.5; nevertheless, throughout the four hours of simulation the mixing lines remain on the "unstable" side of the Randall–Deardorff criterion (i.e., $\kappa > 0.23$), as did the observed cloud, but the cloud layer remains solid in LES-1 and nearly solid in LES-2.

5.2.3 *Large-scale truncation*

The time series of the liquid water field from DYCOMS-II in-cloud flight legs (I. Faloona, personal communication) often show fluctuations on scales larger than the domain size of LES-1 and LES-2. This raises the question as to whether the truncation of scales larger than the numerical domain biases the resultant statistics. To investigate this issue we conducted LES-3, whose domain was a factor of three larger. For the most part the statistics were insensitive to the presence of larger scales (although truly large-scale variations may not have had time to spin-up in this short time period). The cloud field in LES-3 reveals the presence of scales larger than 2.5 km, which compares more favorably with observations, but these larger-scale fluctuations do not change the statistics examined here.

5.3 Sharp-edged framework

Having demonstrated that LES plausibly represents a real flow, we now use it to address specific scientific questions. The first (e.g., Lilly, 2002a) arises in response to lingering criticisms of mixed-layer theory which, as originally formulated (Lilly, 1968), rests on the idealization of the cloud-top interface as a discontinuity in the mean thermodynamic profiles. This so-called zero-order jump condition (see Chapter 4 by Randall and Schubert) has been criticized in part because field measurements and LES (e.g., Betts, 1974; Deardorff, 1979) often show a smooth transition in state variables over a non-negligible depth across the top of the mixed layer. To address this criticism, Lilly (2002a, b) introduced an interface-tracking coordinate which he identified with the local position of the entrainment interface. His idea was that the apparent diffusiveness in averaged entrainment interfaces could be a product of averaging over a locally sharp interface that fluctuates in space and time. By working in a coordinate system following the local interface, Lilly argues that such artifacts can be avoided.

5.3.1 Cloud-top interface

Central to Lilly's argument is the idea that the cloud top constitutes an unambiguous interface. However in practice, nothing is unambiguous. For example, in Fig. 5.6 we plot the location of two interfaces: (1) the liquid-water interface defined as the uppermost level where the liquid-water field changes from non-zero to zero, which is shown as a solid curve and denoted as z_{lwc} and (2) the maximum-θ_l-gradient interface defined as the height where the maximum vertical gradient of θ_l occurs, which is shown as a dotted curve and denoted as z_{grd}. (We also checked the maximum-q_T-gradient interface, and found it is nearly coincident with the z_{grd} interface.) To see the fluctuations in a larger domain, we combine two different horizontal segments of the LES-3 solution (each 7.5 km long) to form a total domain of 15 km.

These two interfaces seldom coincide: z_{grd} is most often above z_{lwc}, and the gap between them becomes wider where z_{lwc} is smaller; this is similar to what was observed by Stevens *et al.* (1999) for the case of a smoke cloud. Here we see that the vertical separation can be more than 100 m. These cloud-top fluctuations reveal some interesting physical processes near the cloud top. The interface (particularly when identified with z_{lwc}) is higher than average above vigorous updrafts. As these updrafts penetrate into the inversion, they squeeze the constant θ_l surfaces aloft intensifying the maximum gradient right above them. Thus, in these segments z_{grd} is likely to be about the same as z_{lwc} and both are near the top of the penetrating updrafts. Adjacent regions tend to have greater separation between z_{grd} and z_{lwc} in part for kinematic reasons. Similar effects are evident in the study of the dry convective boundary layer (CBL), cf. Sullivan *et al.* (1998), and of the smoke cloud, cf. Stevens *et al.* (1999). However in contrast to both the smoke-cloud CBL and the dry CBL, in the stratocumulus-topped PBL the manner in which the top of the layer is affected by mixing depends on how the top of the layer is defined – in large part because cloud top is not a material surface. If some inversion air is entrained and mixed in with these returning eddies, mixtures of clear and cloudy air can be expected to characterize the properties of the air near cloud top. Mixtures with small amounts of inversion air remain saturated, while those with more inversion air totally evaporate and become non-cloudy air. Similarly, in regions of active mixing the θ_l gradients will be reduced and thus the level where the gradient attains its

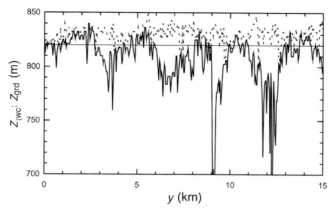

Figure 5.6. Spatial variation of the liquid-water interface (solid curve) and the maximum-θ_l-gradient interface (dotted curve) from LES-3. The horizontal line is the mid-level of the liquid-water interface undulation, which contains approximately equal amounts of clear and cloudy air.

maximum value can be expected to be above these active mixing regions. Hence, this process can simultaneously lower z_{lwc} and raise z_{grd} in mixing zones.

Figure 5.6 also reveals another interesting feature: Lenschow *et al.* (2000) used cloud-top penetration flight legs to study the jump conditions across the cloud-top interface. They purposely flew at a nearly constant height to ensure that flights spent equal time inside and outside the cloud layer (as postulated by the horizontal straight line near 820 m shown in Fig. 5.6). By averaging data collected on either side of the cloud edge (with about 10 m segment on either side) from multiple cloud penetrations, Lenschow *et al.* (2000) constructed composite profiles of the mean temperature, moisture and ozone concentration, and used these profiles to infer the jump condition across the cloud-top interface. They found that these temperature and moisture jumps, though remaining sharp, are "considerably" smaller than those measured from ascending/descending sounding flights. Figure 5.6 can help us understand this apparent discrepancy. Any cloud-top penetration flight leg along a horizontal straight line in the middle of the cloud-top undulations (as shown in Fig. 5.6) would miss most of the maximum $\partial\theta_l/\partial z$ areas and hence produce a considerably smaller mean Θ_l-jump across the interface than that measured from ascending/descending flight legs.

The fluctuating amplitude of the z_{grd} interface is clearly much smaller than that of the z_{lwc} interface, as evidenced from Fig. 5.6. The standard deviation of z_{lwc} is about 20 m computed from LES-1 and LES-3, and about 30 m from LES-2, all of which are commensurate with the observed value of 25 m as derived from downward-looking lidar during RF01. The standard deviation of z_{grd} is only about 7–8 m in all three LESs. The skewness of $z_{lwc}(\equiv \overline{z_{lwc}'^3}/\overline{z_{lwc}'^2}^{3/2})$ is about -2 and that of z_{grd} is around

-0.8 to -0.5, in all LESs. (Note that in these calculations of standard deviation and skewness of z_{lwc} we exclude all "holes" where $z_{lwc} = 0$. If the holes are included, the liquid-water interface would yield a much larger negative skewness.) The large negative skewness of z_{lwc} is consistent with the highly intrusive (into the mixed layer) feature shown in Fig. 5.6.

5.3.2 *Vertical profiles in the sharp-edged coordinate*

One apparent advantage of the sharp-edged top coordinate is its ability to cleanly delineate the transition at the top of the boundary layer. However, even this can be ambiguous; as defined, z_{lwc} can delineate the cloud-top interface, but it may not adequately separate turbulent from non-turbulent air due to entrainment and evaporation at the cloud top. On the other hand, z_{grd} may better delineate the turbulence boundary, but not the cloud boundary as shown in Fig. 5.6.

Both interfaces, particularly z_{lwc}, are highly distorted which makes it difficult to average over using real data or LES flows. For instance, the interface is not guaranteed to remain single valued, and in the case of z_{lwc} it becomes undefined if no cloud exists in a column. For the LES cases studied here we do not experience multi-valued interfaces, and in regions where there is no cloud in a column, we interpolate z_{lwc} based on neighboring points for the purpose of constructing the z/z_{lwc} coordinate. This procedure is admittedly ad hoc but given nearly 100% cloud cover in LES-1, it probably does not significantly bias our subsequent results.

Figure 5.7 compares the mean temperature, moisture and liquid-water profiles averaged along the z/z_{lwc} coordinate with those averaged in the traditional smooth-top framework (i.e., $z/\bar{z}_i$) from LES-1. Note that z_{lwc} varies in x, y, and t, while

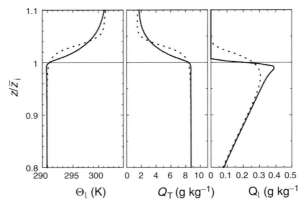

Figure 5.7. Profiles of Θ_l, Q_T and Q_l from LES-1 where solid curves represent averages along the z/z_{lwc} coordinate and dotted curves represent averages along horizontal planes, $\bar{z}_i$, which is the horizontal average of z_{lwc}.

$\overline{z_i}$ varies only in t. The largest difference between the two averaging procedures shows up in the Q_l profile, which is expected given the definition of z_{lwc}. The changed coordinate, however, doesn't significantly affect the profiles of Θ_l and Q_T. The jumps of Θ_l and Q_T across the cloud top are not as sharp as one might have anticipated for a sharp-edged coordinate. (We do not plot the averages on the z/z_{grd} coordinate because they are essentially identical to the traditional smooth-top averages, mainly because the LES case analyzed here has a very sharp and strong capping inversion that is barely resolved even with our fine vertical grid spacing; the z_{grd} interface fluctuates within 1 to 2 vertical levels.)

It would be interesting to compare the flux profiles between the sharp-edged and smooth-top coordinates; as argued by Lilly (2002a, b) the minimum θ_l flux obtained from the smooth-top coordinate is much smaller than that obtained from the sharp-edged coordinate. However, we had trouble in constructing flux profiles for the sharp-edged coordinate from the LES solutions. Constructing the fluxes in such a coordinate requires computing the vertical velocity ω in the z/z_{lwc} coordinate (see Equation (6.1) in Lilly, 2002a), which involves taking the time and space derivatives of the highly distorted surface z_{lwc} and hence is difficult to perform with either LESs or field measurements. In particular, ω becomes ill-defined near cloud-free columns where $z_{lwc} = 0$.

The difference between the entrainment θ_l-flux (denoted as $\overline{w\theta_l}_i$, and by definition the flux averaged along the sharp-edged top) and the minimum θ_l-flux obtained from horizontal averaging was discussed by Lilly (2002a). Based on smoke-cloud LESs from Moeng *et al.* (1999) and assuming a Gaussian distribution for smoke-top fluctuations, Lilly showed that the ratio of these two fluxes $\overline{w\theta_l}_i / \overline{w\theta_l}_{min} \sim \exp^{(4s/\lambda)}$, where s is the standard deviation of the interface fluctuations and λ is the decay length scale of longwave radiation as defined in Section 5.2.1. From LES-1, $s \sim 20$ m using z_{lwc} as the interface and $\lambda \sim 26$ m, which yields $\exp^{(4s/\lambda)} \sim 20$. Figure 5.3(a) shows (dotted line) that $\rho_0 c_p \overline{w\theta_l}_{min} \sim -38$ Wm^{-2}, and since $\rho_0 c_p \overline{w\theta_l}_i \equiv -\rho_0 c_p w_e \Delta\Theta_l \sim -53$ W m^{-2}, their ratio is only about 1.4. Hence we conclude that Lilly's assumption of a Gaussian distribution for the interface fluctuations works only for smoke-cloud cases, not for wet-cloud cases.

5.4 Interface properties and stability

Lilly's second question is particularly relevant to RF01 and also well suited to evaluation by LES. He asks what characterizes the effective stability of the interface. This is essentially the same question which motivates the theoretical discussion presented in Chapter 4 by Randall and Schubert. It arises because for stratocumulus entrainment mixing occurs in conjunction with phase changes of water, and hence buoyancy of the mixtures doesn't depend linearly on mixing fraction. And, because of the phase change of water, the entrainment buoyancy flux (i.e., the buoyancy flux along the cloud-top interface where entrainment mixing occurs) does not simply equal $-w_e \Delta \Theta_v$.

5.4.1 Wetness of cloud-top interface

At the thin cloud-top interface, air is neither completely dry nor completely wet. To characterize the buoyancy flux at this thin interface, Lilly (2002b) defines a "wetness" factor α, which describes the moisture content of the interface, as

$$\overline{w\theta_{vi}} = \alpha \overline{w\theta_{vwi}} + (1 - \alpha)\overline{w\theta_{vdi}}, \tag{5.4}$$

where the dry (hypothetical) buoyancy flux is defined as

$$\overline{w\theta_{vdi}} = a_d \overline{w\theta_{li}} + b_d \overline{wq_{Ti}} \tag{5.5}$$

assuming the interface is completely dry, and the wet (hypothetical) buoyancy flux is defined as

$$\overline{w\theta_{vwi}} = a_w \overline{w\theta_{li}} + b_w \overline{wq_{Ti}} \tag{5.6}$$

assuming the interface is completely wet. The thermodynamic coefficients are $a_d = 1$ and $b_d \sim 175$ K for unsaturated air and $a_w \sim 0.54$ and $b_w \sim 1035$ K for saturated air, following Lilly's thermodynamic approximation. Equation (5.4) makes the entrainment buoyancy flux depend strongly on the wetness factor α. Rewriting (5.4) yields

$$\alpha = \frac{\overline{w\theta_{vdi}} - \overline{w\theta_{vi}}}{\overline{w\theta_{vdi}} - \overline{w\theta_{vwi}}}. \tag{5.7}$$

Lilly related this "wetness" factor to the mixing fraction of dry inversion air at which the mixture is just saturated, denoted as m_*. The concept of mixing fraction and how it modifies the buoyancy of mixtures has been used by many investigators (e.g., Nicholls and Turton, 1986; Kuo and Schubert, 1988; Siems *et al.*, 1990; Lilly, 2002b) to explain interface instability. This instability factor has been explicitly incorporated into the entrainment rate parameterization of Turton and Nicholls (1987). The basic idea is explained with the help of Fig. 5.8, which was adapted from Fig. 3 in Stevens (2002).

The figure illustrates that the buoyancy of mixtures depends linearly on mixing fraction only when the mixtures are completely saturated ($m < m_*$) or completely unsaturated ($m > m_*$); these two linear curves have different slopes. The linear curve on the unsaturated side has a slope that satisfies the clear-air thermodynamic property:

$$\Delta_d \Theta_v \equiv a_d \Delta \Theta_l + b_d \Delta Q_T, \tag{5.8}$$

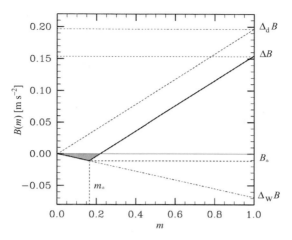

Figure 5.8. Buoyancy of mixtures of boundary-layer air with a mass fraction, m, of above-boundary-layer air. Here the buoyancy, $B = g\delta\Theta_v/\Theta_0$, measured relative to the boundary-layer air, is plotted versus m. See text for further discussion.

while the linear curve on the saturated side has a slope that satisfies the cloudy-air thermodynamic property:

$$\Delta_w\Theta_v \equiv a_w\Delta\Theta_l + b_w\Delta Q_T. \tag{5.9}$$

The curves intersect at m_* where

$$m_* = \frac{\Delta_d\Theta_v - \Delta\Theta_v}{\Delta_d\Theta_v - \Delta_w\Theta_v}. \tag{5.10}$$

(A more detailed derivation of the above equation is given in Moeng *et al.*, 1995.)

There is a similarity between (5.7) and (5.10). Because $\overline{w\theta}_{vdi} = -w_e\Delta_d\Theta_v$ for completely unsaturated air and $\overline{w\theta}_{vwi} = -w_e\Delta_w\Theta_v$ for completely saturated air, (5.7) and (5.10) imply that $\overline{w\theta}_{vi} = -w_e\Delta\Theta_v$ only if $\alpha = m_*$. Based on several LES solutions Lilly (2002b) argues that α is in general larger than m_* according to

$$\alpha = 1 - (1 - m_*)^{8/3}. \tag{5.11}$$

As discussed by Lilly, $\alpha = m_*$ only if no mixing occurs near the cloud top; for a typical stratocumulus where mixing does occur near the cloud top, $\alpha > m_*$.

LES-1 of RF01 allows us to check Lilly's ideas using a case independent of those used by Lilly in calibrating the above relationships. From LES-1, we estimate $\Delta_w\Theta_v \sim -1.5$ K from (5.9) and $\Delta_d\Theta_v \sim 9.7$ K from (5.8) using Lilly's values of a_d, a_w, b_d, and b_w. We also deduce the jump of mean virtual potential temperature from a mixing-line analysis of θ_v and q_T (not shown), which produces $\Delta\Theta_v \sim 9$ K. Substituting these values into (5.10) yields $m_* \sim 0.06$ and in (5.11) gives $\alpha \sim 0.15$.

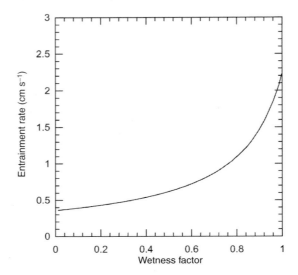

Figure 5.9. Dependence of the entrainment rate on the wetness factor, using Lilly's new entrainment-rate formula (i.e., Equation (2.15) in Lilly, 2002b) and the LES-1 simulation fields.

Alternatively α can be derived by picking the value that yields the best agreement between the entrainment rate obtained from Lilly's new entrainment-rate formula (i.e. Equation (2.15) of Lilly (2002b)) and the simulated rate. We solve this graphically by plotting in Fig. 5.9 the dependence of the entrainment rate on the wetness factor α using the LES-1 results for surface fluxes, F_i, $\Delta_d \Theta_v$, $\Delta_w \Theta_v$, and the cloud-base to cloud-top ratio. Figure 5.9 shows a strong dependence of w_e on α, particularly when α becomes larger, i.e., where the interface becomes wetter. For $w_e \sim 0.48$ cm s^{-1} as in LES-1, Fig. 5.9 yields $\alpha \sim 0.3$. If we assume $\alpha = 1$ (a completely saturated interface), the entrainment rate would have been 4–5 times larger than this predicted value.

In the above, we have discussed the value of α based on a z_{lwc} interface where the layer below is completely wet and the layer above is completely dry. If the sharp-top interface is assumed to be z_{grd}, Fig. 5.6 suggests that the interface would be close to completely dry because z_{grd} is mostly above z_{lwc}.

5.4.2 A new CTEI criterion

With his new entrainment-rate formula, Lilly (2002b) derived a new CTEI criterion, which can be summarized as

$$-\frac{L\Delta Q_T}{c_p \Delta \Theta_l} > F\left(\alpha, \frac{\overline{z_b}}{\overline{z_i}}\right), \tag{5.12}$$

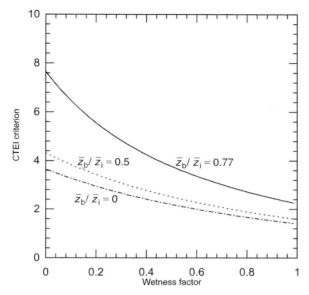

Figure 5.10. Dependence of the CTEI criterion on the wetness factor and the ratio of the cloud base-to-top ratio calculated from Equation (5.2) in Lilly (2002b).

where L is the latent heat of evaporation. The right-hand side depends on the wetness factor and the ratio of the mean cloud-base ($\overline{z_b}$) to cloud-top $\overline{z_i}$ heights. Lilly derived the above formula by setting the denominator of his entrainment-rate formula to zero such that the entrainment rate becomes infinitely large and hence corresponds to instability. We plot this new instability criterion as a function of α in Fig. 5.10 for three different $\overline{z_b}/\overline{z_i}$ values; $\overline{z_b}/\overline{z_i} = 0.77$ represents the DYCOMS-II RF01 case. In all cases this CTEI criterion is more stringent than the Deardorff–Randall criterion [which is 1.28 in terms of $-L\Delta Q_T/c_p\Delta\Theta_l$ and is the smallest value one can find in Fig. 5.10, where the cloud-top interface is assumed to be completely saturated (i.e., $\alpha = 1$) and the cloud base zero]. The reasons Lilly's new criterion is more stringent are two-fold: (i) the energetics of the whole layer, rather than just the cloud-top interfacial layer are considered in the entrainment formula; (ii) the buoyancy flux at the interface is a weighted combination of fluxes due to saturated and unsaturated mixtures rather than obtained by simply assuming all mixtures are saturated as has been done by Deardorff (1980) and Randall (1980).

For the DYCOMS-II RF01 case, $\overline{z_b}/\overline{z_i} \sim 0.77$ and $\alpha \sim 0.15$–0.3, and hence the jump ratio in (5.12) has to be larger than ~ 5 in order for instability (defined here as $w_e \to \infty$) to occur. The mixing line in Fig. 5.5 indicates that the jump ratio in the DYCOMS-II RF01 case is only about 1.6, which appears consistent with the sustenance of the cloud layer. Some other CTEI criteria (e.g., Kuo and

Schubert, 1988; MacVean and Mason, 1990; Siems *et al.*, 1990) might also predict a stable cloud layer for this RF01 case, but here we examine only Lilly's new criterion.

5.5 Summary

Large-eddy simulation provides scientists with an invaluable tool in their efforts to unravel the mysteries of flows beyond the reach of laboratories. The value of this tool depends in large part on the extent to which simulations are insensitive to the assumptions upon which they are based, for instance on the fidelity of the numerical representation of the resolved component of the flow, or the faithfulness with which the parameterized scales (both large and small) are represented. To illustrate these points we start with the construction of benchmark simulations, built around measurements derived from a recent field study. Comparisons between the simulations and the observed cloud evolution suggest that the simulations perform reasonably well. Moreover, sensitivity studies indicate that the LES representation is not markedly sensitive to the truncation of larger scales, and that details of the numerical representation only modestly affect the macroscopic statistics of the LES. The results of this benchmarking encourage the use of LES to investigate questions raised in two recent papers by Lilly (2002a, b).

The first question we address is motivated by Lilly's argument that the idealization of the stratocumulus-topped boundary layer as a well-mixed layer topped by a discontinuity in profiles of averaged state variables is most appropriate from the perspective of a coordinate system following the local cloud-top height. In contrast, the use of the geometrical height as the vertical coordinate artificially smoothes out the sharp jump over a layer equal to the depth of the cloud-top fluctuations. Because it resolves the three-dimensional structure of the cloud-top interface, LES is well suited to investigating questions relating to these possible choices of the vertical coordinate. It is shown that the conceptual simplicity of the interface-following coordinate is partially offset by ambiguity in defining the interface. In general, an interface defined by the cloud tops lies below the interface defined by the maximum gradient in temperature or moisture, and also fluctuates more. For this reason, the averaged statistics depend upon how one defines the interface. As a result we find it difficult to recommend one coordinate framework as clearly superior to the other.

The second question is what determines the effective stability of the cloud-top interface. With his new entrainment-rate formula, Lilly (2002b) derives a new criterion for the stability of the interface, and that criterion depends strongly on an interpolation factor (called wetness because it is related to the moisture content of mixtures) and the ratio of cloud-base to cloud-top heights. From the LES of the DYCOMS-II RF01 case, the wetness is estimated to be between 0.15 to 0.3 and the cloud base-to-top ratio is about 0.77. These values yield a more stringent criterion compared to Deardorff–Randall's and put the RF01 cloud layer in the stable regime with respect to the interface stability.

References

Betts, A. K. (1974). Reply to comment on the paper "Non-precipitating cumulus convection and its parameterization." *Quart. J. Roy. Meteor. Soc.*, **100**, 469–471.

Deardorff, J. W. (1979). Prediction of convective mixed-layer entrainment for realistic capping inversion structure. *J. Atmos. Sci.*, **36**, 424–436.

(1980). Cloud-top entrainment instability. *J. Atmos. Sci.*, **37**, 131–147.

Koren, B. (1993). A robust upwind discretization method for advection, diffusion and source terms. In *Notes on Numerical Fluid Mechanics*, **45**, C.B. Vreugdenhil and B. Koren, eds., Vieweg, 117–138.

Kuo, H. and Schubert, W. H. (1988). Stability of cloud-topped boundary layers. *Quart. J. Roy. Meteor. Soc.*, **114**, 887–917.

Lenschow, D. H., Zhou, M., Zeng, X., Chen, L. and Xu, X. (2000). Measurements of fine-scale structure at the top of marine stratocumulus. *Boundary-Layer Meteorol.*, **97**, 331–357.

Lilly, D. K. (1968). Models of cloud-topped mixed layers under a strong inversion. *Quart. J. Roy. Meteor. Soc.*, **94**, 292–309.

(2002a). Entrainment into mixed layers, Part I: Sharp-edged and smoothed tops. *J. Atmos. Sci.*, **59**, 3340–3352.

(2002b). Entrainment into mixed layers, Part II: A new closure. *J. Atmos. Sci.*, **59**, 3353–3361.

MacVean, M. K. and Mason, P. J. (1990). Cloud-top entrainment instability through small-scale mixing and its parameterization in numerical models. *J. Atmos. Sci.*, **47**, 1012–1030.

Moeng, C.-H. (1986). Large-eddy simulation of a stratus-topped boundary layer. Part I: Structure and budgets. *J. Atmos. Sci.*, **43**, 2886–2900.

(2000). Entrainment rate, cloud fraction, and liquid water path of PBL stratocumulus clouds. *J. Atmos. Sci.*, **57**, 3627–3643.

Moeng, C.-H., Lenschow, D. H. and Randall, D. A. (1995). Numerical investigations of the roles of radiative and evaporative feedbacks in stratocumulus entrainment and breakup. *J. Atmos. Sci.*, **52**, 2869–2883.

Moeng, C.-H., Sullivan, P. P. and Stevens, B. (1999). Including radiative effects in an entrainment-rate formula for buoyancy-driven PBLs. *J. Atmos. Sci.*, **56**, 1031–1049.

Nicholls, S. and Turton, D. J. (1986). An observational study of the structure of stratiform cloud sheets: Part II. Entrainment. *Quart. J. Roy. Meteor. Soc.*, **112**, 461–480.

Randall, D. A. (1980). Conditional instability of the first kind upside-down. *J. Atmos. Sci.*, **37**, 125–130.

Siems, S. T., Bretherton, C. S., Baker, M. B., Shy, S. S. and Breidenthal, R.E. (1990). Buoyancy reversal and cloud-top entrainment instability. *Quart. J. Roy. Meteor. Soc.*, **116**, 705–739.

Stevens, B. (2002). Entrainment in stratocumulus mixed layers. *Quart. J. Roy. Meteor. Soc.*, **128**, 2663–2690.

Stevens, B., Moeng, C.-H. and Sullivan, P. P. (1999). Large-eddy simulations of radiatively driven convection: sensitivities to the representation of small scales. *J. Atmos. Sci.*, **56,** 3963–3984.

Stevens, B., Lenschow, D. H., Vali, G., . . . , Lilly, D. K., *et al.* (2003a). Dynamics and chemistry of marine stratocumulus. *Bull. Amer. Meteor. Soc.*, **84**, 579–593.

Stevens, B., Lenschow, D. H., Faloona, I., *et al.* (2003b). On entrainment rates in nocturnal marine stratocumulus. *Quart. J. Roy. Meteor. Soc.*, **129**, 3469–3493.

Sullivan, P. P., Moeng, C.-H., Stevens, B., Lenschow, D. H. and Mayer, S.D. (1998). Structure of the entrainment zone in the convective atmospheric boundary layer. *J. Atmos. Sci.*, **55**, 3042–3064.

Turton, J. D. and Nicholls, S. (1987). A study of the diurnal variation of stratocumulus using a multiple mixed-layer model. *Quart. J. Roy. Meteor. Soc.*, **113**, 969–1009.

Part II

Mesoscale meteorology

6

Model numerics for convective-storm simulation

Joseph B. Klemp and William C. Skamarock
National Center for Atmospheric Research, Boulder, Colorado

6.1 Introduction

Over the past forty years, the numerical simulation of atmospheric convection has evolved from its infancy in two-dimensional dry thermals to highly sophisticated three-dimensional models used for numerical weather prediction (NWP) at convective scales. This advancement has been feasible because of the enormous growth in computing power, increasing from thousands to billions of calculations per second during this period (Wilhelmson and Wicker, 2001). However, significant advancements in model numerics, physical parameterizations, and data analysis have also been required to capture the complexity of atmospheric convection and convective storms in numerical simulation models. Throughout these decades, Doug Lilly has been a major force in advancing this technology, both in his own research and in motivating the achievements of others.

Lilly (1962) conducted pioneering research on the numerical simulation of buoyant thermals that laid the groundwork for the 3D convective storm models that evolved in subsequent decades. This work included a new approach for grid staggering (Lilly, 1961) and improved techniques for the treatment of subgrid turbulence in an inertial subrange using a nonlinear eddy viscosity proportional to the local shear and modified by buoyancy effects through a Richardson-number dependency. Lilly solved the full 2D compressible equations in flux form, placing strong emphasis on both numerical stability and accurate conservation (analyzed systematically for alternative numerical schemes in Lilly, 1965). Lilly fostered the development of one of the early 3D cloud models, the Klemp–Wilhelmson model, and in founding and directing the Center for Analysis and Prediction of Storms (CAPS), he promoted development of the Atmospheric Research and Prediction System (ARPS).

Throughout his long association with convective storm modeling, Lilly has had an abiding interest in evaluating and improving subgrid turbulence closure techniques for these models (Lilly, 1962, 1966, 1967, 1992; Scotti *et al.*, 1993; Wong and Lilly, 1994). However, the benefit of improved subgrid turbulence schemes will only be realized if the computational damping in the model numerics is small compared to the dissipation produced by the physical parameterization. Cloud models that have been used extensively for storm research [e.g., Klemp–Wilhelmson (Klemp and Wilhelmson, 1978), RAMS (Tripoli and Cotton, 1982), ARPS (Xue *et al.*, 2000)] have integrated the nonhydrostatic equations in advective form (non-conservative for flux quantities) using split-explicit leapfrog time integration techniques to efficiently accommodate acoustic modes, and used numerical filters to maintain stability. In conducting analyses of momentum and energy budgets for supercell storm simulations with the Klemp–Wilhelmson model, Lilly and Jewett (1990) found that these numerical filters can cause excessive damping in the simulated storms, removing more energy than the physically based subgrid turbulence closure. The numerical filters are required in models using split-explicit leapfrog time integration to prevent unrealistic energy buildup near the grid scale, particularly in strongly nonlinear simulations.

In considering alternative numerical approaches in designing a new Weather Research and Forecasting Model (WRF), we have sought techniques that improve the numerical characteristics to mitigate some of the limitations encountered in earlier models. Although several different numerical solvers for the dynamic equations are being developed and tested as part of the WRF project, the discussion here will focus on the Eulerian split-explicit version using the basic equation set and numerical techniques as described by Klemp *et al.* (2000) and Wicker and Skamarock (2002). Further references to the WRF model in this chapter focus on this particular model solver. In this version of the WRF model, we have implemented higher-order numerical techniques that obviate the need for additional numerical filters and significantly reduce artificial damping near the grid scale. Without this artificial energy sink, it has been found that the subgrid turbulence parameterization needs to be re-tuned in order to achieve realistic energy spectra at small scales (Takemi and Rotunno, 2003). The WRF model also returns to flux form integration of the prognostic equations (as Lilly adopted in his early convection model) to ensure conservation of first-order quantities.

In this presentation, we will discuss the characteristics of the numerical techniques used to integrate the dynamic equations in WRF, and compare them to those used in the predecessor Klemp–Wilhelmson model. We will describe how the previous subgrid turbulence closure should be adjusted in light of the improved numerics and how the power spectra in convective simulations are improved. These numerics also appear to provide a realistic representation of the kinetic-energy spectra in NWP applications, and we will illustrate the nature of this behavior. To begin,

(a)

July 1999–2002 Mean Surface Wind (SeaWinds)

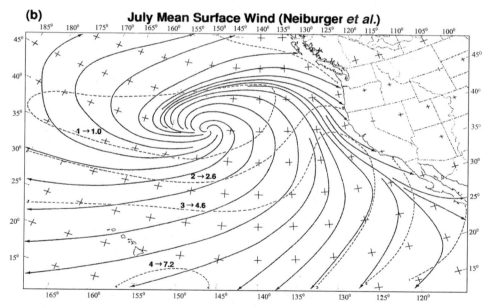

(b)

July Mean Surface Wind (Neiburger *et al.*)

Plate 1 (a) Streamlines and (b) isotachs of the July average surface winds as determined by the SeaWinds scatterometer aboard QuikSCAT and as estimated by Neiburger *et al.* (1961). The QuikSCAT average, shown in (a), is for July of the years 1999–2002 (only July 19–31 was available for 1999) and has 0.5° latitude/longitude resolution. The Neiburger *et al.* average, shown in (b), is actually a redrafted version of McDonald's (1938) Charts 9 and 21, in which the wind speeds are contoured in Beaufort units. We have converted the Beaufort units to m s^{-1} with the arrow notation, e.g., $4 \rightarrow 7.2$ meaning 4 Beaufort units is equivalent to 7.2 m s^{-1}.

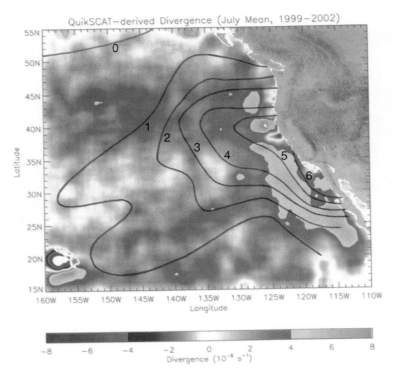

Plate 2 The color analysis shows the divergence field associated with the QuikSCAT winds displayed in Plate 1. For comparison the July mean divergence estimates of Neiburger *et al.* (1961) are shown by the solid black isolines, labeled in units of 10^{-6} s^{-1}. The white areas adjacent to the North American coast, the Hawaiian Islands, the Aleutians, and Guadalupe are regions where the divergence could not be calculated.

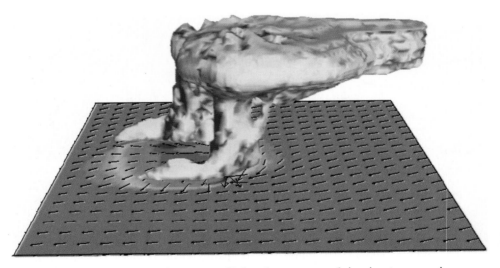

Plate 3 Simulated splitting supercell thunderstorms evolving in strong environmental wind shear, displayed at 2 h. The cloud field is shaded in gray, surface temperature is colored in shades ranging from red (warm) to blue (cold), and surface wind vectors are included at every fourth interval. The model integration employed $\Delta x = \Delta y = 1$ km, $\Delta z = 500$ m, within an $80 \times 80 \times 20$ km domain.

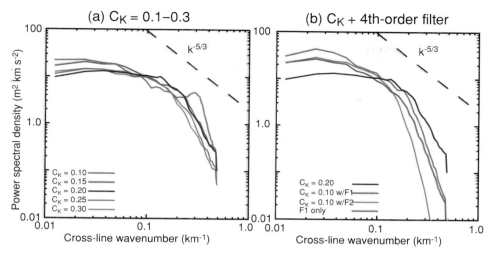

Plate 4 Power spectral density of vertical velocity at a height of 3 km in the cross-line (x) direction, obtained using the TKE scheme for turbulence closure. (a) Influence of C_K ranging from 0.1 to 0.3; (b) Effects of fourth-order horizontal filter with and without the TKE scheme. $C_\epsilon = 0.93$ for all TKE runs. A dashed $k^{-5/3}$ line is also plotted for reference. (From Takemi and Rotunno, 2003.)

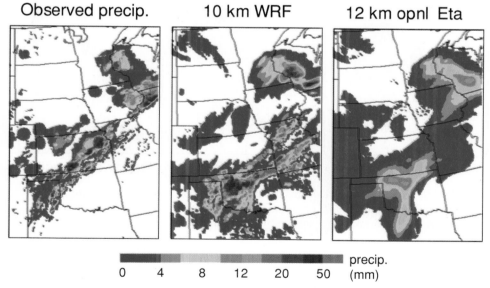

Plate 5 15–18 UTC accumulated precipitation from 10 km WRF and 12 km operational Eta model forecasts initialized at 12 UTC on 4 June 2002. The observed 15–18 UTC precipitation is also displayed. (From Baldwin and Wandishin, 2002.)

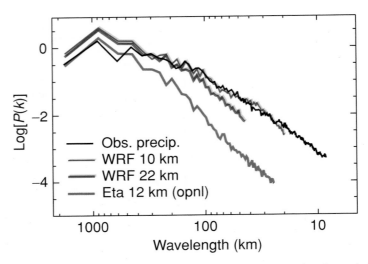

Plate 6 Power spectra for the observed 15–18 UTC accumulated precipitation on 4 June 2002 along with spectra for the forecast precipitation from the 10 km WRF, the 22 km WRF, and the 12 km operational Eta model. (From Baldwin and Wandishin, 2002.)

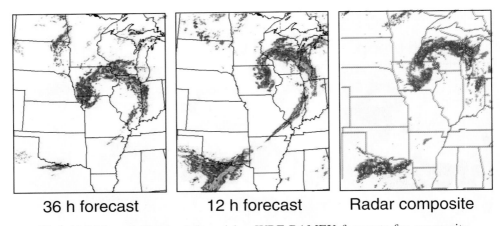

36 h forecast   12 h forecast   Radar composite

Plate 7 36 h and 12 h real-time 4 km WRF BAMEX forecasts for composite reflectivity valid at 12 UTC on 8 June 2003. Observed composite reflectivity from NEXRAD Radar are also displayed for comparison.

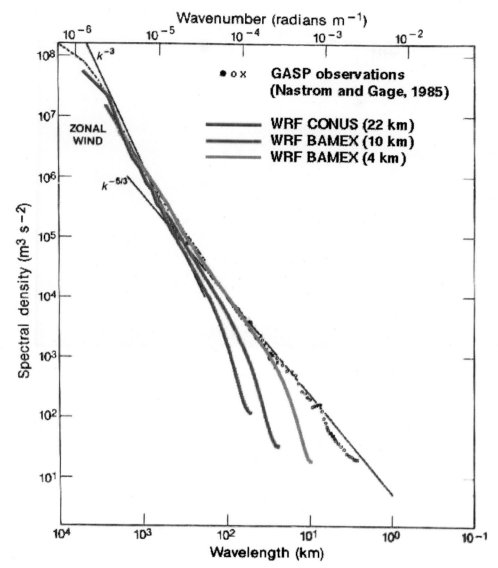

Plate 8 Power spectral density of kinetic energy from experimental WRF 22 km grid CONUS forecasts, 10 km grid BAMEX forecasts, and 4 km grid BAMEX forecasts. Also included is the observational spectral data from commercial aircraft presented by Nastrom and Gage (1985).

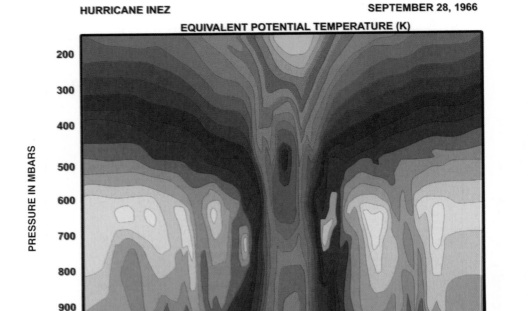

Plate 9 Equivalent potential temperature (K) as a function of pressure and radius from the centre of Hurricane Inez on 28 September 1966, based on aircraft data at 500 m, and at pressures of 750, 650, 500 and 180 mbar. Contours are at intervals of 2 K with a minimum value of 336 K (light blue) and a maximum value of 376 K (yellow). (After Hawkins and Imbembo, 1976.)

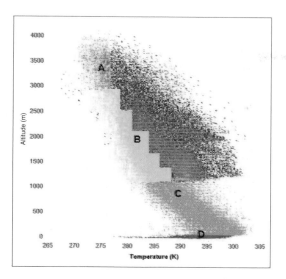

Plate 10 Terrain elevation and temperature of the Sierra Nevada range in California. Temperature map was derived from thermal infrared channel 11 on MODIS instrument on 30 October 2002. The scattergram is partitioned to show four cooler-than-average regions (A, B, C, D) at different altitudes.

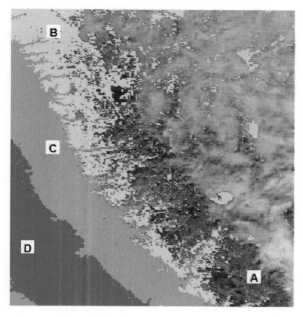

Plate 11 Map view of Sierra Nevada range. Partitions in Plate 10 are shown in the same color and labeled A, B, C, D.

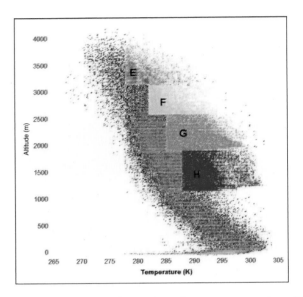

Plate 12 Same as Plate 10, but for warmer-than-average regions (E, F, G, H).

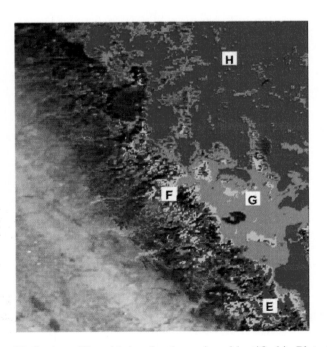

Plate 13 Same as Plate 11, but for the regions identified in Plate 12.

we will briefly summarize the evidence in Lilly and Jewett's (1990) analysis that reveals the significant computational damping in these storm simulations.

6.2 Computational damping in supercell storm simulations

Lilly and Jewett (1990, hereafter referred to as LJ) computed momentum and kinetic-energy budgets for simulated supercell storms to determine their sources, sinks, and transport using model datasets that overcome the problem of incomplete data inherent in observational studies. Supercells are powerful rotating thunderstorms that are often long-lived and produce severe weather such as tornadoes, large hail, and heavy precipitation. Plate III illustrates the nature of supercells developing in a numerical simulation with WRF for environmental conditions similar to those in the cases analyzed by LJ. This simulation depicts splitting supercells forming in strong environmental shear and characterized by single-cell rotating updrafts with cold precipitation-induced downdrafts spreading out to form gust fronts along the flanking lines of the storms. The LJ analysis revealed that the vertical flux of horizontal momentum is consistently down the velocity gradient, and that the mean-flow kinetic energy associated with the vertical shear is a major contributor to the kinetic energy of the storm. This transfer of mean kinetic energy to the storm was found to occur primarily through a gravity wave produced by the cold rear-flank downdraft outflow. Their analysis also revealed that the computational damping in the model had a significant influence on the overall kinetic-energy budget, as illustrated in Fig. 6.1 for the right-moving supercell case simulated by Klemp *et al.* (1981). Here, the shear-stress product term (VSH), representing the

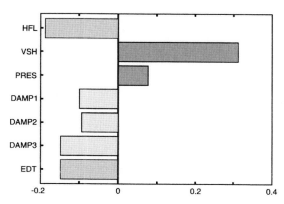

Figure 6.1. Bar-graph representation of the volume-averaged contributions to change of the kinetic energy of disturbance horizontal flow at 2 h in the Klemp *et al.* (1981) supercell simulation. (From Lilly and Jewett, 1990.)

energy transfer from the mean vertical shear, provides the dominant contribution to the horizontal kinetic energy of the storm. The other positive contribution comes from the pressure-induced transfer of vertical kinetic energy into the horizontal (PRES). The sinks for the horizontal kinetic energy arise from the horizontal flux convergence through the lateral boundaries (HFL), and from the damping terms, and produce a net rate of change (EDT) that is negative at this particular time.

To clarify the nature of damping effects, LJ computed the individual damping terms separately: DAMP1 results from the subgrid turbulence closure that solves a turbulence energy equation to calculate a shear- and buoyancy-dependent eddy viscosity; DAMP2 arises from the fourth-order horizontal filter; and DAMP3 is associated with a second-order vertical filter on perturbation quantities. The damping effects on the horizontal kinetic-energy budget from the numerical filters are about twice the size of the damping produced by the physically based turbulence parameterization. This does not necessarily mean that the overall damping is three times too large; with less computational damping, the physical damping would increase as more smaller-scale structure would be present. However, it does mean that the selective physics in the turbulence closure is being compromised by the less discriminating damping in the computational filters. As stated by LJ, the effects of these filters "do not necessarily vitiate the usefulness of the models, since the more important larger scales of the simulations may not be much affected. They do indicate that the modeling techniques used in these simulations need improvement."

The significant influence of the computational filter was also documented for squall-line simulations by Weisman *et al.* (1997). Their examination of the dependence of squall-line structure on horizontal resolution was complicated by the fourth-order computational filter, which for filter coefficients required to control small-scale noise, does not produce the same magnitude of damping as the horizontal resolution is varied. Furthermore, their sensitivity simulations revealed that changes in the filter coefficient had a much larger effect on the precipitation characteristics of the squall lines than comparable changes in coefficients in the turbulence closure scheme.

6.3 Higher-order numerics in WRF

In seeking to improve the numerics of the earlier cloud models, we have placed emphasis on techniques that could provide better conservation properties, increased accuracy, and a more robust behavior. The importance of strict conservation of linear and quadratic integral properties is arguable for many cloud-scale and NWP applications, which have open lateral boundaries, relatively short integration times, and processes that contribute large sources and sinks to the model equations (Lilly, 1962). Although conservation of certain quadratic quantities may improve nonlinear

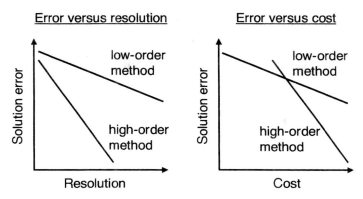

Figure 6.2. Dependence of solution error in low- and high-order numerics on resolution and computational cost.

stability (Phillips, 1959), linear stability characteristics and numerical accuracy may be more important considerations. Also, strict conservation of mass and entropy is important for dynamic models used in atmospheric-chemistry and air-quality applications (Byun, 1999). For WRF, we have chosen to solve the full nonhydrostatic compressible equations in flux form, thereby conserving first-order quantities that have conservation properties. We have developed time-split integration techniques for the flux form of the equations for a terrain-following height coordinate and a terrain-following hydrostatic pressure (mass) coordinate (Klemp *et al.*, 2000). Using the flux form of the equations, we integrate prognostic equations for conserved variables and recover other variables (such as pressure and temperature) from diagnostic relationships. This approach allows exact mass conservation, which is typically not possible when solving a prognostic pressure equation.

In considering the appropriate accuracy for numerical techniques, we desire the most efficient scheme for a particular application. Here, by efficiency, we mean the most accurate solution for a given amount of computer time, or equivalently, a solution of given accuracy in the minimum amount of computer time. Thus, we are faced with the tradeoffs between the benefits of higher-order (more accurate) numerics and their accompanying increased computational costs, as illustrated schematically in Fig. 6.2. Clearly, higher-order numerics should yield smaller error than lower-order techniques for a given resolution. However, since the higher-order schemes are computationally more expensive, there will typically be a crossover point in efficiency between the high- and low-order numerics. At low resolution the low-order schemes may be most efficient, but because the high-order schemes converge more rapidly with increasing resolution, they will at some point become more cost effective. Unfortunately, for complex applications including parameterized physics, it is difficult, if not impossible, to quantify these error profiles, which may vary depending on the specific application. Nevertheless, since the resolution of model simulations will continue to increase as computer power advances, we believe that higher-order techniques will become increasingly advantageous.

For multi-dimensional models, higher-order numerics also tend to be more efficient because of the strong dependence of the number of computations on the grid size. In a three-dimensional model, the number of calculations required to integrate over a specified time interval is inversely proportional to the fourth power of the grid size (i.e., halving the grid interval in three dimensions plus halving the time step increases the computations by a factor of 16). As a result, a significant increase in the number of calculations at each grid point that might be needed for higher-order schemes can be offset by a small increase in grid size. For example, if a high-order scheme requires twice as many computations per grid point as a lower-order scheme, the high-order model can run in the same amount of computer time as the low-order version by using a grid that is only 19% coarser than low-order grid.

The time integration scheme chosen for our WRF solver is a time-split third-order Runga–Kutta (RK3) method proposed and evaluated by Wicker and Skamarock (2002). The basic algorithm for the integration of the equation $\phi_t = L(\phi)$ from time t to $t + \Delta t$ is as follows:

$$\phi^* = \phi^t + (\Delta t/3)L(\phi^t) \tag{6.1}$$

$$\phi^{**} = \phi^t + (\Delta t/2)L(\phi^*) \tag{6.2}$$

$$\phi^{t+\Delta t} = \phi^t + \Delta t L(\phi^{**}). \tag{6.3}$$

Although leapfrog time integration occurs in a single step, it is almost always coupled with a time-smoother to stabilize the computational mode (uncoupling of the odd and even time steps) that arises because it is a three time level scheme (cf. Durran, 1999, pp. 60–64). Thus, the time-smoothed leapfrog integration has the form:

$$\phi^{*t+\Delta t} = \phi^{t-\Delta t} + 2\Delta t L(\phi^{*t}) \tag{6.4}$$

$$\phi^t = \phi^{*t} + \alpha(\phi^{*t+\Delta t} - 2\phi^{*t} + \phi^{t-\Delta t}), \tag{6.5}$$

where in cloud models typically $\alpha \simeq 0.1 - 0.2$.

In the time-split implementation of the RK3 method, the terms involving the fast-moving (acoustic) modes L_f are separated from the slower modes L_s of meteorological interest, and $\phi_t = L_f(\phi) + L_s(\phi)$ is stepped forward using a series of smaller forward–backward time steps to update the L_f terms. This technique is depicted in Fig. 6.3 for the original time-split leapfrog integration developed by Klemp and Wilhelmson (1978), and for the RK3 integration analyzed by Wicker and Skamarock (2002). Notice that the number of small time steps in the three-stage RK3 integration is less than the number of small time steps in the

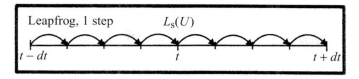

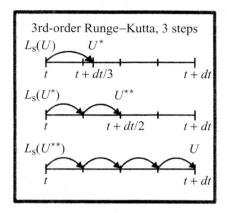

Figure 6.3. Schematic illustrating the time-split leapfrog integration developed by Klemp and Wilhelmson (1978) and the time-split third-order Runge–Kutta technique proposed by Wicker and Skamarock (2002), shown for the case of 4 small time steps per large time step.

one-stage leapfrog scheme because the leapfrog step must span a time interval of $2\Delta t$.

The improved response characteristics of the RK3 scheme over the leapfrog approach are demonstrated for the simple oscillation equation $\phi_t = i\omega_0\phi$ in Fig. 6.4. The RK3 integration remains stable for $\omega_0\Delta t \leq 1.73$ (Durran, 1999, pp. 68–69), which is nearly double the time-step limit for leapfrog ($\omega_0\Delta t \leq 0.90$) with a time filter having a coefficient $\alpha = 0.1$. While the pure leapfrog scheme is non-dissipative, including the time filter makes it more dissipative than the RK3 scheme. Notice also that the relative phase errors in the RK3 integration are significantly less than those arising in the leapfrog differencing.

The choice of numerics used for the advection terms affects both the accuracy and dissipation of the model simulations. This is illustrated in Fig. 6.5 for simple numerical integrations of the one-dimensional linear advection equation $\phi_t = -U\phi_x$ for an advecting top-hat profile. These integrations are carried out in a periodic domain for a top hat of width $15\Delta x$, using small time steps (Courant number $\lambda = U\Delta t/\Delta x \to 0$) such that temporal errors are negligible. Clearly, the higher-order schemes reduce the advective phase errors, but notice also that the odd-ordered schemes produce smoother profiles with significantly less small-scale "noise." This occurs because the even-ordered schemes (centered differences) are non-dissipative

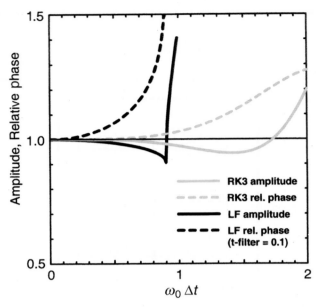

Figure 6.4. Phase and amplitude errors for leapfrog (LF) and third-order Runge–Kutta (RK3) integration of the oscillation equation. Representing $\phi = \exp(i\omega_r t)\exp(-\omega_i t)$, ω_r/ω_0 is the relative phase propagation, and the amplitude $\exp(-\omega_i t)$ is the amplification per time step.

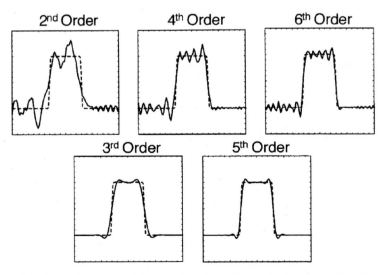

Figure 6.5. One-dimensional linear advection of a top-hat profile using 2nd–6th-order differencing for the advection term. The integrations were conducted for a top-hat width $L_h = 15\Delta x$ in a periodic domain and are displayed at a dimensionless time of $Ut/L_h = 13.33$. The dashed line denotes the correct position of the top hat at this time.

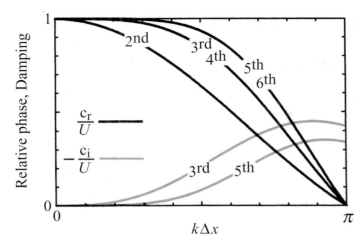

Figure 6.6. Real (c_r) and imaginary (c_i) components of the phase speed in the dispersion equation for the linear advection equation for 2nd–6th-order differencing of the advection term for $\lambda \rightarrow 0$, displayed a function of horizontal wavenumber k. Here c_r/U represents the relative accuracy of the phase propagation, while $-c_i/U$ is a measure of the relative damping associated with the advection operator.

while the odd-ordered ones (upwind differences) contain inherent numerical dissipation. This behavior is quantified in Fig. 6.6, which displays the real (c_r) and imaginary (c_i) parts of the phase speed from the linear dispersion equation for one-dimensional 2nd–6th-order advection (assuming perfect time resolution). As expected, the errors in phase propagation (c_r) are reduced in the higher-order advection schemes. Interestingly, the phase speeds for the odd-order schemes are identical to those of the next-higher even-order scheme. In addition, the odd-order schemes contain dissipation (the imaginary portion of the phase speed, c_i) that is increasingly high wavenumber specific as the order increases. Although the WRF model is coded to allow selectable 2nd–6th-order advection, the 5th-order upwind scheme is recommended as providing the best overall response characteristics in combination with RK3 (Wicker and Skamarock, 2002). Writing the advection terms as flux divergences for use in the flux form integration of the prognostic equations,

$$\frac{\partial(U\phi)}{\partial x} = \frac{1}{\Delta x}\left[F_{i+\frac{1}{2}}(U\phi) - F_{i-\frac{1}{2}}(U\phi)\right] \tag{6.6}$$

the 5th-order representation of the flux becomes:

$$F_{i+\frac{1}{2}}(U\phi) = \frac{1}{60}U_{i+\frac{1}{2}}\{37(\phi_{i+1} + \phi_i) - 8(\phi_{i+2} + \phi_{i-1}) + (\phi_{i+3} + \phi_{i-2})$$
$$- \text{sgn}(U_{i+\frac{1}{2}})[10(\phi_{i+1} - \phi_i) - 5(\phi_{i+2} - \phi_{i-1}) + (\phi_{i+3} - \phi_{i-2})]\}$$
$$\tag{6.7}$$

where the i subscripts denote the locations of variables on a staggered C grid.

Table 6.1. *Maximum stable Courant number* $\lambda = U \Delta t / \Delta x$ *for linear 2nd–6th-order advection for leapfrog and both second- and third-order Runge–Kutta time integration. Unstable configurations are denoted by "uns." (Adapted and revised from Wicker and Skamarock, 2002.)*

Time integration	Advection scheme				
	2nd	3rd	4th	5th	6th
Leapfrog ($\alpha = 0.1$)	0.91	uns.	0.66	uns.	0.57
RK2	uns.	0.90	uns.	0.39	uns.
RK3	1.73	1.63	1.26	1.43	1.09

For constant flow, it can be shown that the odd-order flux divergence schemes are equivalent to the next-higher even-order flux divergence scheme plus a dissipation term of the higher even order with a coefficient proportional to the Courant number (Wicker and Skamarock, 2002). Thus, for constant (positive) U, the 5th-order flux divergence tendency becomes:

$$\Delta t \nabla (U\phi)\Big|_{5\text{th}} = \Delta t \nabla (U\phi)\Big|_{6\text{th}} - \lambda \frac{\Delta x^6}{60} \nabla^6 \phi. \tag{6.8}$$

The stability limits for leapfrog and Runge–Kutta time integration of the one-dimensional advection equation are listed in Table 6.1 for the 2nd–6th-order advection schemes. Again, the RK3 scheme exhibits a stability envelope that is nearly double the limit for leapfrog with a time filter of $\alpha = 0.1$ for the even-order advection operators. Leapfrog integrations are unstable with odd-order advection, so upwind advection schemes cannot be used. For comparison, the stability limits are also shown for a second-order Runge–Kutta (RK2) time integration, which was proposed as an alternative to leapfrog for time-split integrations by Wicker and Skamarock (1998). The RK2 scheme is stable for odd-order advection operators (though unstable for the even-order ones), but has a maximum Courant number that becomes quite restrictive for higher-order advection. Wicker and Skamarock (2002) proposed the time-split RK3 integration scheme as an improvement over both leapfrog and the RK2 approaches: RK3 is stable for both centered and upwind advection operators and allows significantly larger time steps within the limits of linear stability. Although the RK3 scheme requires three evaluations of the advection term per time step compared to only one for leapfrog, much of this increased computational burden is offset by the significantly larger time steps that can be taken in the RK3 integration. These larger time steps are justified since the RK3 scheme is

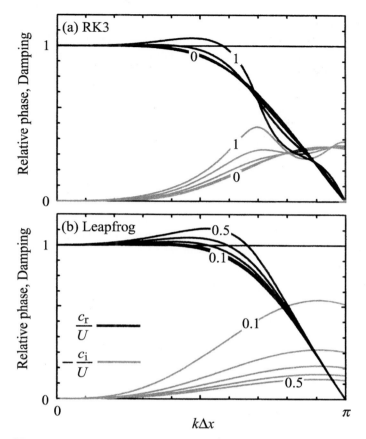

Figure 6.7. As in Figure 6.6 except for numerical integrations at finite Courant number. (a) RK3 time integration with 5th-order upwind advection for Courant numbers (labeled) ranging from 0 to 1 in increments of 0.2. (b) 6th leapfrog integration (without time-smoother) with 6th-order advection and fourth-order filter for $\beta = 0.01$ for Courant numbers (labeled) increasing from 0.1 to 0.5 in increments of 0.1.

3rd-order accurate for linear disturbances (2nd-order for nonlinear forcing), while leapfrog (with the time filter) is formally only accurate to first order.

The behavior of the RK3 scheme for the one-dimensional linear advection equation with 5th-order advection over a range of stable Courant numbers is displayed in Fig. 6.7(a). At the higher Courant numbers the normalized phase propagation increases somewhat above unity over the lower half of the wave spectrum, and the damping increases in the higher-wavenumber regime. Because the effective dissipation is proportional to the Courant number as indicated in (6.8), the damping associated with the upwind differencing does not change in direct response to the

Courant number; doubling the Courant number doubles the damping coefficient, but in doubling the time step, the damping terms are applied only half as many times in integrating out to a specified time. Thus, changes with Courant number are due solely to the changing truncation errors in the RK3 scheme.

As mentioned above, time-split leapfrog integration models typically incorporate a spatial filter to stabilize the overall scheme. Typically, a fourth-order horizontal filter is used with a dimensionless coefficient $\beta = \nu \Delta t / \Delta x^4$ set equal to a constant. For the supercell storm simulations analyzed by LJ, the filter coefficient was $\beta = 0.01$. It turns out that by defining the filter in this manner, the damping is strongly dependent on the Courant number and, in fact, increases as the Courant number is reduced. This occurs because the damping per time step remains proportional to the filter coefficient, but decreasing the Courant number increases the number of time steps needed to integrate over the same time period. This behavior is demonstrated in Fig. 6.7(b), showing the response for the linear advection equation for a leapfrog integration with no time filter for 6th-order advection, and including a fourth-order filter with $\beta = 0.01$, displayed for several Courant numbers over the range of stability ($\lambda \leq 0.58$). Consequently, if the top-hat simulations depicted in Fig. 6.5 were recomputed at small Courant number including this fourth-order filter, the resulting response would be highly damped. The second-order vertical filter on perturbation variables in the Klemp–Wilhelmson (1978) model also exhibits this inverse dependence on Courant number.

In cloud- and mesoscale simulations, the wind speed typically varies widely over the model domain. Thus, for a given model time step, the local Courant number will vary correspondingly. With upwind differencing, the internal computational damping is largest in regions of higher wind speed where the local Courant numbers are larger. With a constant coefficient computational filter, all regions within the model domain are damped by the same amount. The selectivity of the damping inherent in higher-order upwind advection schemes to high wavenumbers and higher-Courant-number locales appears to improve the robustness of model simulations with a lower overall amount of computational damping. Within the context of 5th-order upwind advection, the coefficient multiplying the 6th-order damping (the last term in (6.8)) can be altered to change the damping characteristics while still maintaining the 5th-order accuracy. We have experimented with alternative coefficients and have found it difficult to improve upon the form shown in (6.8).

6.4 Re-tuning the subgrid turbulence closure

In applying the WRF model using RK3 integration with 5th-order upwind advection to simulations of three-dimensional squall lines, Takemi and Rotunno (2003, hereafter referred to as TR) confirmed that good results could be achieved without the use of any additional computational filters. However, they found that without these added filters, unrealistic cell structure tended to develop at small scales. By re-tuning the coefficients in the turbulence closure, they restored realistic behavior in the simulations without resorting to additional computational damping.

The turbulence closure in WRF is based on subgrid eddy mixing, with the eddy-mixing coefficient derived either from a Smagorinsky-type formulation or from the integration of a prognostic turbulence kinetic-energy (TKE) equation. The Smagorinsky approach uses a modified version, developed by Lilly (1962), that includes buoyancy effects in computing the eddy mixing coefficient K_m:

$$K_m = (C_S \Delta)^2 |D| \sqrt{1 - \frac{Ri}{Pr}}, \tag{6.9}$$

where D is the deformation:

$$D^2 = \frac{\partial u_j}{\partial x_i} \left(\frac{\partial u_i}{\partial x_j} + \frac{\partial u_j}{\partial x_i} \right), \tag{6.10}$$

Ri is the Richardson number based on moist stability, and $Pr = K_m/K_h$ is the eddy Prandtl number (specified to be 1/3). From theoretical considerations, Lilly (1966) estimated the Smagorinsky constant $C_S \simeq 0.23$, while Mason (1994) determined that $C_S \simeq 0.2$ provided the most realistic treatment of small scales in large-eddy simulations.

With the TKE approach, proposed by Lilly (1966, 1967), the prognostic TKE equation is integrated, including terms for advection, heat flux sources, Reynolds stress sources, diffusion and dissipation. The eddy mixing coefficient is then related to the turbulence kinetic energy E_t through the expression:

$$K_m = C_K E_t^{1/2} \ell \tag{6.11}$$

where ℓ is a measure of the grid scale. The dissipation ϵ is also expressed in terms of the turbulence energy:

$$\epsilon = \frac{C_\epsilon E_t^{3/2}}{\ell} \tag{6.12}$$

Lilly's (1966) theoretical derivation led to $C_K \simeq 0.12$ and $C_\epsilon \simeq 0.68$, while Moeng and Wyngaard (1988) proposed $C_K \simeq 0.1$ and $C_\epsilon \simeq 0.93$ based on comparisons of simulated spectra with theory.

To evaluate these turbulence closure techniques in WRF, TR conducted three-dimensional squall-line simulations in a 300×80 km domain with periodic boundary conditions in the along-line (north–south) direction. No wind shear was present in the initial environment to promote the evolution of less-organized convective cells along the outflow boundary beneath the squall line. Simulations were evaluated at

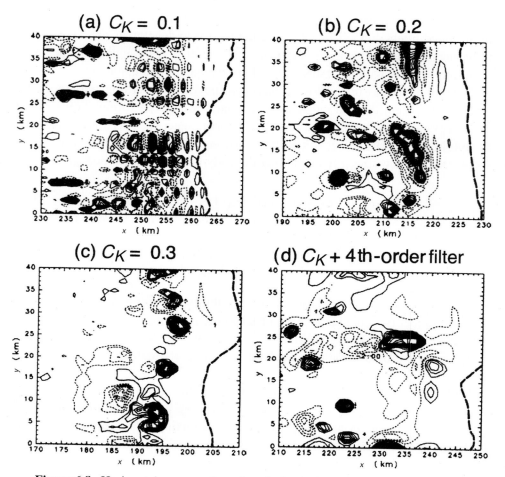

Figure 6.8. Horizontal cross-section of vertical velocity at 4 h at a height of 3 km, with a contour interval of 0.1 m s^{-1}, for simulations using the TKE closure with $C_\epsilon = 0.93$ and (a) $C_K = 0.1$, (b) $C_K = 0.2$, (c) $C_K = 0.3$, and (d) $C_K = 0.1$, but including a weak fourth-order filter ($\beta = 0.0025$). The heavy dashed line indicates the position of the right-moving gust front within the displayed 40 × 40 km window (which differs for each case). (From Takemi and Rotunno, 2003.)

four hours for runs with the Smagorinsky closure for varying C_S, with the TKE scheme for different combinations of C_K and C_ϵ, and for runs including fourth-order horizontal damping.

For the TKE scheme, TR used the coefficients $C_K = 0.1$ and $C_\epsilon = 0.93$ suggested by Moeng and Wyngaard (1988) as reference values, which produced significant cell structure near the grid scale, indicating too little energy removal at high wavenumbers (Fig. 6.8(a)). Increasing C_K to 0.2, the convective cells appear better represented and no longer display any grid-scale features (Fig. 6.8(b)). Further increasing C_K to 0.3, the cells are smoother yet, and confined to the zone of strongest lifting behind the gust front (Fig. 6.8(c)). In this case, the squall line produces less precipitation, a weaker cold pool and, consequently, a slower rightward propagation of the gust front.

To gain a better perspective of the effect of physical and numerical dissipation across the range of horizontal scales present in the simulations, TR computed the power spectral density for the vertical velocity at a height of 3 km in both the across-line (x) and along-line (y) directions. These spectra were calculated within an 80×80 km window positioned with its right boundary approximately 20 km ahead of the gust front, and averaged over the last hour of the simulation. Plate IV displays the spectra for simulations with the TKE scheme for values of C_K ranging from 0.1 to 0.3. For $C_K = 0.1$, there is a significant accumulation of energy at small scales, reflecting the convective structure near the grid scale apparent in Fig. 6.8(a). For $C_K = 0.15$ and 0.20, this small-scale energy buildup disappears and the two coefficients produce very similar spectra. Further increasing C_K, the spectra begin to drop off more rapidly at the high-wavenumber end due to the increasing dissipation. Based on these and other results, TR concluded that $C_K = 0.2$ was the best choice for the TKE scheme in this model system. The spectral slope decreases more rapidly than a $-5/3$ decay at small scales; this may be due to the influence of the implicit filtering in the 5th-order advection or to the fact that these smaller scales are still well outside the inertial subrange.

TR obtained similar behavior with the Smagorinsky turbulence closure. With the reference value of $C_S = 0.2$, noticeable cell structure and and energy buildup was evident at small scales. Increasing C_S into the range 0.25–0.30 produced realistic cell structure without spurious energy buildup at small scales, and TR endorsed these values for use in the model. With further increases in C_S, the convective cells began to appear over-damped.

To assess the comparative effects of added computational damping, TR included a fourth-order horizontal filter with $\beta = 0.0025$ (labeled F1) with the TKE scheme. Using the reference coefficients, the grid-scale features are also removed, although the resulting convective-cell structure appears somewhat over-damped (Fig. 6.8(d)). This is confirmed in the plot of the power spectral density in Plate IV(b), which exhibits a much more rapid energy decay at small scales than the run with $C_K = 0.20$ and no filter. From simulations with the fourth-order filter and no TKE scheme, TR further demonstrated that the filter itself had a significantly greater effect on smoothing the convective cells than the TKE scheme with the reference coefficients. With the F1 fourth-order filter, the power spectra in Plate IV(b) for runs with and without the TKE scheme are nearly the same. Increasing the fourth-order filter coefficient to $\beta = 0.012$ (labeled F2), the spectral drop off at high wavenumber is much more rapid. These results explain why previous models using added numerical filters have not noticed much sensitivity to the choice of coefficients in the subgrid turbulence closure schemes.

6.5 Resolved spatial scales in WRF forecasts

In considering WRF for numerical weather prediction applications, the model is being evaluated in real-time experimental forecasts over scales ranging from convective to synoptic. As part of the evaluation of these experiments, it is of interest to determine how well the spectral distribution of predicted model fields compares to the real atmosphere, particularly near the grid scale where artificial dissipation might produce unphysical behavior. Although the verification of model spectra does not, in itself, assess the accuracy or skill of forecasts, it does shed light on how realistically the model represents the spatial variability observed in nature.

One assessment of the spatial variability in WRF has been provided by Baldwin and Wandishin (2002, hereafter referred to as BW), who examined the Fourier power spectra for forecast precipitation fields. As part of this study, BW compared the variability of the forecast precipitation for one event using the WRF model with 10 km and 22 km horizontal grids, and the operational Eta model with a 12 km grid. Forecasts over the continental United States (CONUS) were initialized at 12 UTC on 4 June 2002 from data interpolated from the Eta analysis. At this time, a cold front was moving to the southeast across Kansas, northern Oklahoma, and the northern tip of Texas, producing heavy rainfall in the vicinity of the front. The accumulated precipitation from 15–18 UTC, derived from the NCEP gage and radar data, depicts this band of frontal precipitation (Plate V). The 3–6 h forecast precipitation fields for the 10 km WRF model and the 12 km Eta model are also displayed in Plate V. While both models forecast significant precipitation in the general vicinity of the cold front, there are noticeable differences in the spatial variability of the precipitation. These visual differences are confirmed in the Fourier power spectra BW computed for the observed and forecast precipitation that are displayed in Plate VI. The spectrum for the 10 km WRF precipitation forecast is quite similar to the observed profile, with the slope beginning decline from the observed profile at scales below about 30–40 km. For the 22 km WRF, the spectrum agrees with observations for scales down to about 100 km, and then begins to drop off more rapidly at smaller scales. In contrast, the 12 km Eta model spectrum diverges significantly from the observed profile for scales less than several hundred kilometers, and contains much less spatial variability at the smaller scales represented in the model.

The WRF and Eta models contain numerous differences in model physics as well as numerics that merit further analysis to clarify the reasons for this disparate model behavior. However, both the precipitation fields in Plate V and their spectral decomposition in Plate VI suggest that WRF is producing more realistic spatial variability in the predicted precipitation.

Harris *et al.* (2001) conducted a similar analysis, computing spectra for model-derived reflectivity in convective forecasts with the ARPS model using a 3 km horizontal grid for comparison with reflectivity spectra from Doppler radar. The model spectrum agreed well with observations for scales greater than about 15 km ($5\Delta x$); at smaller scales, the model variability dropped off rapidly in comparison to the observations. Harris *et al.* attributed this reduced variability at smaller scales to the physically based subgrid-scale mixing and the fourth-order computational damping included for model stability.

BW emphasized that guidance on the likely mode of convection from NWP models exhibiting realistic precipitation patterns, even with errors in timing and location, is of significant value to forecasters. To test the current capabilities in WRF to provide improved convective forecast information, real-time WRF forecasts on a 4 km grid were conducted in support of the Bow echo And Mesoscale EXperiment (BAMEX) in the central US from mid May to early July 2003. The 36 h forecasts initiated at 00 UTC provided guidance for field operations on the following day. These forecasts rely on the explicit treatment of convection, and no cumulus parameterization is employed. Plate VII shows an example of a 36 h and the subsequent 12 h forecast for the model-derived composite reflectivity valid at 12 UTC on 8 June 2003, along with the composite reflectivity from the NEXRAD Radar. At this time a strong baroclinic cyclone is centered in northeastern Iowa, producing heavy precipitation along the cold front spiraling around the center of circulation. The 36 h forecast captures the structure of the cyclone and the strong convection along the frontal boundary, but the position of the system is displaced to the southwest. The next forecast, verifying at 12 h, corrects the position of the frontal convection and also captures the convective system moving through northern Texas. The earlier forecast also depicted strong isolated cells along the frontal line at 00 UTC on 8 June where the line of tornadic supercells actually occurred.

From daily evaluation of the BAMEX forecasts, Done *et al.* (2004) concluded that the 4 km WRF simulations, in comparison with coarser-grid forecasts, provided a much better indication of the likely mode of convection as well as the timing and location of convective initiation. Bow-echo structures were frequently predicted in the vicinity of their actual occurrence. Although the quantitative prediction of precipitation did not improve noticeably in the high-resolution forecasts (possibly due to deficiencies in the microphysics and a tendency for the model convection to decay too slowly), the improved realism of the forecasts provided significant value in support of the field operations, even in the absence of improved quantitative accuracy.

Having conducted WRF real-time forecasts over a range of horizontal grid sizes, it is of interest to see how the distribution of the model kinetic energy as a function of horizontal scale compares to observed energy spectra. Analysis of aircraft measurements by Lilly and Peterson (1983), Nastrom and Gage (1983, 1985), and others, have documented that at large scales the spectral slope is approximately -3, consistent with the behavior of quasi-geostrophic turbulence. However, at scales of several hundred kilometers and below, the atmospheric kinetic energy exhibits a significantly shallower $-5/3$ slope characteristic of the mesoscale. This mesoscale variability has been attributed to either the spectrum of internal gravity and inertia-gravity waves in the atmosphere (Dewan, 1979), or to the upscale transfer of energy through quasi-two-dimensional stratified turbulence (Gage, 1979; Lilly, 1983). (These mechanisms explaining the mesoscale variability are addressed in depth by Gage in Chapter 10.) Nastrom and Gage (1985) further demonstrated that velocity and temperature spectra have essentially the same universal profile, which is largely independent of latitude, season, and location in the troposphere or stratosphere. A preliminary comparison of the model-derived kinetic-energy spectra with the observed profile is displayed in Plate VIII for three experimental WRF forecast applications: CONUS forecasts with a 22 km horizontal grid; BAMEX forecasts over the central US with a 4 km grid, and BAMEX forecasts with a 10 km grid over a domain intermediate between the other two forecasts. For each forecast application, kinetic energy spectra are computed in the east–west direction (removing grid points in the immediate vicinity of the lateral boundaries) at several model levels and averaged over the 24–48 h forecast interval (12–36 h for the 4 km forecasts) for three forecast days in early June 2003.

While the 22 km grid forecasts are fully within the hydrostatic regime, the 4 km forecasts, running without cumulus parameterization, are beginning to explicitly resolve the nonhydrostatic organized convection. Although there is a slight shift of the spectra to higher energy with increasing model resolution, the overall model spectra appear quite similar, and consistent with observed data as plotted by Nastrom and Gage (1985). The nearly -3 slope at large scales transitions to a shallower slope approaching $-5/3$ for scales smaller than several hundred kilometers. At the small-scale end of each model spectrum, the curve tails downward, reflecting the influence of increased computational dissipation near the grid scale. This effect begins to appear in the model spectra at horizontal scales smaller than about $8\Delta x$. Although further analysis of these kinetic-energy spectra is still being conducted (W. C. Skamarock, unpublished paper), these preliminary results suggest that the WRF model is producing a realistic distribution of energy over the broad range of the scales resolved in the model.

6.6 Summary

Lilly (1990) challenged the convective-storm research community to apply the advances in this field to the development of new capabilities in storm-scale numerical weather prediction. Lilly addressed the sobering limitations in the predictability of convective storms, but also identified situations, such as isolated organized storms, that might have significantly enhanced predictability. In seeking to improve storm-scale NWP, Lilly emphasized the need for further advances in utilizing Doppler radar data to capture storm-scale information in model initializations, in developing adaptive-mesh techniques to provide selective resolution enhancement for highly intermittent phenomena, and in refining model physics to accommodate more of the real-world complexity of these processes. In discussing current practices, Lilly lamented that "subgrid-scale turbulence parameterizations are always applied, but tend to be mixed somewhat haphazardly with computational damping." Although the ambiguity between physical and computational damping may never be fully resolved in NWP models, we believe the WRF model has made significant strides in this area.

The numerics of the new WRF model have been designed with the intent to enhance the accuracy and stability of simulations over a diverse spectrum of applications. Integration of the conservative prognostic equations is achieved using a two time level, split-explicit, third-order Runge–Kutta scheme along with fifth-order upwind advection. With these numerics, the model time steps can be about double those used in earlier leapfrog integrations, and without the addition of explicit computational filters required with leapfrog. The implicit damping associated with the RK3 and fifth-order advection appears to have a more selective influence on high wavenumbers at large Courant numbers than other computational filters, thus producing less computational damping on resolved modes of physical interest. Without the presence of added computational damping, the subgrid turbulence closure used in the model plays a more prominent role in selectively removing energy near the grid scale, and the coefficients in these closure schemes needed to be retuned to achieve the desired behavior. In NWP applications, the WRF numerics appear well suited for high-resolution, convection-resolving forecasts, producing realistic precipitation and kinetic-energy spectra across a broad range of scales.

References

Baldwin, M. E. and Wandishin, M. S. (2002). Determining the resolved spatial scales of Eta model precipitation forecasts. In *Proc. 15th Conference on Numerical Weather Prediction, 12–16 August 2002, San Antonio, TX*, American Meteorological Society, 85–88.

Byun, D. W. (1999). Dynamically consistent formulations in meteorological and air quality models for multiscale atmospheric studies. Part I: Governing equations in a generalized coordinate system. *J. Atmos. Sci.*, **56**, 3789–3807.

Dewan, E. M. (1979). Stratospheric spectra resembling turbulence. *Science*, **204**, 832–835.

Done, J., Davis, C. A., and Weisman, M. L. (2004). The next generation of NWP: explicit forecasts of convection using the Weather Research and Forecast (WRF) Model. *Atmos. Sci. Lett.*, to appear.

Durran, D. R. (1999). *Numerical Methods for Wave Equations in Geophysical Fluid Dynamics*, New York: Springer Verlag.

Gage, K. S. (1979). Evidence for a $k^{-5/3}$ law inertial range in mesoscale two-dimensional turbulence. *J. Atmos. Sci.*, **36**, 1950–1954.

Harris, D., Foufoula-Georgiou, E., Droegemeier, K. K. and Levit, J. J. (2001). Multiscale statistical properties of a high-resolution precipitation forecast. *J. Hydrometeorol.*, **2**, 406–418.

Klemp, J. B. and Wilhelmson, R. B. (1978). The simulation of three-dimensional convective storm dynamics. *J. Atmos. Sci.*, **35**, 1070–1096.

Klemp, J. B., Wilhelmson, R. B. and Ray, P. S. (1981). Observational and numerically simulated structure of a mature supercell thunderstorm. *J. Atmos. Sci.*, **38**, 1558–1580.

Klemp, J. B., Skamarock, W. C. and Dudhia, J. (2000). Conservative split-explicit time integration methods for the compressible nonhydrostatic equations. Available at: www.mmm.ucar.edu/individual/skamarock/wrf_equations_eulerian.pdf

Lilly, D. K. (1961). A proposed staggered-grid system for numerical integration of dynamic equations. *Mon. Wea. Rev.*, **89**, 59–66.

    (1962). On the numerical simulation of buoyant convection. *Tellus*, **14**, 148–172.

    (1965). On the computational stability of numerical solutions of time-dependent non-linear geophysical fluid dynamics problems. *Mon. Wea. Rev.*, **93**, 11–26.

    (1966). On the application of the eddy viscosity concept in the inertial sub-range of turbulence. NCAR Manuscript No. 123, Boulder, CO: National Center for Atmospheric Research.

    (1967). The representation of small-scale turbulence in numerical simulation experiments. In *Proc. IBM Scientific Computing Symposium on Environmental Sciences, November 14–16, 1966*, Thomas J. Watson Research Center, Yorktown Heights, NY, H. H. Goldstein, ed., IBM Form No. 320–1951, 195–210.

    (1983). Stratified turbulence and the mesoscale variability of the atmosphere. *J. Atmos. Sci.*, **40**, 749–761.

Lilly, D. K. and Peterson, E. (1983). Aircraft measurements of atmospheric energy spectra. *Tellus*, **35A**, 379–382.

Lilly, D. K. (1990). Numerical prediction of thunderstorms – has its time come? *Quart. J. Roy. Meteor. Soc.*, **116**, 779–798.

Lilly, D. K. and Jewett, B. F. (1990). Momentum and kinetic energy budgets of simulated supercell thunderstorms. *J. Atmos. Sci.*, **47**, 707–726.

Lilly, D. K. (1992). A proposed modification of the Germano subgrid-scale closure method. *Phys. Fluids A*, **4**(3), 633–635.

Mason, P. J. (1994). Large-eddy simulation: A critical review of the technique. *Quart. J. Roy. Meteor. Soc.*, **120**, 1–26.

Moeng, C.-H. and Wyngaard, J. C. (1998). Spectral analysis of large-eddy simulations of the convective boundary layer. *J. Atmos. Sci.*, **45**, 3573–3587.

Nastrom, G. D. and Gage, K. S. (1983). A first look at wavenumber spectra from GASP data. *Tellus*, **35A**, 383–388.

    (1985). A climatology of atmospheric wavenumber spectra of wind and temperature observed by commercial aircraft. *J. Atmos. Sci.*, **42**, 950–960.

Phillips, N. A. (1959). An example of non-linear computational instability. In *The Atmosphere and Sea in Motion*, New York: Rockefeller Institute Press in association with Oxford University Press, 501–504.

Scotti, A., Meneveau, A. C. and Lilly, D. K. (1993). Generalized Smagorinsky model for anisotropic grids. *Phys. Fluids A*, **5**(9), 2306–2308.

Takemi, T. and Rotunno, R. (2003). The effects of subgrid model mixing and numerical filtering in simulations of mesoscale cloud systems. *Mon. Wea. Rev.*, **131**, 2085–2101.

Tripoli, G. J. and Cotton, W. R. (1982). The Colorado State University three-dimensional model – 1982. Part I: General theoretical framework and sensitivity experiments. *J. de Rech. Atmos.*, **16**, 185–220.

Weisman, M. L., Skamarock, W. C. and Klemp, J. B. (1997). The resolution dependence of explicitly modeled convective systems. *Mon. Wea. Rev.*, **125**, 527–548.

Wicker, L. J. and Skamarock, W. C. (1998). A time-splitting scheme for the elastic equations incorporating second-order Runge–Kutta time differencing. *Mon. Wea. Rev.*, **126**, 1992–1999.

  (2002). Time-splitting methods using forward time schemes. *Mon. Wea. Rev.*, **130**, 2088–2097.

Wilhelmson, R. B. and Wicker, L. J. (2001). Numerical modeling of severe storms. In *Severe Convective Storms*, Meteorological Monographs, Boston: American Meteorological Society, 123–166.

Wong, V. C. and Lilly, D. K. (1994). A comparison of two dynamic subgrid closure methods for turbulent thermal convection. *Phys. Fluids A*, **6**(2), 1016–1023.

Xue, M., Droegemeier, K. K. and Wong, V. (2000). The Advanced Regional Prediction System (ARPS) – A multi-scale nonhydrostatic atmospheric simulation and prediction model. Part I: Model dynamics and verification. *Meteorol. Atmos. Phys.*, **75**, 161–193.

7

Numerical prediction of thunderstorms: fourteen years later

Juanzhen Sun

National Center for Atmospheric Research, Boulder, USA

7.1 Introduction

Numerical weather prediction (NWP) has been an essential part of large-scale weather forecasting since the1950s. Although the steady increase of computer power has pushed operational numerical models to higher resolution and greater levels of sophistication, until the 1990s, the numerical simulation of convective clouds and storms was conducted only for basic research, without much thought towards forecast application. In 1990, Douglas Lilly wrote an article entitled "Numerical prediction of thunderstorms – has its time come?", stating that it was time for convective-storm scientists to apply their knowledge to the purpose of weather prediction. Lilly (1990) argued that continued support and vigor of convective-storm modeling research depends on identifying an applied objective, and weather prediction is the principal reason for the support that we are given by our fellow citizens.

Since the first three-dimensional cloud simulations were attempted in the 1970s (see, e.g., Miller and Pearce, 1974; Schlesinger, 1975; Klemp and Wilhelmson, 1978), active research has been conducted in this area and almost all of the studies have focused on the understanding of the dynamics of convective clouds. Due to the lack of detailed data on convective clouds and storms for initialization and comparison, and computational constraints, numerical simulations of convection using cloud models have been started from composite soundings and artificial thermal bubbles. Although some of the simulations were compared with reflectivity observations (Klemp *et al.*, 1981; Wilhelmson and Klemp, 1981), the comparison was made only in a qualitative sense. Lilly (1990) argued that, with the availability of the nationwide radar WSR-88D (Weather Surveillance Radar – 1988 Doppler) observational network and the rapid increase in computer power, it was time to launch a new research endeavor, which would examine whether storm-scale NWP was a realistic goal. With this mission in mind, a new program "Center for Analysis and Prediction of Storms" (CAPS) was proposed by Lilly in 1989 and funded by the National Science Foundation.

There is no doubt that numerical forecasting of convective storms possesses considerable societal and economic significance. It also poses a number of challenges. Among them are the predictability of small-scale flows, the initialization of cloud-scale models, computing, data and networking, and quantifying forecast skills (Droegemeier, 2000). Although research has been conducted to cope with each of these challenges, more attention has been given to the initialization of cloud-scale numerical models due to its critical role in NWP. In Section 7.2, the progress in dealing with the challenges posed to the numerical prediction of convective storms is briefly reviewed. Initialization methods for explicit forecast of storms are described in Section 7.3. In Section 7.4, results are presented from a recent case study of numerical prediction of a supercell storm to demonstrate our current ability in the numerical prediction of thunderstorms. In Section 7.5, conclusions are drawn and future directions are discussed.

7.2　Progress in the last fourteen years

7.2.1　Initialization of storm-scale numerical models

One of the crucial issues for explicit prediction of thunderstorm evolution is how to initialize a storm-scale prediction model. Since the WSR-88D network is the key observing system capable of sampling the four-dimensional structure of storm-scale flow and this network is able to provide only single Doppler observations, many of the studies in the initialization of storm-scale NWP have focused on obtaining the model initial fields from the limited observations of a single Doppler radar. Unlike the large-scale forecast problem, in which all variables necessary to initialize a forecast model (except for vertical velocity) can be obtained directly from the radiosonde, on the convective scale, a single Doppler radar will only provide observations of radial velocity and reflectivity (intensity of precipitation). The variables required to initialize a cloud model, such as three-dimensional wind, temperature, pressure, and water-substance fields, within the storm must therefore be retrieved in the initialization or data-assimilation process. Because no simple balances or approximations apply, it seems necessary to use the prognostic equations in the initialization process for the convective scale.

Early efforts in the storm-scale initialization were devoted to the retrieval of the 3D wind from single-Doppler clear-air observations. When the reflectivity signal is due primarily to clear-air scatterers (as distinct from precipitation), the reflectivity conservation equation should be valid. This assumption forms the basis for a number of single-Doppler wind retrieval techniques (e.g., Qiu and Xu, 1992; Laroche and Zawadzki, 1995; Shapiro *et al.*, 1995). These techniques showed acceptable accuracy when compared with dual-Doppler analysis and can be combined with techniques for retrieving thermodynamic and microphysical fields (Gal-Chen, 1978; Roux, 1985; Ziegler, 1985) to initialize storm-scale numerical models. Weygandt *et al.* (2002a, b) applied the Shapiro *et al.* (1995) single-Doppler retrieval technique and a thermodynamic retrieval technique, and followed by a moisture-specification step in an initialization and forecasting experiment of a supercell storm that occurred in Arcadia, OK.

Techniques were also developed to obtain the initial conditions of all the prognostic variables in a storm-scale numerical model in a single step with the aid of a numerical model. The four-dimensional variational assimilation (4D-VAR) method was first applied to the convective scale by Lilly's graduate student D. Wolfsberg (1987) using a Boussinesq boundary-layer model. The results were critical to the NSF proposal that established CAPS. The technique was further developed at the University of Oklahoma (Sun *et al.*, 1991) and later tested using real data at NCAR (Sun and Crook, 1994). The prediction model along with the adjoint model[1] in the 4D-VAR system were further extended to include microphysics such that the system could be used to perform initialization and prediction of localized convection (Sun and Crook, 1997, 1998).

The two initialization methodologies will be explained in more detail in Section 7.3. In Section 7.4, results from a recent study of initialization and prediction of a supercell storm using the 4D-VAR technique will be presented.

7.2.2 Practical predictability experiments

The classic analysis by Lorenz (1969) laid the foundation for the theoretical study of predictability. However, since the real atmosphere is much more complex than the Lorenz model, predictability experiments using real atmospheric models are necessary. A number of predictability experiments have been conducted using cloud-resolving models since the 1990s. These experiments were based on either idealized simulations (e.g., McPherson and Droegemeier, 1991; Brooks, 1992; Droegemeier and Levit, 1993) or numerical simulations using more realistic models and initial conditions. For example, real-time predictability experiments were conducted using the Advanced Regional Prediction System (ARPS) (Droegemeier *et al.*, 1996a, b; Xue *et al.*, 1996a, b). More recently, three cloud-resolving numerical models [WRF (Weather Research and Forecasting model), RAMS (Regional

---

[1] An adjoint model is a model composed of adjoint equations that maps the gradient vector of a cost function from a forecast time to an initial time of the integration of a prediction model.

Atmospheric Modeling System), and MM5 (PennState/NCAR mesoscale model)] were run at a 4 km grid during BAMEX (Bow-echo and MCV Experiment) to assess the predictability of mesoscale convective systems. A number of practical predictability studies assimilated high-resolution Doppler-radar observations into cloud-scale numerical models to provide the detailed storm-scale flow structure (Crook and Sun, 2001; Montmerle *et al.*, 2001; Sun and Crook, 2001a; Weygandt, 2002a, b). Although results from the studies cited above are generally encouraging, large sensitivity of the storm-scale prediction with respect to changes in initial conditions, environmental condition, and model physics are reported by several authors (Crook, 1996; Sun and Crook, 2001b; Hu and Xue, 2002). The issue of predictability for convective weather has been and remains a topic of debate (Brooks *et al.*, 1992). Important questions concerning the impact on forecast quality of various observations, model physics, and error in the initial conditions are being investigated. More research and operational tests are needed before the operational storm-scale NWP becomes a reality.

7.2.3 Computing, networking, and data management

Numerical prediction of storms presents an enormous challenge to computing, networking and data management because the storm-scale NWP requires a high-resolution numerical model along with an advanced data-assimilation system. For example, the short time scale and thus the rapid perishability of convective-storm predictions demands almost instant transmission of observations and output. In the last decade, however, computing power has increased steadily, allowing operational models to run at much higher resolution and to produce output of shorter-term prediction. Moreover, the project CRAFT (Collaborative Radar Acquisition Field Test, Droegemeier *et al.*, 2002) has successfully demonstrated that the real-time compression and internet-based transmission of WSR-88D Level II data from multiple radars is feasible.[2] Currently, the NCDC and CAPS are receiving Level II data from 62 radars. The CRAFT concept will be applied to the entire WSR-88D network in the near future. Plans have been made at NWS to deliver near real-time base data to customers through NWSTG (NWS Telecommunication Gateway, Crum *et al.*, 2003).

7.2.4 Measuring the quality of convective weather forecasts

Traditional verification approaches based on simple grid overlays between forecast and observation are generally inadequate for convective precipitation forecasts. For

---

[2] WSR-88D Level II data consist of reflectivity, mean radial velocity, and spectral width from the NWS WSR-88D Radar Data Acquisition (RDA) processor. These data are located at spherical coordinates from the radar. There has been a threshold on signal-to-noise ratio, and second trip echoes have been removed.

example, small phase errors in the forecast of a small-scale feature can lead to zero overlap with the observed feature and thus a zero skill score. However, this forecast may still have value to forecasters. It would be useful in such a case to determine the accuracy of the precipitation forecast if there were no position error. A number of efforts are underway toward improving verification approaches for convective and quantitative precipitation forecasts (Brooks *et al.*, 1998; Ebert and McBride, 2000; Baldwin *et al.*, 2001; Brown *et al.*, 2002). These approaches specifically attempt to evaluate errors in location, intensity, and sometimes the shape of convective or precipitation areas. It is clear that further efforts are needed to develop techniques that can take into account the intensity, area coverage, location, timing, and scale of convective precipitation.

7.3 Methods of initialization for the convective scale

Initialization of a cloud-resolving numerical model using high-resolution observations such as those from a Doppler-radar network is crucial to storm-scale NWP. Therefore, active research has been conducted in the last decade to develop techniques that are able to provide initial conditions to cloud-scale numerical models. Two basic methodologies have been employed in the past: one that determines all unobserved fields simultaneously; and one that retrieves the three-dimensional wind first, followed by a retrieval or specification of the thermodynamic and microphysical fields. For simplicity, we will hereafter refer to the first methodology as "simultaneous initialization" and to the second one as "sequential initialization".

7.3.1 Simultaneous initialization

A data-assimilation system that retrieves all unobserved model variables simultaneously requires the use of the prognostic model equations. The four-dimensional variational (4D-VAR) technique is the usual approach for retrieving all fields simultaneously with the aid of a cloud-scale numerical model. If a numerical model represents the atmospheric motion without error and the initial conditions and boundary conditions of the model are known, the predicted trajectory should match observations to their measurement accuracy. Based on this concept, the 4D-VAR technique seeks to determine the initial conditions of the model prognostic variables by iteratively minimizing a cost function. Model errors are usually neglected and boundary conditions are assumed known from larger-scale analyses. The cost function consists of two major terms: the observation term; and the background term. The observation term is defined by the discrepancy between the output, from the forward integration of the numerical model, and the observations within a specified assimilation window (the time period in which data assimilation is performed before a forecast). The background term measures the discrepancy between the initial conditions and a forecast background or a priori analysis obtained using data from an observation network excluding radar observations. The assimilation window is usually determined such that it covers three observation updates (10–12 min for WSR-88D observations). In most storm-scale applications, we have found this window not wide enough to propagate the information from the observed variables to the unobserved variables. However, because of the nonlinearity of the numerical model, a wider assimilation window tends to make the cost function behave more nonlinearly and, as a result, the minimization algorithm does not perform efficiently. To circumvent this problem, a cycling procedure is employed in which a 4D-VAR data-assimilation period is followed by a forecast period of similar duration and then by another 4D-VAR period. The output at the end of the final assimilation period is used as the model initial conditions from which the model forecast commences.

The National Center for Atmospheric Research's VDRAS (Variational Doppler Radar Analysis System) was designed for the assimilation of high-resolution Doppler-radar observations using the 4D-VAR data-assimilation method. The constraining numerical model in VDRAS is a cloud-scale model with a bulk warm-rain parameterization. The reader is referred to Sun and Crook (1997) for a detailed description of the numerical model. The prognostic variables in this model include the three wind components (u, v, w), the liquid-water potential temperature (θ_l), the total liquid-water mixing ratio (q_t), and the rain-water mixing ratio (q_r). The cloud-water mixing ratio (q_c) and temperature (T) are diagnosed from the prognostic variables by assuming that all vapor in excess of the saturation value is converted to cloud water. The water-vapor mixing ratio (q_v) and the pressure (p) are also prognostic variables. Once q_t, q_r, and q_c are known, q_v is obtained because q_t is the sum of q_c, q_r, and q_v. The pressure is obtained by solving a Poisson equation. The cost function (its exact form will be given and described in Section 7.4) is minimized by iteratively adjusting the initial conditions. The gradient of the cost function is computed by the backward integration of the so-called adjoint model.

Figure 7.1 provides a flow chart that shows the procedure of 4D-VAR data assimilation in VDRAS. For generality of numerical models, the thermodynamic variable T (instead of Θ_l) and the microphysical variables q_r, q_c, and q_v (instead of q_t) are used in Fig. 7.1 and (later) in Fig. 7.2. The first guess of the initial conditions is provided and the numerical model is integrated forward. The cost function is computed to evaluate the discrepancy between the observations and the model output. If the value is greater than a given criterion, the adjoint model is integrated backward to find out how to adjust the initial conditions such that the observation and model prediction would be in better agreement. This process is repeated until a satisfactory solution is obtained.

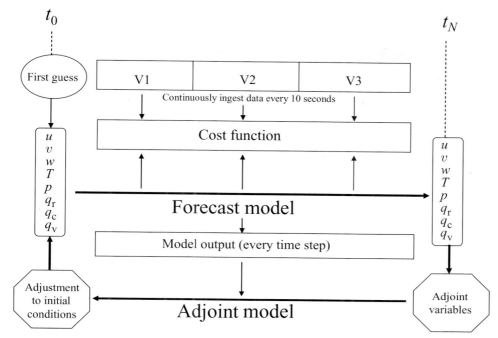

Figure 7.1. Flow chart illustrating the optimization procedure in VDRAS. The three volume scans of radar data denoted by V1, V2, and V3 are continuously assimilated into the numerical model as the model is integrated forward. Values t_0 and t_N represent the beginning and ending times of the data-assimilation window. Variables are defined in text.

Another technique that can retrieve the model initial conditions simultaneously is the ensemble Kalman filter (EnKF, Evensen, 1994). Recently, this technique has been applied to convective-scale data assimilation using simulated data (Snyder and Zhang, 2003; Zhang *et al.*, 2003). Similar to 4D-VAR, the EnKF aims at minimizing a cost function consisting of an observation term and a background term. However, the observational term in EnKF is defined at a single time in contrast to the 4D-VAR in which it is defined over a specified data-assimilation window. The EnKF attempts to retrieve the unobserved prognostic variables by computing the error covariance between the variables. The error covariance is estimated using the deviation of each forecast from the ensemble mean calculated from an ensemble of forecasts. As the forecasts in the ensemble are carried forward by the numerical model, the error covariance is propagated forward for analysis at the next observation time. When the model forecast converges to the observations, the unobserved model variables are recovered through the dynamic relation represented by the numerical model. Application of the EnKF to simulated radar data experiments by Snyder and Zhang (2003) and Zhang *et al.* (2003) have shown encouraging results.

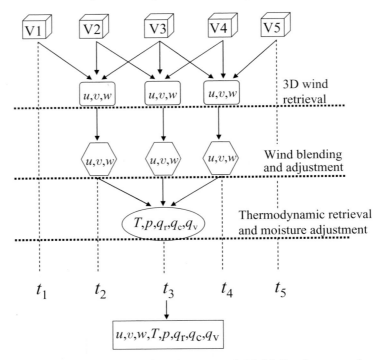

Figure 7.2. Flow chart illustrating the sequential initialization procedure as in Weygandt *et al.* (2002). Five successive volume scans, V_1 to V_5, are used to obtain 3D wind fields at three successive times. The wind is then blended with a background wind. Finally, thermodynamic fields are retrieved for the middle time level and the moisture fields are specified.

7.3.2 Sequential initialization

Both the 4D-VAR and the EnKF techniques are computationally expensive, preventing them from being widely applied to complex and high-resolution numerical models. An alternative methodology that can be used for initialization of cloud-scale numerical models involves the sequential application of a three-dimensional single-Doppler wind retrieval followed by a thermodynamic retrieval and a moisture adjustment. An example of a sequential single-Doppler retrieval procedure is described by Weygandt *et al.* (2002a, b). The three-dimensional wind retrieval algorithm used in their study is that developed by Shapiro *et al.* (1995). The retrieval assumes that the three-dimensional distribution of two conserved scalars (one is the reflectivity and the other is derived from the reflectivity conservation by imposing a temporal constraint on the velocity field) and the radial velocity are known. Figure 7.2 shows a schematic diagram of the sequential initialization procedure. Note that a total of five successive volume scans are needed to produce an initial state of the model dynamic variables. A separate step is necessary to blend the retrieved wind with the background wind from the model forecast or a large-scale analysis. After the 3D wind is obtained, a Gal-Chen-type thermodynamic retrieval procedure (Gal-Chen, 1978) is applied to determine the thermodynamic fields (temperature and pressure). As a final step in the retrieval process, the water vapor mixing ratio is adjusted to saturation in specified regions.

Other 3D wind retrieval methods, for instance the three-dimensional variational technique (Gao *et al.*, 2001), can be used to obtain the 3D wind in the sequential initialization procedure. The variational wind retrieval technique is able to blend the single-Doppler retrieval with a background wind field in a single step, thus eliminating the second step in Fig. 7.2.

The major drawback of sequential initialization is that it is unable to determine the thermodynamic and microphysical fields in a dynamically consistent manner. Consequently, some of the features that play an important role in storm evolution (for instance, the low-level cold pool) may not be well retrieved, and thus the forecast of the storm could be significantly impaired. In addition, sequential initialization has to assume that the 3D volume of radar observations is collected at a single time, neglecting the time difference associated with the sequential scanning of the volumetric observation. As it will be described in the next section, the 4D-VAR technique is able to assimilate each radar data sample at its actual observation time.

7.4 Initialization and forecast experiments of a supercell storm

The objective in this section is to demonstrate our current ability in numerical forecasting of thunderstorms when the numerical model is initialized using an advanced data-assimilation system. In order to give the reader a broader picture, a number of sensitivity experiments with respect to environmental conditions are also presented to show how the numerical prediction depends on the environment.

7.4.1 Description of the 29 June 2000 supercell case

The supercell storm studied here occurred on 29 June 2000 during STEPS (Severe Thunderstorm Electrification and Precipitation Study) near Bird City, Kansas. This supercell storm was observed by the WSR-88D radar located at Goodland, Kansas and two research radars. It appeared to have formed along an advancing surface boundary propagating to the southeast. The first echo appeared on radar around 2130 UTC. The pre-storm environmental sounding observed at 2022 UTC from the NCAR Mobile GPS/Loran Sounding System (MGLASS) is interpolated to the model levels and shown in Fig. 7.3(a). The sounding indicates a southerly component to the low-level flow with veering winds up to the tropopause and the CAPE (Convective Available Potential Energy) of the environment is 1350 J kg^{-1}.

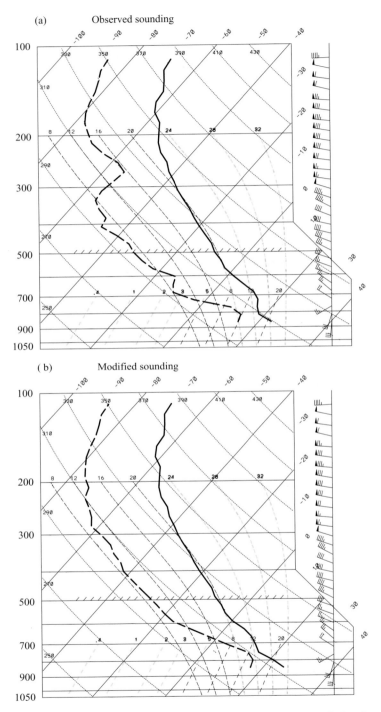

Figure 7.3. (a) Observed MGLASS sounding at 2022 UTC, plotted after interpolation to the model levels. (b) Modified sounding used in the control experiments.

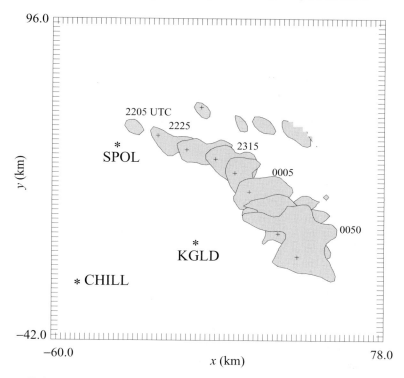

Figure 7.4. Storm positions indicated by the area greater than 40 dBZ at $z = 0.75$ km with a temporal interval of 20 min. The locations of the three radars are marked by *. The x and y distances are relative to the KGLD radar.

The storm's track is illustrated by the 40 dBZ contour line plotted every 20 min starting from 2205 UTC in Fig. 7.4. The position of the three radars is also indicated in Fig. 7.4. The storm propagated southeastward, from about 295° at a speed of about 9.7 m s^{-1}, before ~2325 UTC. It then turned right from 295° to 330° and moved with a velocity of 8.9 m s^{-1}. An F1 tornado was reported at 2328 UTC. Reflectivities near 65 dBZ, which are believed to represent large hail, developed around 2230 UTC.

7.4.2 *Experiments and results*

The supercell storm described above was initialized at 2235 UTC using the observations from the WSR-88D radar KGLD located at Goodland, Kansas. NCAR's VDRAS was used for the initialization experiments. A two-hour numerical forecast was performed after the initialization using the cloud model in VDRAS. There is no ice physics in the cloud model and a bulk warm-rain parameterization scheme is

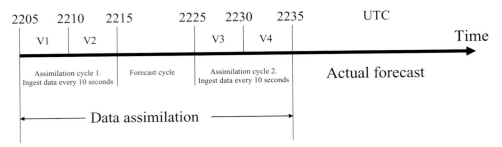

Figure 7.5. Illustration of data-assimilation and forecast cycles.

employed. The numerical domain covers an area of 140 km × 140 km and extends to a height of 15 km; the resolution is 2 km in the horizontal and 500 m in the vertical; the temporal resolution is 5 s. A cycling procedure is implemented in VDRAS and shown in Fig. 7.5. The numbers on top of the figure indicate the times in UTC. The entire assimilation period of 30 min includes two assimilation cycles and one forecast cycle. Each assimilation cycle assimilates two volumes of data. These data are interpolated to a Cartesian grid of the same resolution as the numerical model. At every second time step, a portion of data, whose observation time is within 10 s of that time step, is assimilated. The data assimilation can be performed using only one assimilation cycle starting at 2225 UTC or the assimilation–forecast–assimilation procedure that spans a period of 30 min. We have found that the cycling procedure improves the accuracy of the initial conditions and hence the prediction.

In each assimilation cycle, a trajectory, that optimally fits the observations distributed in the assimilation window of 10 min, is sought by minimizing the following cost function:

$$J = (x_0 - x_b)^T B^{-1}(x_0 - x_b) + \sum_{\sigma,t} \left[\eta_v \left(v_r - v_r^0\right)^2 + \eta_q \left(q_r - q_r^0\right)^2 \right] + J_p, \quad (7.1)$$

where x_0 represents the model state at the beginning of the assimilation window and x_b the previous forecast for the second assimilation cycle or a large-scale background for the first assimilation cycle. The symbol B denotes the background covariance matrix and is assumed diagonal and constant in this study. The variable v_r is the radial velocity computed from the model velocity components; v_r^0 is the observed radial velocity; q_r is the rain-water mixing ratio from the model; and q_r^0 is the rain-water mixing ratio estimated from the reflectivity observation using the formula:

$$Z = 43.1 + 17.5 \log(\rho q_r), \quad (7.2)$$

where Z denotes the reflectivity factor and ρ the air density. The quantities η_v and η_q in (7.1) are weighting coefficients for radial velocity and reflectivity, respectively.

The summation is over space (denoted by σ) and time t. The symbol J_p denotes the spatial and temporal smoothness penalty term. The function of the smoothness penalty term is to ensure a smooth fit to the observations. Its exact form can be found in Sun and Crook (2001c). Since the reduction of the cost function slows down significantly after 70 iterations, the minimization is terminated then.

Data quality control and pre-processing are performed using NCAR's software SPRINT (Sorted Position Radar INTerpolation, Mohr and Vaughan, 1979; Miller *et al.*, 1986) and CEDRIC (Custom Editing and Display of Reduced Information in Cartesian space, Mohr *et al.*, 1986). Three major pre-processing steps are carried out before the data are used in the assimilation: (1) interpolation from the data spherical coordinates to the model Cartesian coordinates; (2) data filtering to reduce random noise; and (3) partial data filling using a least-squares technique. Since the data with high reflectivity values are associated with hail and cannot be easily quantified, and the numerical model does not have ice physics, we truncated the reflectivity data at 55 dBZ.

An assimilation experiment using the observed sounding at 2022 UTC (Fig. 7.3(a)) produced an initial storm with an updraft of 9 m s^{-1} and a positive temperature perturbation of about 2 K. However, this initial storm in the analysis dissipated rapidly during the forecast. When comparing the low-level temperature and dew-point temperature from the sounding (Fig. 7.3(a)) with those from surface mesonet observations and the wind with a VAD (Velocity Azimuth Display) analysis (obtained using the KGLD radar radial velocity observations at 2130 UTC), we found there were significant differences. Therefore a composite sounding was made by first replacing the surface temperature and dew-point temperature in the sounding with the surface mesonet observations and then extending the values up to the top of the boundary layer by assuming a well-mixed boundary layer. In addition, smoothing is applied to the dew-point profile. As a result, the environmental CAPE in the modified sounding increased to the value of 3647 J kg^{-1}. The wind profile is replaced by the VAD analysis at low level (below 1.75 km) and by the average wind between the observations at 2022 UTC and at high level 2338 UTC (above 4.75 km). A cubic-spline interpolation was used to determine the wind between 1.75 km and 4.75 km. The modified sounding is shown in Fig. 7.3(b) and the horizontal wind components before and after the modification is plotted in Fig. 7.6. The main difference of the wind before and after the modification is in the mid-level where the northwest wind is increased after the modification. The shear in the north–south direction is reduced according to the VAD analysis.

The performance of the forecasts is verified by computing the three-dimensional relative correlation coefficient between the forecast rain-water mixing ratio and the rain-water mixing ratio estimated from the reflectivity observation using (7.2). It should be noted that this correlation coefficient has a dependence on the conversion

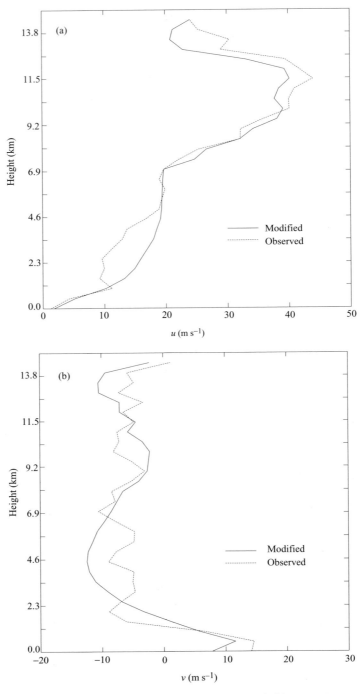

Figure 7.6. Vertical profile of (a) x-component, u, and (b) y-component, v, of velocity. The solid lines show the profiles after the modification and the dotted lines from the observation.

Table 7.1. *Summary of experiments.*

Description of experiment	Experiment notation	Initiated storm
Control sounding.	4DV_CTR35	Yes
Model initialized at 2235 UTC	BUB_CTR35	Yes
Control sounding.	4DV_CTR55	Yes
Model initialized at 2255 UTC	BUB_CTR55	Yes
Observed wind profile.	4DV_OBV	Yes
Model initialized at 2235 UTC	BUB_OBV	Yes
Observed water vapor.	4DV_OBQV	Yes
Model initialized at 2235 UTC	BUB_OBQV	No
Observed temperature.	4DV_OBT	Yes
Model initialized at 2235 UTC	BUB_OBT	No

formula. However, we believe that it should still be able to provide a reasonable measure of the performance of the forecast.

A number of experiments are conducted to demonstrate the impact of radar data assimilation on thunderstorm prediction and the sensitivity of that prediction to the environmental conditions. These experiments are summarized in Table 7.1. The last column of the table indicates whether the model initiated convection. For each experiment, two runs are performed: one using VDRAS to initialize the model and one using a warm thermal bubble. For the warm-bubble experiment, we have varied the size, magnitude, and height of the thermal bubble in order to obtain the best forecast, which is presented here. This experiment is initiated by perturbing the liquid-water potential temperature by a warm bubble of 4 °C placed at the observed location of the storm. The size of the bubble is 20 km in the horizontal and 3 km in the vertical.

The initial conditions retrieved from the VDRAS 4D-VAR assimilation are shown in Fig. 7.7 by a vertical cross-section through the center of the storm from southwest to northeast at 2235 UTC. The magnitude of the updraft (Fig. 7.7(a)) is about 15 m s^{-1}. There is a positive temperature perturbation of over 2 K in the mid-level and a weak cold pool near the surface (Fig. 7.7(b)). The maximum cloud-water mixing ratio is a little over 2 g kg^{-1}. The reflectivity field shown in Fig. 7.7(a) by the shaded areas is converted from the analysis rain-water mixing ratio using (7.2).

Two-hour forecasts of the thunderstorm are performed for all of the experiments starting from the VDRAS analysis or the thermal bubble initialization. As mentioned previously, both VDRAS and the warm-bubble runs failed to generate convection when the observed sounding (Fig. 7.3(a)) was employed. On the

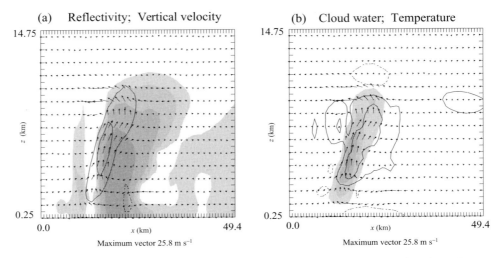

Figure 7.7. Analysis fields at 2235 UTC shown in a vertical cross-section through center of the storm. (a) Reflectivity (shaded with a 20 dBZ increment starting from 20 dBZ); vertical velocity (contours of −2.5 (dash line), 2.5, and 5.0 m s^{-1} are shown); and velocity vector. (b) Cloud-water mixing ratio (shaded with a 0.5 g kg^{-1} increment starting from 0.5 g kg^{-1}); perturbation temperature (contours of −1 (dash lines), +1, and +2K are shown); and velocity vector.

contrary, both runs initiated convection with the modified sounding (Fig. 7.3(b)). The effect of 4D-VAR initialization on subsequent prediction is shown in Fig. 7.8 by comparing the rain-water correlation coefficient of the two-hour prediction between the two runs. The results of the two-hour prediction initialized at 2235 UTC and 2255 UTC are shown by Figs. 7.8(a) and 7.8(b), respectively. At both times, the rain-water correlation from the prediction that is initialized by the VDRAS 4D-VAR data assimilation (4DV_CTR35 and 4DV_CTR55) is much higher than that initialized by a thermal bubble (BUB_CTR35 and BUB_CTR55), just as expected. It is interesting to note that in the last half-hour the correlation of BUB_CTR35 begins to rise and reaches the value of 4DV_CTR35, suggesting that when the environment is favorable for convection, the supercell storm can be initiated with a thermal bubble and the developed storm will have some agreement with the observations. Comparison of the structure of the storms from BUB_CTR35 and 4DV_CTR35 (not shown) reveals that both the motion and the precipitation fields show similarities at the end of the prediction period. The 4D-VAR data assimilation of radar observations provides initial conditions of 3D wind, thermodynamic, and microphysical fields, and hence results in better prediction throughout most of the two-hour period. However, the environmental forcing may have played a more important role than the initial conditions in the latter part of the simulation. When the model is initialized at 2255 UTC, the correlation from the warm-bubble experiment

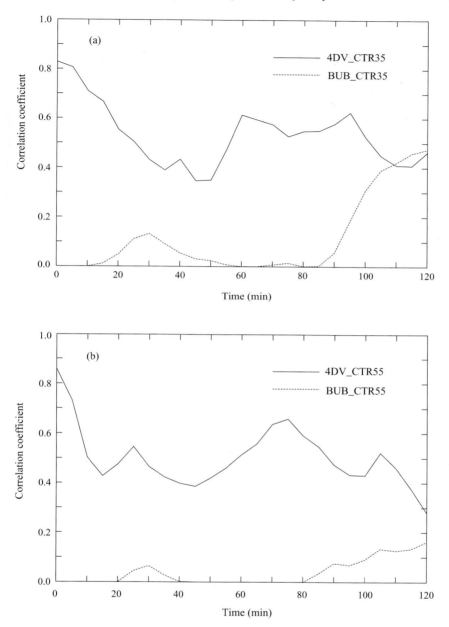

Figure 7.8. Rain-water correlation coefficient for (a) experiments 4DV_CTR35 and BUB_CTR35; and (b) experiments 4DV_CTR55 and BUB_CTR55.

BUB_CTR55 (dashed curve in Fig. 7.8(b)) is lower than the 4D-VAR experiment 4DV_CTR55 (solid curve in Fig. 7.8(b)) throughout the two-hour forecast period, suggesting that the data assimilation provides greater benefit to the forecast if the model is initialized at the later time when the storm is further developed.

The storm tracks from the experiments 4DV_CTR35 and BUB_CTR35 are displayed in Fig. 7.9 by plotting the 40 dBZ contour lines every 20 min and compared with that of the observed storm. The + sign indicates the location of the maximum reflectivity at each selected time. The predicted storm track from the experiment 4DV_CTR35 (Fig. 7.9(b)) shows good agreement with the observations. The storm made a right turn at about the same time as it did in the observations. In contrast to the experiment 4DV_CTR35, the experiment initiated by the thermal bubble produces a storm that initially moves to the east before the precipitation is produced and it then made a greater right turn and propagates to the location of the observed storm near the end of the prediction.

We next present results from a set of experiments that investigate the sensitivity of the prediction with respect to changes of the low-level environmental conditions. A previous study by Crook (1996) on convective initiation showed that variations in boundary-layer temperature and moisture that are within typical observation variability (1 K and 1 g kg^{-1}, respectively) can make the difference between no initiation and intense convection. The objective here is to examine the sensitivity of predicted convection to the environment when the model is initialized with better initial conditions through the 4D-VAR data assimilation and compare it with the simple initialization using a thermal bubble. In the following experiments, the vertical profile of temperature, dew-point temperature, and wind in the control sounding (Fig. 7.3(b)) is alternatively set back to that in the observed sounding (Fig. 7.3(a)) while the other profiles remain the same as in the control sounding. In experiments 4DV_OBV and BUB_OBV, the wind profile in the control sounding is replaced by the observed wind profile at 2022 UTC. In the experiments 4DV_OBQV and BUB_OBQV, the dew-point temperature profile is replaced by the observation. In the last two experiments, 4DV_OBT and BUB_OBT, the temperature profile is replaced.

When the observed wind profile is used, both VDRAS and the thermal bubble runs initiated convection, but the tracks of the simulated storms are significantly different. The storm track from the experiment 4DV_OBV is shown in Fig. 7.10(a). Comparing it with Fig. 7.9(a), we note that the storm does not propagate to the southeast in the first 40 min as in the observation and it moves more toward the south after becoming a supercell. The simulated storm from the experiment BUB_OBV (not shown) moves much slower than the simulated storm from 4DV_OBV and the observed storm, and as a result its rain-water correlation is near zero throughout the two-hour prediction period. The rain-water correlation from 4DV_OBV is shown

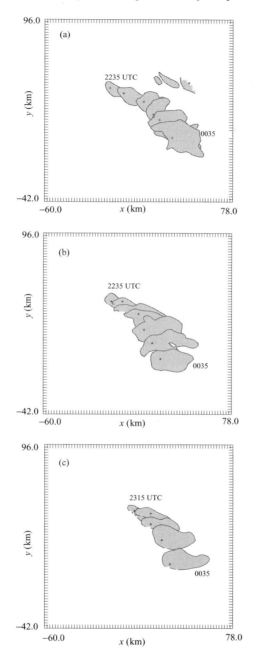

Figure 7.9. Storm positions indicated by the area greater than 40 dBZ at $z = 0.75$ km with a temporal interval of 20 minutes. The x and y distances are relative to the KGLD radar. The results shown are from (a) observation, (b) 4DV_CTR35, and (c) BUB_CTR35.

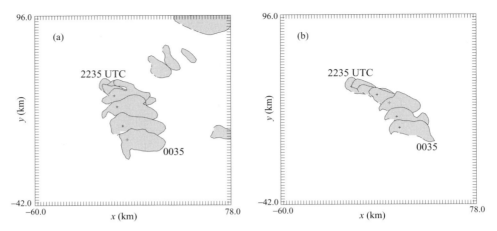

Figure 7.10. Same as Fig. 7.9 but for experiments (a) 4DV_OBV and (b) 4DV_OBT.

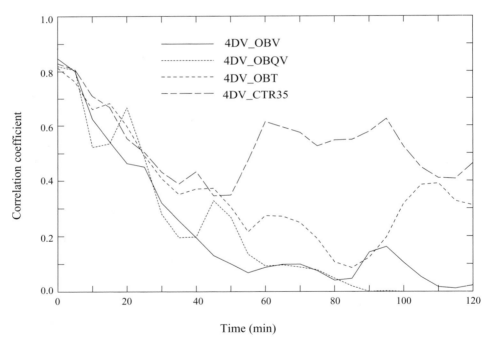

Figure 7.11. Rain-water correlation coefficient for experiments 4DV_OBV, 4DV_OBQV, 4DV_OBT, and 4DV_CTR35.

by the solid line in Fig. 7.11. The correlation is substantially reduced from the control simulation (4DV_CTR35) in the second hour.

When either the profile of the temperature or the dew-point temperature is set back to the observed values, the two experiments initialized by the thermal bubble,

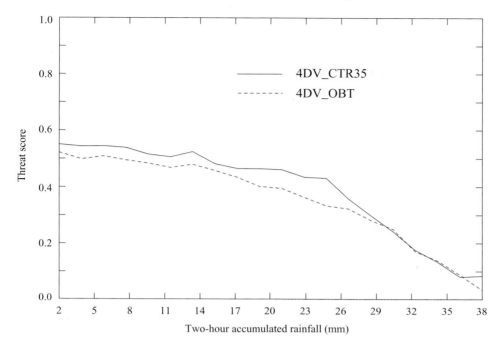

Figure 7.12. Threat score with respect to two-hour accumulated rainfall amount for experiments 4DV_CTR35 and 4DV_OBT.

BUB_OBQV and BUB_OBT, are unable to initiate convection as indicated in Table 7.1. In the 4D-VAR experiment 4DV_OBQV, an updraft of about 13 m s^{-1} is obtained from the 4D-VAR analysis. This updraft develops further in the first 30 min, reaching a maximum of 25 m s^{-1}. The storm then dissipates due to the lack of CAPE. The storm simulated by the experiment 4DV_OBT in which the low-level temperature is set back to the observations, bears the closest resemblance to that in 4DV_CTR35 that uses the control sounding. This is indicated by the plot of the 40 dBZ contours in Fig. 7.10(b) and the rain-water correlation in Fig. 7.11 (dashed line). The rain-water correlation from the experiment 4DV_OBQV is also shown in Fig. 7.11 by the dotted curve. Since the storm dissipates after one hour in 4DV_OBQV, the correlation goes down to zero after 90 min. In contrast, the correlation rises up to 0.4 in 4DV_OBT. Comparing the threat score[3] of the two-hour accumulated rainfall from 4DV_OBT with that from the control experiment, it is seen from Fig. 7.12 that the two are very close. It should be noted that in 4DV_OBT and BUB_OBT the CAPE is not reduced as much as in 4DV_OBQV and BUB_OBQV. The experiment 4DV_OBT seems to suggest that the amount

---

[3] Threat score is a verification measure of precipitation forecast performance equal to the total number of grid points of correct forecast divided by the number of observations plus the number of misses.

of CAPE in the environment plays a major role in determining the accumulated rainfall, and a good initialization as well as an accurate estimate of CAPE are the two key factors in producing a good forecast.

In summary, the prediction of the supercell storm is sensitive to the low-level water vapor but not as much as it is to the low-level temperature. The change to the wind profile produces a storm that has a different propagation track. Better initialization of the cloud model using Doppler radar observations and the 4D-VAR technique results in less sensitivity to the environmental conditions.

7.5 Conclusions and future directions

Numerical prediction of thunderstorms has drawn considerable attention in the last decade. Early efforts have demonstrated some promising results, but there remain great challenges. In this chapter, I briefly reviewed research progress in numerical prediction of convection over the last 14 years. I then described some of the techniques that were used for initialization of cloud-scale model. A recent case study of a supercell storm was presented to show the feasibility of initializing thunderstorms using high-resolution radar data and its impact on the subsequent prediction. Sensitivity of the prediction of the supercell storm with respect to the environmental conditions was investigated. The objective is to examine how the performance of the prediction depends on the large-scale condition with and without data assimilation. From this case study, the following conclusions can be drawn.

- Data assimilation using the 4D-VAR technique and high-resolution radar data is able to provide initial conditions for all of the prognostic variables of a cloud-scale numerical model simultaneously.
- A two-hour prediction of the supercell storm of 29 June 2000 starting from the 4D-VAR analysis using VDRAS showed good agreement with the observations.
- Forecasts are sensitive to low-level environmental conditions, especially to the low-level moisture. VAD wind analysis can be used to adjust the observed large-scale wind and result in better prediction.
- Data assimilation reduces the sensitivity of the initiation of moist convection to variations of the environmental condition. In particular, with the initialization of radar observations, the predicted storm is not very sensitive to a reduction in the boundary-layer temperature of 3 K.

Although considerable progress has been made in the last 14 years in the research of numerical prediction of thunderstorms, further effort toward the improvement of initialization and modeling are necessary before the operational NWP can become a reality. Although the capability of 4D-VAR data assimilation has been demonstrated in storm-scale initialization, it has been applied only to limited domains and observations from a single radar. To expand the effort to mesoscale convective systems, it requires that observations from a number of WSR-88D radars be assimilated into a mesoscale model. Given the current computation capability, we may need to resort to a less costly technique such as 3D-VAR. In comparison with the 4D-VAR technique, the 3D-VAR has its difficulty in thermodynamic and microphysical retrieval. Assumptions and approximations of dynamic relations must be made in order to initialize all the model variables.

The high-resolution radar data have to be integrated with other types of data (satellite, GPS, surface network) as well as other aspects of the physical systems (soil type and soil cover) in order to produce longer-range forecasts of convection and tackle the problem of convective initiation. The optimal integration of different types of observations requires sophisticated data-assimilation techniques that can retain as much information as each data type contains and meanwhile maintain smoothness of the analysis.

Ensemble forecasting will play an important role in storm-scale NWP since the storm-scale numerical prediction is sensitive to uncertainties in initial conditions, large-scale variations, and physical processes and structures that are not well observed or represented in current numerical models. Issues should be addressed concerning the generation of initial conditions for various ensemble members. The ensemble members should include not only variations in initial conditions but also variations in model parameters.

The last issue concerns the operational implementation of convective-scale data assimilation and weather prediction. Given that most weather systems on the convective scale are often driven by highly local effects, it is probably most economical to conduct the convective-scale NWP, and its data assimilation, in a distributed and "on demand" manner. That means local forecast offices guide the execution of their own customized version of a unified model (e.g., WRF, see Chapter 6 by Klemp and Skamarock) and target areas of particular active weather. This scenario might prove effective for both forecast and data assimilation. Research and operational tests are needed to determine the most effective strategy for operational convective-scale NWP.

References

Baldwin, M. E., Lakshmivarahan, S. and Kain, J. S. (2001). Verification of mesoscale features in NWP models. Preprint, *9th Conf. on Mesoscale Processes, 30 July–2 August, Ft. Lauderdale, FL*, Amer. Meteor. Soc., 255–258.
Brooks, H. E. (1992). Operational implications of the sensitivity of modelled thunderstorms to thermal perturbations. Preprint, *4th AES/CMOS Workshop on Operational Forecasting. Weather and Forecasting*, **7**, 120–132.

Brooks, H. E., Kay, M. and Hart, J. A. (1998). Objective limits on forecasting skill of rare events. Preprint, *19th Conf. on Severe Local Storms, Minneapolis, MN*, Amer. Meteor. Soc., 552–555.

Brown, B. G., Mahoney, J. L., Davis, C. A., Bullock, R. and Mueller, C. K. (2002). Improved approaches for measuring the quality of convective weather forecasts. Preprint, *16th Conf. on Probability and Statistics in the Atmospheric Science, 13–17 January, Orlando, FL*, 20–25.

Crook, N. A. (1996). Sensitivity of moist convection forced by boundary layer processes to low-level thermodynamic fields. *Mon. Wea. Rev.*, **124**, 1767–1785.

Crook, N. A. and Sun, J. (2001). Assimilation and forecasting experiments on supercell storms: Part II: experiments with WSR-88D data. Preprint, *14th Conference on Numerical Weather Prediction, Ft. Lauderdale, FL*, Amer. Meteor. Soc., 147–150.

Crum, T., Evancho, D., Horvat, C., Istok, M. and Blanchard, W. (2003). An update on NEXRAD program plans for collecting and distributing WSR-88D base data in near real time. Preprint, *19th Int. Conf. on Interactive Information Processing System (IIPS) for Meteorology, Oceanography, and Hydrology, 9–13 February, Long Beach, CA*, Amer. Meteor. Soc., Paper 14.2.

Droegemeier, K. and Levit, J. (1993). The sensitivity of numerically-simulated storm evolution to initial conditions. Preprint, *17th Conf. On Severe Local Storms, St. Louis, MO*, Amer. Meteor. Soc., 431–435.

Droegemeier, K., Xue, M., Sathye, A., *et al.* (1996a). Real-time numerical prediction of storm-scale weather during VORTEX '95, Part I: Goals and methodology. Preprint, *18th Conf. On Severe Local Storms, 15–20 January, San Francisco, CA*, Amer. Meteor. Soc., 6–10.

Droegemeier, K., Xue, M., Brewster, K., *et al.* (1996b). The 1996 CAPS spring operational forecasting period – Real-time storm-scale NWP, Part I: Goals and methodology. Preprint, *11th Conf. on Numerical Weather Prediction, 19–23 August, Norfolk, VA*, Amer. Meteor. Soc., 294–296.

Droegemeier, K. K. (2000). The numerical prediction of thunderstorms: challenges, potential benefits and results from real-time operational tests. *World Meteor. Org. Bull.*, **46**, 324–336.

Droegemeier, K. K., Kelleher, K., Crum, T., *et al.* (2002). Project CRAFT: A test bed for demonstrating the real-time acquisition and archival of WSR-88D level II data. Preprint, *19th Int. Conf. on Interactive Information Processing System (IIPS) for Meteorology, Oceanography, and Hydrology, 13–17 January, Orlando, FL*, Amer. Meteor. Soc., 136–139.

Ebert, E. E. and McBride, J. L. (2000). Verification of precipitation in weather system: determination of systematic errors. *J. Hydrology*, **239**, 179–202.

Evenson, G. (1994). Sequential data assimilation with a nonlinear quasigeostrophic model using Monte Carlo methods to forecast error statistics. *J. Geophys. Res.*, **99**(C5), 10143–10162.

Gal-Chen, T. (1978). A method for the initialization of the anelastic equations: Implications for matching models with observations. *Mon. Wea. Rev.*, **106**, 587–606.

Gao, J., Xue, M., Shapiro, A. and Droegemeier, K. K. (2001). Three-dimensional simple adjoint velocity retrievals from single-Doppler radar. *J. Atmos. Ocean. Tech.*, **18**, 26–38.

Hu, M. and Xue, M. (2002). Sensitivity of model thunderstorms to modifications to the environmental conditions by a nearby thunderstorm in the prediction of 2000 Fort Worth tornado case. Preprint, *15th Conf. on Numerical Weather Prediction, San Antonio, TX*, Amer. Meteor. Soc., 19–22.

Klemp, J. B. and Wilhelmson, R. B. (1978). The simulation of three-dimensional convective storm dynamics. *J. Atmos. Sci.*, **35**, 1070–1096.

Klemp, J. B., Wilhelmson, R. B. and Ray, P. S. (1981). Observed and numerically simulated structure of a mature supercell thunderstorm. *J. Atmos. Sci.*, **38**, 1558–1580.

Laroche, S. and Zawadzki, I. (1995). A variational analysis method for the retrieval of three-dimensional wind field from single-Doppler data. *J. Atmos. Sci.*, **51**, 2664–2682.

Lilly, D. K. (1990). Numerical prediction of thunderstorms – has its time come? *Quart. J. Roy. Meteor. Soc.*, **116**, 779–798.

Lorenz, E. N. (1969). The predictability of a flow which possesses many scales of motion. *Tellus*, **21**, 289–307.

McPherson, R. A. and Droegemeier, K. K. (1991). Numerical predictability experiments of the 20 May 1977 Del City, OK supercell storm. Preprint, *9th Conf. On Numerical Weather Prediction, Denver, CO*, Amer. Meteor. Soc., 734–738.

Miller, M. J. and Pearce, R. P. (1974). A three-dimensional primitive equation model of cumulonimbus convection. *Quart. J. Roy. Meteor. Soc.*, **100**, 133–154.

Miller, L. J., Mohr, C. G. and Weinheimer, A. J. (1986). The simple rectification to Cartesian space of folded radial velocities from Doppler radar sampling. *J. Atmos. Oceanic Technol.*, **3**, 162–174.

Mohr, C. G. and Vaughan, R. L. (1979). An economical procedure for Cartesian interpolation and display of reflectivity data in three-dimensional space. *J. Appl. Meteor.*, **18**, 661–670.

Mohr, C. G., Vaughan, R. L. and Frank, H. W. (1986). The merger of mesoscale datasets into a common Cartesian format for efficient and systematic analyses. *J. Atmos. Oceanic Technol.*, **3**, 144–161.

Montmerle, T., Caya, A. and Zawadzki, I. (2001). Simulation of a midlatitude convective storm initialized with bistatic Doppler radar data. *Mon. Wea. Rev.*, **129**, 1949–1967.

Qiu, C. and Xu, Q. (1992). A simple adjoint method of wind analysis for single-Doppler data. *J. Atmos. Oceanic Technol.*, **9**, 588–598.

Roux, F. (1985). Retrieval of thermodynamic fields from multiple Doppler radar data, using the equations of motion and the thermodynamic equation. *Mon. Wea. Rev.*, **113**, 2142–2157.

Schlesinger, R. E. (1975). A three-dimensional numerical model of an isolated deep convective cloud: Preliminary results. *J. Atmos. Sci.*, **32**, 934–957.

Shapiro, S., Ellis, S. and Shaw, J. (1995). Single-Doppler velocity retrievals with Phoenix II data: Clear air and microburst wind retrievals in the planetary boundary layer. *J. Atmos. Sci.*, **52**, 1265–1287.

Snyder, C. and Zhang, F. (2003). Assimilation of simulated radar observations with an ensemble Kalman filter. *Mon. Wea. Rev.*, 131, 1663–1677.

Sun, J., Flicker, D. W. and Lilly, D. K. (1991). Recovery of three-dimensional wind and temperature fields from single-Doppler radar data. *J. Atmos. Sci.*, **48**, 876–890.

Sun, J. and Crook, N. A. (1994). Wind and microphysical retrieval from single-Doppler measurements of a gust front observed during Phoenix II. *Mon. Wea. Rev.*, **122**, 1075–1091.

  (1997). Dynamical and microphysical retrieval from Doppler radar observations using a cloud model and its adjoint: Part I. Model development and simulated data experiments. *J. Atmos. Sci.*, **54**, 1642–1661.

  (1998). Dynamical and microphysical retrieval from Doppler radar observations using a cloud model and its adjoint: Part II. Retrieval experiments of an observed Florida convective storm. *J. Atmos. Sci.*, **55**, 835–852.

(2001a). Assimilation and forecasting of a supercell storm: Simulated and observed data experiments. Preprint, *30th Int. Conf. on Radar Meteorology, Munich, Germany*, Amer. Meteor. Soc., 188–190.

(2001b). Assimilation and forecasting experiments on supercell storms: Part I: experiments with simulated data. Preprint, *14th Conf. on Numerical Weather Prediction. Ft. Lauderdale, FL*, Amer. Meteor. Soc., 142–146.

(2001c). Real-time low-level wind and temperature analysis using single WSR-88D data, *Weather Forecasting*, **16**, 117–132.

Weygandt, S., Shapiro, A. and Droegemeier, K. (2002a). Retrieval of model initial fields from single-Doppler observations of a supercell thunderstorm. Part I: Single-Doppler velocity retrieval. *Mon. Wea. Rev.*, **130**, 433–453.

(2002b). Retrieval of model initial fields from single-Doppler observations of a supercell thunderstorm. Part II: Thermodynamic retrieval and numerical prediction. *Mon. Wea. Rev.*, **130**, 454–476.

Wilhelmson, R. B. and Klemp, J. B. (1981). Three-dimensional numerical simulation of splitting severe storms on 3 April 1964. *J. Atmos. Sci.*, **38**, 1558–1580.

Wolfsberg, D. (1987). Retrieval of three-dimensional wind and temperature fields from single-Doppler radar data. CIMMS Report No. 84, Cooperative Institute for Mesoscale Meteorological Studies.

Xue, M., Brewster, K., Droegemeier, K. K., *et al.* (1996a). Real-time numerical prediction of storm-scale weather during VORTEX '95, Part II: Operations summary and example predictions. Preprint, *18th Conf. On Severe Local Storms*, 19–23 February, San Francisco, CA, Amer. Meteor. Soc., 178–182.

Xue, M., Brewster, K., Droegemeier, K. K., *et al.* (1996b). The 1996 CAPS spring operational forecasting period – real-time storm-scale NWP, Part II: Operational summary and sample cases. Preprint, *11th Conf. on Numerical Weather Prediction, 19–23 August, Norfolk, VA*, Amer. Meteor. Soc., 297–300.

Zhang, F., Snyder, C. and Sun, J. (2003). Convective-scale data assimilation with an ensemble Kalman filter: impact of initial estimate and available observations. *Mon. Wea. Rev.*, **130**, 1913–1924.

Ziegler, C. (1985). Retrieval of thermal and microphysical variables in observed convective storms, Part. 1: Model development and preliminary testing. *J. Atmos. Sci.*, **42**, 1487–1509.

8

Tropical cyclone energetics and structure

Kerry Emanuel

Program in Atmospheres, Oceans, and Climate; Massachusetts Institute of Technology,
Cambridge, USA

8.1 Introduction

Aircraft measurements that commenced during World War II allowed scientists
of that era to paint the first reasonably detailed picture of the wind and thermal
structure of tropical cyclones. This led to the first attempts to quantify the energy cycle of these storms and to understand the physical control of their structure. In this contribution, I review the history of research on the energy cycle
and structure of tropical cyclones and offer a revised interpretation of their
structure.

8.2 Energetics

The first reasonably accurate description of the energy cycle of tropical cyclones
appeared in a paper by Herbert Riehl (1950). To the best of my knowledge, this is the
first paper in which it is explicitly recognized that the energy source of hurricanes
arises from the *in-situ* evaporation of ocean water.[1] By the next year, another German
scientist, Ernst Kleinschmidt, could take it for granted that "the heat removed from
the sea by the storm is the basic energy source of the typhoon" (Kleinschmidt,
1951). Kleinschmidt also showed that thermal wind balance in a hurricane-like
vortex, coupled with assumed moist adiabatic lapse rates on angular momentum
surfaces, implies a particular shape of such surfaces. He assumed that a specified
fraction, q_f, of the azimuthal velocity that would obtain if angular momentum were
conserved in the inflow, is left by the time the air reaches the eyewall, and derived

---

[1] Byers (1944) recognized that the observation of nearly constant temperature following air flowing down the
pressure gradient near the surface implies a sensible heat source from the ocean. The existence of isothermal
inflow has been called into question by more recent observations.

an expression for the maximum wind speed:

$$v_{max}^2 = 2E \frac{q_f^2}{1 - q_f^2}, \qquad (8.1)$$

where E is the potential energy found from a tephigram, assuming that air ascending in the eyewall has acquired some additional enthalpy from the ocean. Kleinschmidt did not provide a specific method for estimating this enthalpy increase, and (8.1) is sensitive to the arbitrary value of q_f specified.

In his widely circulated textbook, now regarded as a classic, Riehl (1954) described hurricanes as heat engines and showed that for air ascending in the eyewall to be appreciably warmer than that of the distant environment, a condition for conversion of potential to kinetic energy, the inflowing air had to acquire enthalpy from the underlying surface.

The work of Riehl and his colleagues, most notably Joanne Malkus, culminated in the publication of two papers in the early 1960s: Malkus and Riehl (1960) and Riehl (1963). The first of these once again emphasized that the horizontal temperature gradients that sustain tropical cyclones arise from heat transfer from the ocean. Making use of the observation that the horizontal pressure gradient is very weak at the top of the storm, that temperature lapse rates are very nearly moist adiabatic in the eyewall, and that the temperature of lifted parcels is a function of their boundary-layer equivalent potential temperature θ_{eb}, Malkus and Riehl (1960) used the hydrostatic relation to calculate a relationship between the surface pressure fall from the environment to the inner edge of the eyewall:

$$\delta p_s = -2.5 \, \delta \theta_{eb}, \qquad (8.2)$$

where δp_s is the surface pressure drop in millibars, and $\delta \theta_{eb}$ is the increase in boundary-layer equivalent potential temperature, in kelvin. In deriving this, the horizontal isobaric height gradient was assumed to vanish at 100 mbar. This is a simple quantitative relationship showing explicitly the relationship between a measure of hurricane intensity and the increase in boundary-layer entropy necessarily arising from sea–air enthalpy transfer. Riehl (1963) showed that (8.2) is well verified in observations of actual storms (with a best-fit coefficient of 2.56) and extended the Malkus and Riehl work in several ways. First, he made use of the Riehl and Malkus (1961) argument that outside the eyewall, where latent heat release is weak, conservation of potential vorticity integrated over a volume capped by an isentropic surface above the boundary layer leads to the conclusion that the curl of the surface stress must vanish, which for an axisymmetric vortex gives

$$r \tau_{z_\theta} = const, \qquad (8.3)$$

where r is the radius outward from the storm center, and $\tau_{z\theta}$ is the azimuthal component of the surface stress. We will test this proposition in Section 8.3. Given that the stress varies nearly as the square of the wind speed, (8.3) implies that

$$v_\theta \sim r^{-1/2}, \tag{8.4}$$

where v_θ is the azimuthal wind speed. Using (8.4) and assuming cyclostrophic balance gives an approximate expression for the pressure drop from some outer radius, r_o, (at which the wind speed is assumed to become small) to the radius of maximum azimuthal winds, r_m:

$$v_{max}^2 \simeq -\rho \delta p_s, \tag{8.5}$$

where v_{max} is the maximum azimuthal wind speed and ρ is a mean air density in the boundary layer. Eliminating δp_s between (8.5) and (8.2), and using an estimate of ρ gives

$$v_{max} \simeq 14.1(\delta\theta_{eb})^{1/2}, \tag{8.6}$$

where v_{max} is in m s^{-1}.

In the next step, Riehl estimated $\delta\theta_{eb}$ from conservation of entropy and angular momentum in the inflow. I will slightly abbreviate and generalize his derivation here. Assuming that both entropy (proportional to the logarithm of equivalent potential temperature, θ_e) and angular momentum (M) are vertically uniform in the boundary layer, integration of the conservation equations for entropy and angular momentum through the depth of the boundary layer gives

$$\psi \frac{\partial\theta_e}{\partial r} = C_k(\theta_{es}^* - \theta_e) r|V|, \tag{8.7}$$

and

$$\psi \frac{\partial M}{\partial r} = -r^2 \tau_{z\theta}, \tag{8.8}$$

where ψ is the mass streamfunction of the flow in the r–z plane evaluated at the top of the boundary layer, C_k is an enthalpy transfer coefficient, θ_{es}^* is the saturation equivalent potential temperature of the sea surface, $|V|$ is a surface wind speed, and M is the absolute angular momentum per unit mass, given by

$$M = rv_\theta + \frac{1}{2} fr^2. \tag{8.9}$$

Eliminating ψ between (8.7) and (8.8) gives

$$\frac{\partial\theta_e}{\partial r} = -\frac{C_k(\theta_{es}^* - \theta_e)}{r\tau_{z\theta}}|V|\frac{\partial M}{\partial r}. \tag{8.10}$$

Recall from (8.3) that $r\tau_{z\theta}$ is assumed constant. Also assuming that $|V| \simeq v_\theta$, we use (8.4) to express v_θ as

$$v_\theta = v_{max} \left({}^{r_m}/r \right)^{1/2},$$

where r_m is the radius of maximum winds. Using (8.9) for M, we can integrate (8.10) from r_m to some arbitrary radius, r_a, to get

$$\delta\theta_{eb} \simeq \frac{C_k(\theta_{es}^* - \theta_e)}{2r\tau_{z\theta}} v_{max} \left[v_{max} r_m \ln\frac{r_a}{r_m} + \frac{4}{3} f r_m^{1/2} \left(r_a^{3/2} - r_m^{3/2} \right) \right], \quad (8.11)$$

where we have assumed that $(\theta_{es}^* - \theta_e)$ does not vary with radius. Now using $r\tau_{z\theta} \simeq r_m C_D v_{max}^2$, where C_D is the drag coefficient, we can write (8.11) as

$$\delta\theta_{eb} \simeq \frac{C_k(\theta_{es}^* - \theta_e)}{2\,C_D} \left\{ \ln\frac{r_a}{r_m} + \frac{4}{3}\frac{f r_m}{v_{max}} \left[\left(\frac{r_a}{r_m}\right)^{3/2} - 1 \right] \right\}. \quad (8.12)$$

Noting that, from (8.3) $r_m v_{max}^2 = r_a v_a^2$, where v_a is the wind speed at radius r_a, making the approximation that $r_m \ll r_a$, and substituting (8.12) into (8.6) gives

$$v_{max}^2 \simeq 100\frac{C_k}{C_D}(\theta_{es}^* - \theta_e)\left[\ln\left(\frac{r_a}{r_m}\right) + \frac{4}{3}\frac{f r_a}{v_a} \right], \quad (8.13)$$

which is equation (27) from Riehl (1963), except that Riehl assumed that $C_D = C_k$.

Note that this, together with $r_m v_{max}^2 = r_a v_a^2$ [from (8.3)], gives a transcendental equation for the maximum wind speed as a function of the degree of thermodynamic disequilibrium between the ocean and atmosphere, the Coriolis parameter, and the wind velocity at some specified radius. (Riehl goes on to make what in my view is a somewhat circular argument that there is another dynamic limit on the relationship between v_{max} and r_m which, together with (8.13), determines the radius of maximum wind and an outer radius at the same time.) I shall show later that Riehl comes very close, in (8.13), to an energetic limit on hurricane intensity.[2]

---

[2] Malkus and Riehl (1960) came even closer. Their Equation (33) invokes conservation of θ_e along a boundary-layer streamline, yielding

$$v\frac{\partial\theta_e}{\partial\ell} = C_k v \frac{\theta_{es}^* - \theta_{eb}}{h},$$

where θ_{es}^* is the saturation θ_e of the sea surface, θ_{eb} is the θ_e of the ambient boundary-layer air, h is the boundary-layer depth, and the differentiation is along a streamline. Note that I have changed the notation for consistency, and that Malkus and Riehl unintentionally omitted the factor h. Combining this with (8.2) gives

$$-v\frac{\partial p}{\partial\ell} = 2.5 C_k v \frac{\theta_{es}^* - \theta_{eb}}{h}. \quad (a)$$

This is essentially the unnumbered equation after (33) in Malkus and Riehl. They also wrote down an expression (their Equation (35)) for conservation of energy along a streamline in the boundary layer:

$$-v\frac{\partial p}{\partial\ell} = C_D \rho \frac{v^3}{h}. \quad (b)$$

Several years earlier, Miller (1958) had developed a theory for the minimum central pressure in hurricanes. Miller also started by assuming a moist adiabatic eyewall, but explicitly ignored any increase in entropy from the outer region into the eyewall, opting instead to assume that the eyewall air starts out at sea surface temperature and with a relative humidity of 85%. Miller then estimated a vertical profile of temperature in the eye itself by assuming dry adiabatic descent modified by mixing with the eyewall air, along the line of reasoning explored by Malkus (1958). Once the eye temperature profile was constructed, the central surface pressure was calculated hydrostatically, assuming a level of zero horizontal pressure gradient at the standard pressure level nearest the level of neutral buoyancy for undilute pseudo-adiabatic ascent in the environment. The calculated central pressures were in good agreement with the minimum pressures recorded in a limited sample of intense hurricanes.

It is important to note here that Miller's work departs in a significant way from the line of reasoning adopted by Riehl and Malkus. The latter had emphasized the crucial importance of enthalpy transfer from the ocean, while Miller regarded the hurricane as resulting from the release of conditional instability of the ambient atmosphere, requiring no enhanced air–sea enthalpy flux. He quotes Byers' (1944) statement that compared the hurricane to "one huge parcel of ascending air" and states in his opening sentence that "the principal source of energy of the tropical storm is the release of the latent heat of condensation," a statement rather precisely analogous to a claim that elevators are driven upward by the downward force on the counterweights: both statements are true but miss the point. In hindsight, Miller's estimate of the maximum intensity of hurricanes is energetically inconsistent. As the eyewall entropy is no larger than that of its environment, there can be no conversion of potential to kinetic energy by the overturning circulation of the storm; at the same time, the eye itself contains descending air with high temperature, a process that converts kinetic to potential energy. Thus, the net effect of Miller's energy cycle is absorbtion rather than production of kinetic energy and so cannot maintain the system against dissipation. In an important sense, Miller's analysis presaged the CISK[3] thinking that became the dominant paradigm for tropical cyclone physics after the publication of Charney and Eliassen (1964). This thinking emphasizes the interaction between cumulus convection and the cyclone circulation rather than enthalpy flow from the ocean.

Meanwhile, the development of new observational tools and techniques continued apace. By the late 1960s, the axisymmetric structure of mature hurricanes had

---

Had they eliminated pressure between (a) and (b), they would have obtained the correct expression for maximum wind speed, my (8.21), albeit with fixed thermodynamic efficiency. Instead, they combined (b) with a balance equation for sensible heat along a boundary-layer streamline, to obtain a peculiar relationship between maximum wind and air–sea temperature difference (their Equation (36)).

[3] CISK = conditional instability of the second kind.

been well determined by aircraft and dropsonde observations. It had been known for some time that hurricanes are warm-core vortices; the aircraft data showed that much of the horizontal temperature gradient is concentrated in the eye and eyewall and that in the upper troposphere, the eye temperature can be 15 K warmer than its environment at the same pressure. Analyses of the entropy distribution (Plate IX) tended to confirm the Riehl–Kleinschmidt–Malkus view of the energy cycle, with a pronounced inward increase of equivalent potential temperature near the storm's eyewall. These observations made it clear that there is a strong surface entropy source under the eyewall.

At about the same time, Ooyama (1969) published the results of the first successful numerical simulation of a tropical cyclone, showing among other things that intensification of such storms indeed relies crucially on surface enthalpy fluxes. A decade later, Rosenthal published the results of a numerical simulation in which he had accidentally omitted the cumulus parameterization; the simulated storm had no difficulty intensifying into a mature tropical cyclone (Rosenthal, 1978). Influenced by the Rosenthal and Ooyama results, Douglas Lilly started work on a steady-state model based on conservation of certain key quantities along streamlines emanating from the boundary layer. As he was not satisfied with certain properties of his model, Lilly put this work aside until 1984 when he learned of research on the same subject being carried out by myself and Richard Rotunno. The three of us conducted a lively correspondence over the following year, with the intention of publishing our results in two or three papers. Although we wrote two conference preprint papers together (Emanuel *et al.*, 1985; Lilly and Emanuel, 1985) and my own work was written up (Emanuel, 1986), Lilly never formally published his own work on the steady-state hurricane model. As there are some interesting features of this work and because it departs in certain substantial ways from Emanuel (1986), it is worth reviewing here.

Assuming a steady, circularly symmetric vortex with reversible adiabatic flow above the boundary layer, Lilly first derived the differential relationship

$$T\,ds + \frac{M}{r^2}dM - \frac{1}{\rho r}\xi\,d\psi = d\left[E + \frac{1}{2}fM\right], \tag{8.14}$$

where M is the angular momentum per unit mass, s is the specific (moist) entropy, ξ is the azimuthal component of the vorticity, ψ is the mass streamfunction, f is the Coriolis parameter (assumed constant over the diameter of the storm), r is the radius from the storm center, ρ is the air density and E is the energy content per unit mass, defined as

$$E \equiv \frac{1}{2}|\mathbf{V}|^2 + c_p T + L_v q + gz, \tag{8.15}$$

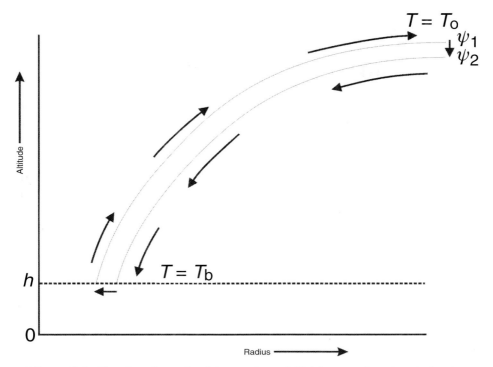

Figure 8.1. Showing the path of integration of (8.14) around a closed circuit consisting of a pair of adjacent streamlines emanating from a boundary layer of depth h.

where $\mathbf{V}$ is the three-dimensional velocity vector, c_p is the heat capacity at constant pressure, L_v is the latent heat of vaporization, q is the specific humidity, g is the acceleration due to gravity and z is the altitude.

Lilly pointed out that the last term on the left-hand side of (8.14) vanishes if hydrostatic and gradient wind balances are assumed. Ignoring that term and integrating around a closed circuit consisting of adjacent streamlines, as illustrated in Fig. 8.1, gives

$$\frac{1}{r_b^2} = \frac{1}{r_o^2} - 2(T_b - T_o)\frac{ds}{dM^2}, \tag{8.16}$$

where r_b and T_b are the radius and absolute temperature, respectively, at the top of the boundary layer, r_o is the radius that the streamline passing through the maximum surface wind attains at the point where the tangential wind vanishes, and T_o is the absolute temperature at that point.

The expression (8.16) was derived on quite different grounds by Kleinschmidt (1951) and later by Emanuel (1986). They assumed hydrostatic and gradient wind balance from the start, and simply integrated the thermal wind equation upward along angular momentum surfaces assuming that the saturation entropy, s^*, is constant on angular-momentum surfaces. Kleinschmidt argued, as did Lilly, that air ascending in the core would be saturated and would preserve its values of both entropy and angular momentum as it ascended; thus entropy would be invariant along angular-momentum surfaces. I used a different (and I think more general) argument: even outside the core, where the air is not saturated on the vortex scale, slantwise moist convection should adjust the saturation entropy to be constant on angular-momentum surfaces; this is just the condition for neutrality to slantwise convection. This condition of slantwise neutrality has been well verified in simulations using a nonhydrostatic model (Rotunno and Emanuel, 1987). Thus a more general form of (8.16) is

$$\frac{1}{r_b^2} = \frac{1}{r_o^2} - 2(T_b - T_o)\frac{ds^*}{dM^2}. \tag{8.17}$$

We note here that Lilly's approach has the advantage that neither hydrostatic nor gradient wind balance has to be assumed; the approach of Kleinschmidt and Emanuel has the advantage that there is no need to invoke energy conservation or to assume that streamlines are along angular-momentum surfaces. Thus the approach based on thermal wind balance is equally applicable to a non-steady vortex, as long as the evolution of the vortex is slow enough that thermal wind balance still applies. I argue that in contrast to (8.16), the relation (8.17) is valid everywhere that moist convection occurs; in a mature hurricane, this is most everywhere, except in the eye. Shutts (1981) also derived an expression similar to (8.16), except that he assumed that dry entropy (potential temperature) rather than moist entropy or saturation moist entropy is invariant on angular-momentum surfaces.

The relation (8.17) strongly constrains the structure of the hurricane vortex, a fact we shall exploit in the next section. When coupled with relations governing sources and sinks of entropy and angular momentum, this relation also places strong constraints on the maximum wind speed of the hurricane. To demonstrate this, we first put (8.16) in a form that makes explicit its reliance on entropy and angular-momentum sources. Assuming that the radii r_o to which angular-momentum surfaces flair near the top of the storm are very much larger than the radii (r_b) that they have at the top of the boundary layer, we can express (8.16) as

$$(T_b - T_o)\frac{ds}{dt} + \frac{M}{r^2}\frac{dM}{dt} = 0, \tag{8.18}$$

in which it is understood that we shall be evaluating the sources of entropy and angular momentum at the top of the boundary layer. Lilly took the top of the boundary layer to be the top of the shallow convective layer, near the level where entropy (equivalent potential temperature) reaches a minimum value. In practice, this is 3–4 km above the surface. I prefer to take the top of the boundary layer to be the top of the well-mixed, subcloud layer, at an altitude of about 500 m (Emanuel 1986). In either case, the equations for the total derivatives of entropy and angular momentum, integrated through a boundary layer of depth h, may be written

$$h\frac{d\bar{s}}{dt} = \frac{1}{T_s}[C_k|\mathbf{V}|(k_s^* - k) + C_D|\mathbf{V}|^3 + F_b], \tag{8.19}$$

and

$$h\frac{d\overline{M}}{dt} = -C_D r|\mathbf{V}|V, \tag{8.20}$$

where $\bar{s}$ and $\overline{M}$ are the entropy and angular momentum averaged through the depth of the boundary layer, $|\mathbf{V}|$ is a near-surface wind speed, V is the azimuthal velocity of air near the surface, C_k and C_D are surface exchange coefficients for enthalpy and momentum (drag), respectively, k_s^* is the specific enthalpy of air at saturation at sea surface temperature and pressure, k is the specific enthalpy of boundary-layer air, and F_b is the enthalpy flux through the top of the boundary layer. I have used the classical bulk formulae for the surface fluxes of enthalpy and momentum, and assumed that there is little turbulent flux of angular momentum through the top of the boundary layer, owing to the very weak vertical gradients of angular momentum found at lower levels in hurricanes. The first term inside the square bracket on the right-hand side of (8.19) is the surface enthalpy flux; the second term is the entropy source owing to dissipative heating, and the final term is the entropy source (usually a sink) owing to enthalpy fluxes through the top of the boundary layer. Both Lilly and Emanuel neglected to include the dissipative heating term, which Bister and Emanuel (1998) later found to be of first-order importance.

Lilly's approach, taking the boundary-layer depth to be that of the shallow cumulus layer, has the advantage that it is plausible to assume that the enthalpy flux through the top of the boundary layer, F_b, vanishes. On the other hand, since moist entropy itself varies significantly with altitude within this layer, the relationship between $\bar{s}$ and the saturation entropy at the top of the boundary layer is problematic. In my approach, taking the boundary layer to be the well-mixed subcloud layer, the entropy should be well mixed in the vertical, while convective neutrality would argue that $\bar{s}$ should be nearly equal to the saturation entropy above the top of the boundary layer; on the other hand, F_b will usually be significant. In the eyewall of a well-developed storm, however, both the vertical entropy gradient in the boundary layer and F_b should be very small in the eyewall. Thus, in the eyewall, we may assume that $F_b \approx 0$ and $\bar{s} \approx s$. Also approximating M by rV in this region and taking $|\mathbf{V}| \cong V$ allows one to derive, by substituting (8.19) and (8.20) into (8.18), the relation

$$v_{\max}^2 = \frac{T_s - T_o}{T_o} \frac{C_k}{C_D} (k_s^* - k), \tag{8.21}$$

in which we have assumed that the wind speed so computed represents an upper bound, given that we have neglected enthalpy fluxes through the top of the boundary layer, which are almost always negative. The expression derived by both Lilly and Emanuel (1985) differs from (8.21) in that T_s rather than T_o appears in the denominator, as a consequence of neglecting dissipative heating. I interpreted (8.21) in terms of a Carnot cycle, in which enthalpy is added to the system at the high temperature of the ocean and removed at the low temperature of the storm's outflow near the tropopause.

Equation (8.21) is in many respects similar to (8.13) from Riehl (1963), with the same dependence on the ratio of the exchange coefficients and the ambient thermodynamic disequilibrium between the ocean and atmosphere (though here expressed in terms of enthalpy rather than entropy). But there are two differences: unlike (8.13), (8.21) has no explicit dependence on outer radius, radius of maximum wind, or wind speed at some particular radius; and Riehl's factor of 100 is replaced by a modified thermodynamic efficiency. Riehl's assumption that parcels become neutrally buoyant at 100 mbar has been replaced by an explicit dependence on out-flow temperature, which depends on the level of neutral buoyancy of air ascending in the eyewall. Also, Riehl's use of a power-law dependence of wind on radius has been replaced by the assumption of thermal wind balance and slantwise neutrality (or, equivalently, by an assumption of energy equilibrium); this gets rid of the factor in square brackets in (8.13).

The relation (8.21) suggests a strong sensitivity of hurricane intensity to those boundary-layer processes that determine the exchange of enthalpy and momentum with the ocean, and ocean temperature near the eyewall, which can strongly affect $k_s^* - k$. It is sometimes remarked that (8.21) is especially sensitive to assumptions about the value of the enthalpy (k) under the eyewall (Holland, 1997). But k is not a free parameter. According to the subcloud-layer equilibrium hypothesis (Raymond, 1995), air in the boundary layer is very nearly neutral to adiabatic displacement to a position just above the top of the boundary layer. This may be expressed as $k = h_{b+}^*$, where h_{b+}^* is the saturated moist static energy just above the top of the boundary layer. But h_{b+}^* is not arbitrary: through the thermal wind relation (8.17) it has a specific relationship to the unperturbed saturation moist static energy of the environment. Since angular momentum increases outward, the saturation entropy (and the saturation moist static energy) must increase inward, so that h_{b+}^* (and therefore k) is greater than the value it has in the unperturbed environment. In (8.21), this offsets the inward increase in k_s^* that arises from decreasing surface pressure. These effects are quantified in the Appendix at the end of this chapter. Emanuel (1986, 1995b) simplified the calculation of k at the radius of maximum winds by assuming that the boundary-layer relative humidity is constant outside the radius of maximum winds.

The predictions of (8.21) are in good accord with numerical experiments, beginning with those by Ooyama (1969) and Rosenthal (1971) and continuing with many others in the 1990s, in which the exchange coefficients are simply specified. Unfortunately, little is known about how these coefficients behave at high wind speeds in nature. As is apparent in Plate IX, most of the entropy increase in the inflow occurs very near the eyewall; it is here that hurricanes are sensitive to the exchange coefficients. (For this reason the centers of hurricanes can approach very near to land before their intensity begins to diminish.) Measurements at low to moderate wind speeds suggest that the drag coefficient increases with wind speed, because of increased surface roughness, but the enthalpy exchange coefficient remains approximately constant (Large and Pond, 1982); when extrapolated to hurricane wind speeds, this would yield a ratio C_k/C_D too small to explain the observed intensity of hurricanes (Emanuel, 1995b). This suggests that other physical processes must come into play to enhance the enthalpy exchange and/or diminish drag. Andreas and Emanuel (1999) suggested that the relevant mechanism is re-entrant sea spray, which transfers significant amounts of enthalpy to the air. Recent estimates of the exchange coefficients from wind-wave tank measurements (Alamaro et al., 2004), from measurements of the ocean current response to tropical cyclones (Shay, 1999) and from wind profiles measured using dropwindsondes (Powell et al., 2003) suggests that their ratio in high winds is not too different from unity. A field experiment that took place in the summer of 2003 was designed to make measurements that could help understand the behavior of surface exchanges at extreme wind speeds. These data are now being analyzed.

The calculation of T_o is straightforward in principle. Since tropical cyclones are subcritical vortices – internal waves can propagate inward against the outflow at upper levels – the outflow temperature represents that environmental temperature to which the entropy surface arising at the radius of maximum winds asymptotes at large radius. It can be calculated given an environmental temperature sounding. The saturation enthalpy of the ocean surface, k_s^*, is a function of surface pressure as well as ocean temperature and must be calculated iteratively, using a second relationship between pressure and wind speed. This is discussed in Emanuel (1986) and Emanuel (1995b). Finally, the actual enthalpy of the boundary-layer air, k, must be estimated using a boundary-layer model or by making an assumption about its radial distribution outside the radius of maximum winds.

In spite of these limitations, calculations of the maximum wind speed made using (8.21) are in good agreement with those attained in numerical simulations, reviewed above, in which the ocean temperature is fixed and for which, of course, the exchange coefficients are known since they are specified. Real hurricanes are never observed to exceed the limit given by (8.21) with $C_k/C_D = 1$, but the vast majority fall well short of this limit (Emanuel, 2000). This is probably due in part to the fact that the ocean temperature cools as a hurricane passes over, owing to strong upward mixing of cold water, but also to disruption of the energy cycle by atmospheric interactions which serve, among other things, to import low-entropy air into the storm's core.

The sensitivity of (8.21) to local perturbations of sea surface temperature can be seen by noting that under average tropical conditions, a local decrease of sea surface temperature of only 2.5 K suffices to bring $k_s^* - k$ to zero. (But note that large-scale gradients of sea surface temperature are associated with similar gradients in k, so that $k_s^* - k$ may remain approximately constant over large areas of undisturbed ocean.) This would suggest that the observed ocean cooling of order 1 K under the storm core could have a significant feedback on hurricane intensity. But the first simulation of a hurricane using a coupled ocean–atmosphere model, by Chang and Anthes (1979), showed little effect of the ocean feedback on storm intensity, leading to a period of roughly two decades during which ocean feedback was regarded as unimportant, except perhaps for storms crossing the wakes of previous storms. (In hindsight, the model used by Chang and Anthes had too coarse a resolution and was integrated for too short a period to see appreciable effects from ocean feedback.) Interest in ocean feedback was renewed after publications by Sutyrin and Khain (1984), Gallacher *et al.* (1989), Khain and Ginis (1991), Bender *et al.* (1993), and Schade and Emanuel (1999), all of whom used advanced coupled models to demonstrate that ocean feedback has a first-order effect on hurricane intensity. Emanuel (1999) demonstrated that the intensity of many hurricanes can be accurately predicted using even a very simple atmospheric model coupled to an essentially one-dimensional ocean model (Schade, 1997), as long as storms remain unmolested by adverse atmospheric influences such as environmental wind shear, which has been shown to be a statistically significant predictor of intensity change.

8.3 Physical constraints on hurricane structure

The derivation of (8.21) relies on the assumptions that the boundary-layer entropy is equal to the saturation entropy above the boundary layer (i.e., convective neutrality), that the angular momentum is dominated by rV and that we can neglect the turbulent flux of enthalpy through the top of the boundary layer at the radius of maximum winds. If these assumptions truly applied everywhere, then (8.21) would be valid everywhere; clearly this is not the case as the right-hand side of (8.21) has only a very weak dependence on radius. Here we argue that the main features that

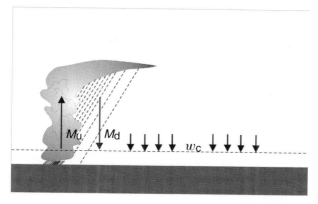

Figure 8.2. Partition of vertical motions at the top of the subcloud layer into convective updrafts M_u, convective downdrafts M_d, and clear-air subsidence (vertical velocity w_c).

determine the radial structure of the hurricane vortex are radial variations in the enthalpy flux through the top of the boundary layer (F_b) and, in the far outer region, the decoupling of the boundary layer from the free troposphere in regions that are stable to convection.

Observations (e.g., Powell, 1990a, b) and numerical simulations (e.g., Rotunno and Emanuel, 1987) reveal that the main mechanisms for evacuating enthalpy from the boundary layer are low-entropy convective downdrafts and turbulent entrainment. In the spirit of simplicity, we represent these processes using a simple convective scheme based on Raymond's (1995) boundary-layer quasi-equilibrium hypothesis. This scheme is described in detail in Emanuel (1995a). As shown in Fig. 8.2, we represent convective updraft volume flux by M_u, convective downdraft volume flux by M_d, clear-air vertical velocity by w_c, and total vertical velocity by w. For convenience, we define M_d and w_c to be positive downward. The flux of low-enthalpy air into the top of the boundary layer is then just

$$F_b = -(w_c + M_d)(h_b - h_{b+}), \qquad (8.22)$$

where h_b and h_{b+} are the moist static energies of air in the boundary layer and just above the top of the boundary layer, respectively. We have assumed that both convective downdrafts and clear-air descent advect the same characteristic value of moist static energy into the boundary layer, and that $w_c > 0$, i.e., that the clear air is actually sinking.

At the same time, mass continuity demands that

$$M_u - M_d - w_c = w, \qquad (8.23)$$

i.e., that the three components add up to the total vertical velocity. Using (8.23) in (8.22) gives

$$F_b = -(M_u - w)(h_b - h_{b^+}). \tag{8.24}$$

Using this in the boundary-layer entropy equation (8.19), and expanding the total derivative of entropy in angular-momentum coordinates:

$$\frac{ds}{dt} = \frac{\partial s}{\partial \tau} + \frac{dM}{dt}\frac{\partial s}{\partial M}$$

allows us to write (8.19) as

$$hT_s\frac{\partial s}{\partial \tau} = C_k|\mathbf{V}|(k_s^* - k) + C_D|\mathbf{V}|^3 - (M_u - w)(h_b - h_{b^+}) - hT_s\frac{dM}{dt}\frac{\partial s}{\partial M}. \tag{8.25}$$

Finally, using (8.20) for the boundary-layer sink of angular momentum in (8.25) gives

$$hT_s\frac{\partial s}{\partial \tau} = C_k|\mathbf{V}|(k_s^* - k) + C_D|\mathbf{V}|^3 - (M_u - w)(h_b - h_{b^+}) + T_sC_Dr|\mathbf{V}|V\frac{\partial s}{\partial M}. \tag{8.26}$$

We have assumed here that both entropy and angular momentum are well mixed in the vertical within the boundary layer.

We are going to use (8.26) in two different ways, depending on whether convection is present or absent. Where convection is absent, it is assumed that the boundary-layer entropy is decoupled from the saturation entropy aloft. We can then use (8.26), with $M_u = 0$, to calculate the radial distribution of entropy in the steady state. But there is little incentive to actually carry out the calculation, since the boundary-layer entropy will then have no control over the vortex structure as a whole.

Where convection is present, we invoke boundary-layer quasi-equilibrium, which sets the left-hand side of (8.26) to zero, and use it as a closure for M_u. We also assume that the saturation entropy above the boundary layer, s^*, is equal to the boundary-layer entropy, s, when convection is active. Then, with the help of the thermal wind balance (8.17) and once again neglecting $1/r_o^2$, we can write the last term of (8.26) as

$$T_sC_Dr|\mathbf{V}|V\frac{\partial s}{\partial M} \simeq -C_D|\mathbf{V}|V\frac{T_s}{T_s - T_o}\frac{M}{r}.$$

Approximating M/r by V, and $|\mathbf{V}|$ also by V, and using this in (8.26) gives the closure for the convective updraft mass flux:

$$M_\mathrm{u} = w + \frac{1}{h_\mathrm{b} - h_\mathrm{b+}} \left[C_k V(k_\mathrm{s}^* - k) - C_D V^3 \frac{T_\mathrm{o}}{T_\mathrm{s} - T_\mathrm{o}} \right]. \qquad (8.27)$$

Using boundary-layer quasi-equilibrium has allowed us to close on the convective mass flux, but in the process we have lost the prediction of boundary-layer entropy. The missing ingredient is the thermodynamic balance above the boundary layer. Along an angular-momentum surface (also a surface of constant s^* by the assumption of slantwise convective neutrality), the temperature (equivalently s^*) is controlled by convection and radiation:

$$\frac{\partial s^*}{\partial \tau} = \frac{\Gamma_\mathrm{d}}{\Gamma_\mathrm{m}} \left[(M_\mathrm{u} - M_\mathrm{d} - w) \frac{\partial s_\mathrm{d}}{\partial z} + \frac{\dot{Q}_\mathrm{rad}}{T} \right], \qquad (8.28)$$

where Γ_d and Γ_m are the dry and moist adiabatic lapse rates, s_d is the entropy of dry air, and $\dot{Q}_\mathrm{rad}$ is the radiative heating. We relate the downdraft mass flux to the updraft mass flux by

$$M_\mathrm{d} = (1 - \varepsilon) M_\mathrm{u}, \qquad (8.29)$$

where ε is a bulk precipitation efficiency. When it is unity, there is no downdraft, while when it is zero the updraft and downdraft mass fluxes are equal. Using this in (8.28) and assuming a steady state gives

$$w = -w_\mathrm{rad} + \varepsilon M_\mathrm{u}, \qquad (8.30)$$

where

$$w_\mathrm{rad} \equiv -\dot{Q}_\mathrm{rad} \Big/ \left(T \frac{\partial s_\mathrm{d}}{\partial z} \right).$$

Since the tropical troposphere is usually cooling radiatively, w_rad is usually positive. It is the rate at which air subsides in the troposphere under the influence of radiative cooling. In this simple model, we shall just take it to be a constant.

In summary, in nonconvective regions in which (8.27) gives a zero or negative value for the convective updraft mass flux, we have, from (8.30), that

$$w = \frac{1}{r} \frac{\partial \psi}{\partial r} = -w_\mathrm{rad} \quad \text{when} \quad M_\mathrm{u} = 0, \qquad (8.31)$$

where ψ is the mass streamfunction at the top of the boundary layer. But where convection is active [i.e., when (8.27) yields a positive convective updraft mass flux], substitution of (8.27) into (8.30) gives

$$(1 - \varepsilon)\frac{1}{r}\frac{\partial \psi}{\partial r} = -w_{\text{rad}} + \frac{\varepsilon}{h_{\text{b}} - h_{\text{b}^+}}$$

$$\times \left[C_k V(k_{\text{s}}^* - k) - C_D V^3 \frac{T_{\text{o}}}{T_{\text{s}} - T_{\text{o}}} \right], \quad \text{when} \quad M_{\text{u}} > 0. \quad (8.32)$$

To close the system, we use the steady-state form of the boundary-layer angular-momentum equation, (8.20):

$$\psi \frac{\partial M}{\partial r} = C_D r^2 V^2,$$

or equivalently,

$$\frac{\partial (rV)}{\partial r} = \frac{C_D r^2 V^2}{\psi} - fr. \quad (8.33)$$

Thus the closed steady-state system consists of (8.33) together with either (8.31) or (8.32), depending on the sign of M_{u} determined from (8.27).

This system of equations is appropriate to the outer region of the storm and to the outer part of its eyewall, where we expect a match between the vertical motion in the free troposphere and that demanded by Ekman dynamics at the top of the boundary layer. However, it is not applicable at the inner edge of the eyewall, where radial diffusion is necessary to balance the strong frontogenetical tendencies, or in the eye where Ekman pumping produces upflow only through a shallow layer, while inward turbulent fluxes of angular momentum drive an axial downflow above the boundary layer (Emanuel, 1997). Thus we terminate integration of the equations near the radius of maximum winds and do not use them to derive a maximum wind speed.

I next proceed to a simple numerical solution of (8.33) with either (8.31) or (8.32). Before doing this, we can absorb most of the parameter dependence of these equations into scaling of the dependent and independent variables. I replace the variables as follows:

$$V \rightarrow v_{\text{max}} V,$$

$$r \rightarrow \frac{v_{\text{max}}}{f} r,$$

$$\psi \rightarrow C_D \frac{v_{\text{max}}^3}{f^2} \psi,$$

$$w_{\text{rad}} \rightarrow C_D v_{\text{max}} w_Q,$$

$$M_{\text{u}} \rightarrow C_D v_{\text{max}} M_{\text{u}},$$

where v_{max} is defined by (8.21). With these substitutions, our system becomes

$$\frac{\partial(rV)}{\partial r} = \frac{r^2 V^2}{\psi} - r, \tag{8.34}$$

$$\frac{1}{r}\frac{\partial \psi}{\partial r} = -w_Q \quad \text{for} \quad M_{\text{u}} = 0, \tag{8.35}$$

$$\frac{1-\varepsilon}{r}\frac{\partial \psi}{\partial r} = -w_Q + \varepsilon \Lambda(V - V^3) \quad \text{for} \quad M_{\text{u}} > 0, \tag{8.36}$$

with

$$M_{\text{u}} = \frac{1}{1-\varepsilon}[-w_Q + \Lambda(V - V^3)] \tag{8.37}$$

Here the additional non-dimensional parameter is defined:

$$\Lambda \equiv \frac{C_k}{C_D} \frac{k_{\text{s}}^* - k}{h_{\text{b}} - h_{\text{b+}}}. \tag{8.38}$$

Although Λ must vary with radius, since all of its components do, we shall take it to be a constant here for simplicity; likewise, we shall neglect radial variations of the bulk precipitation efficiency, ε.

The boundary condition for this system is that ψ vanishes at some outer radius r_{o}. From (8.34) V must vanish there as well. Although the system is second order in r, we do not apply a second boundary condition since we terminate integration at or outside the radius of maximum winds. The control parameters are then Λ, ε, w_Q and r_{o}.

Before turning to numerical integrations, it is instructive to look at approximate analytic solutions in the far outer region, where it will turn out that $M_{\text{u}} = 0$. In that case, we can integrate (8.35) directly and, applying the boundary conditions, we get

$$\psi = \frac{1}{2}w_Q(r_{\text{o}}^2 - r^2).$$

Substituting this into (8.34) gives

$$\frac{\partial(rV)}{\partial r} = \frac{2r^2 V^2}{w_Q(r_{\text{o}}^2 - r^2)} - r.$$

For $w_Q \ll 1$, the dominant balance of the above gives

$$V^2 \approx \frac{1}{2}w_Q \frac{r_{\text{o}}^2 - r^2}{r}. \tag{8.39}$$

At $r \ll r_{\text{o}}$, this gives the same $r^{-1/2}$ dependence of V derived by Riehl (1963) on the somewhat questionable premise that potential vorticity is conserved in the outer region.

Closer in towards the center, but still well outside the radius of maximum winds, another approximate solution presents itself. Here we assume that convection is active, so that both (8.34) and (8.36) apply. We also assume that $\varepsilon \Lambda V \gg w_Q$ but $V^2 \ll 1$, so that (8.36) may be approximated by

$$\frac{\partial \psi}{\partial r} \simeq \frac{\varepsilon \Lambda}{1 - \varepsilon} rV, \tag{8.40}$$

while at the same time, the last term in (8.34) may be neglected (i.e., the relative vorticity is much greater than the Coriolis parameter), giving

$$\frac{\partial}{\partial r}(rV) \simeq \frac{r^2 V^2}{\psi}. \tag{8.41}$$

The system comprised of (8.40) and (8.41) has the power-law solution

$$V \approx r^{-n}, \tag{8.42}$$

where

$$n \equiv \frac{\varepsilon \Lambda - 2(1 - \varepsilon)}{\varepsilon \Lambda - (1 - \varepsilon)}. \tag{8.43}$$

Realistic solutions are thus obtainable only if

$$\Lambda > 2\frac{1 - \varepsilon}{\varepsilon}. \tag{8.44}$$

Thus the bulk precipitation efficiency has to be relatively large and/or the relative air–sea thermodynamic disequilibrium has to be large. Note that for the parameters used in the numerical solution discussed presently, $n = 2/3$.

The numerical solution of (8.34)–(8.36) is straightforward. We start at $r = r_0$ and proceed inward, using a radial step of 0.001. A particular solution for the non-dimensional azimuthal wind velocity is shown in Fig. 8.3 and compared to the profile obtained by running the model of Rotunno and Emanuel (1987) into a statistical steady state and re-scaling the velocity and radius to map into the non-dimensional coordinates. The agreement is quite good, especially considering the crude approximation of neglecting any radial variations of ε and Λ. The longer tail in the Rotunno–Emanuel model is likely owing to the fact that in that model the radiative cooling is proportional to the temperature perturbation rather than being a fixed constant. Thus the dry descent is weaker and must extend over a broader area to carry the same mass flux. Corresponding solutions are shown for the non-dimensional radial velocity (Fig. 8.4), and the total vertical velocity and convective updraft mass flux (Fig. 8.5). Note that convection is absent in the far outer region.

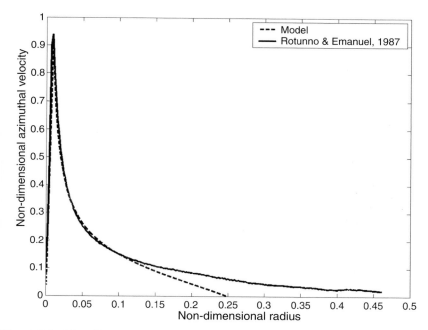

Figure 8.3. Non-dimensional azimuthal velocity as a function of non-dimensional radius as a solution of the steady-state model (dashed line), with $\Lambda = 1$, $\varepsilon = 0.8$, $w_Q = 0.1$ and $r_o = 0.25$. For plotting purposes, we let the azimuthal velocity decrease linearly with radius to zero inside the terminal radius of the integration. Shown for comparison (solid line) is the quasi-steady-state velocity profile from a simulation using the nonhydrostatic model of Rotunno and Emanuel (1987), scaled to these coordinates.

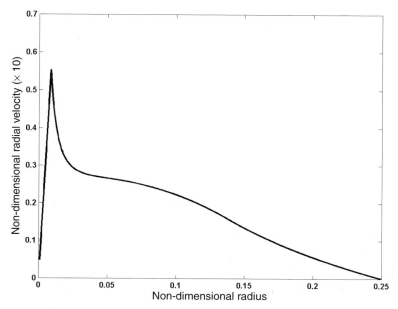

Figure 8.4. Same as in Fig. 8.3 but showing the non-dimensional radial velocity in the boundary layer (with positive values inward).

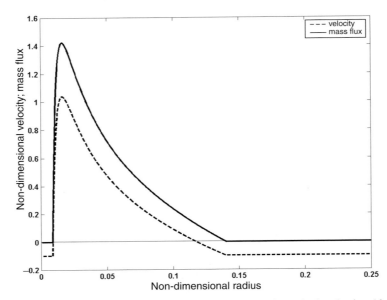

Figure 8.5. Same as in Fig. 8.3 but showing the total vertical velocity (dashed line) and the convective updraft mass flux (solid line).

Having obtained asymptotic solutions for the azimuthal velocity for large and small V (but still well outside the radius of maximum winds), we can attempt to patch these together to form a distribution approximately valid for the whole range of radius outside the radius of maximum winds. At the same time, we can build in an asymptotic limit for the wind profile in the eye, to get a distribution for the whole storm. I have attempted to do this while at the same time altering the large-r asymptotic limit to better fit the Rotunno and Emanuel numerical solution. The result of this exercise is

$$V^2 = V_{\text{max}}^2 \left(\frac{r_0 - r}{r_0 - r_m}\right)^2 \left(\frac{r}{r_m}\right)^{2m} \left[\frac{(1-b)(n+m)}{n+m\left(\dfrac{r}{r_m}\right)^{2(n+m)}} + \frac{b(1+2m)}{1+2m\left(\dfrac{r}{r_m}\right)^{2m+1}}\right],$$

(8.45)

where n is given by (8.43), m is an exponent governing the wind profile in the eye, V_{max} is the maximum wind speed, r_m is the radius of maximum winds, and b is a weighting parameter that governs the transition between the two asymptotic regimes. Note that (8.45) is valid as well using dimensional values of the radius, since all the radii are normalized anyway. Also note that unless r_0 is unreasonably small, the absolute angular momentum implied by (8.45) will be a monotonically increasing function of radius.

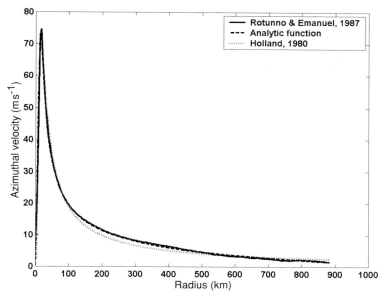

Figure 8.6. Radial profiles of azimuthal velocity from an integration of the Rotunno and Emanuel (1987) model (solid line), from (8.45) (dashed line) and from the analytic model of Holland (1980) (dotted line). See text for parameter values used.

A solution to (8.45) taking $V_{\max} = 74\,\mathrm{m\,s^{-1}}$ and $r_m = 14\,\mathrm{km}$ from the Rotunno and Emanuel (R&E) numerical simulation, and using $n = 0.9$, $m = 1.6$, $b = 0.25$ and $r_o = 1200\,\mathrm{km}$ is compared to the R&E simulation results in Fig. 8.6. The fit is quite good. Also shown is the best fit of Holland's (1980) wind profile, taking his b parameter to be 1.9. The Holland profile is a little too flat in the outer region but quite good in the inner region. It has the advantage, though, of having a simpler form than (8.45).

Once V has been calculated, the boundary-layer streamfunction can be obtained from (8.34), whence the radial velocity in the boundary layer and the vertical velocity at its top may be derived. These are shown for the same solution in Figs. 8.7 and 8.8, respectively. Note that while the vertical velocity is indeed negative in the outer region, it is too small to distinguish from zero in the graph.

We have made one initial attempt to compare (8.45) with observed wind profiles, using data collected from a NOAA WP-3D aircraft and made available by NOAA's Hurricane Research Division. Figure 8.9 compares a profile of azimuthal wind at 3 km altitude from a single radial aircraft pass to the profile in (8.45) using the same parameters as before, except taking $V_{\max} = 60\,\mathrm{m\,s^{-1}}$, $r_m = 32\,\mathrm{km}$ and using $n = 0.8$. Thus this observed profile is a little flatter than the numerically simulated profile just outside the radius of maximum winds.

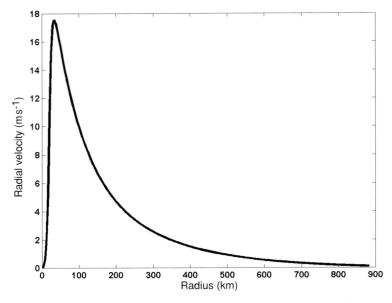

Figure 8.7. Magnitude of the inward radial velocity derived from (8.34) using the azimuthal velocity obtained from (8.45).

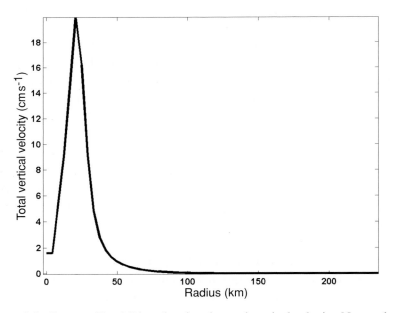

Figure 8.8. Same as Fig. 8.7 but showing the total vertical velocity. Note reduced radial scale.

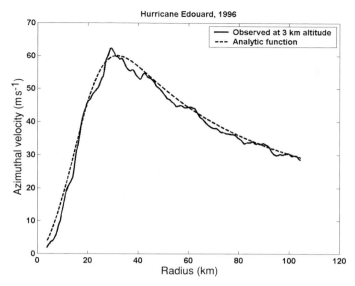

Figure 8.9. Comparison of (8.45) to profile of azimuthal wind observed in Hurricane Edouard of 1996. See text for parameter values.

8.4 Summary

Early work on tropical cyclone energetics by Riehl, Kleinschmidt, and by Riehl and Malkus recognized that such storms are powered by enthalpy transfer from the ocean. The latter authors came close to developing a correct closed-form expression for the maximum sustainable wind speed. Subsequent work by Lilly and by the present author established such an expression, (8.21), although only in the last decade was the importance of dissipative heating recognized. The energy cycle makes clear that tropical cyclones are highly susceptible to small ocean cooling under their eyewalls, and also emphasizes the importance of the outflow temperature, which is governed by the entropy of the air ascending in the eyewall and the ambient temperature profile.

The intensification of tropical cyclones requires a rapid variation of downdraft enthalpy flux across the eyewall (Emanuel, 1997); this process also determines the radial profile of pressure and wind in this region. The wind profile in the eye itself represents a balance between Ekman pumping, which acts to spin down the circulation above the boundary layer, and inward turbulent diffusion of angular velocity from the eyewall, which acts in the opposite sense. Outside the eye, the surface wind controls the surface fluxes which, through the boundary-layer quasi-equilibrium postulate, control the convective flux of enthalpy out the top of the boundary layer. On the other hand, there must be enough upward motion to balance, by adiabatic cooling, the sum of the convective heating and the radiative cooling. But this upward motion must match that demanded by Ekman pumping, which is determined by the radial variation of azimuthal wind. This requirement strongly constrains the radial wind variation outside the core. I here developed a set of nonlinear ODEs that govern this profile and found asymptotic solutions in certain limits. By patching such solutions together, I derived a uniformly valid wind profile, (8.45), which replicates that found in a numerical simulation using a non-hydrostatic, axisymmetric model. While not as elegant as the simple analytic wind profile proposed by Holland (1980), it does depend explicitly on environmental parameters. In particular, it predicts a steeper decline of wind with radius just outside the core when the mid-level environment is moist, the air–sea thermodynamic disequilibrium is large, and/or the bulk precipitation efficiency is large.

Appendix

The expression (8.21) for the maximum wind speed is not closed, because both k_s^* and k vary with radius. The author (Emanuel, 1986; Emanuel, 1995b) developed a closed-form expression by assuming that the boundary-layer relative humidity under the eyewall is the same as that of the unperturbed environment. Here we point out that the radial variation of both k_s^* and k depend on the outer vortex structure. First, and without loss of generality, we write (8.21) in terms of moist static energy rather than enthalpy:

$$v_{\max}^2 = \frac{T_s - T_o}{T_o} \frac{C_k}{C_D} (h_s^* - h). \tag{8.A1}$$

Second, we make use of the boundary-layer quasi-equilibrium postulate, setting h in (8.A1) to h_{b+}^*, the saturation moist static energy just above the top of the boundary layer. Variations in h_{b+}^* at constant altitude are related to variations in the saturation entropy by the first law of thermodynamics:

$$\delta h_{b+}^* = T_b \delta s^* + R_d T_b \delta \ln p, \tag{8.A2}$$

where T_b is the absolute temperature at the top of the boundary layer, R_d is the gas constant for dry air, and we have neglected the difference between total pressure and the partial pressure of dry air. On the other hand, thermal wind balance, as given by (8.17), relates radial variations of saturation entropy to radial variations of angular momentum. In the limit of very large r_o, (8.17) may be written

$$(T_b - T_o) \delta s^* = -\frac{M}{r^2} \delta M. \tag{8.A3}$$

Now using the definition of angular momentum, (8.9), and the gradient balance equation,

$$R_{\mathrm{d}}T\frac{\partial \ln p}{\partial r} = \frac{v^2}{r} + fv,$$

we can re-write (8.A3) as

$$(T_{\mathrm{b}} - T_{\mathrm{o}})\delta s^* = -\left[\delta\left(\frac{1}{2}v^2 + \frac{1}{2}frv\right) + \frac{1}{2}f^2 r\,\delta r + \left(\frac{v^2}{r} + fv\right)\delta r\right]$$

$$= -\delta\left[\frac{1}{2}v^2 + \frac{1}{2}frv + \frac{1}{4}f^2 r^2 + R_{\mathrm{d}}T\ln p\right]. \qquad (8.A4)$$

Substituting (8.A4) into (8.A2) and neglecting any radial variations of T_{b} or T_{o}, we can integrate the result between the radius of maximum winds and the outer limit of the vortex, where, by definition, $v = 0$, to obtain

$$h_{\mathrm{b}+}^* = h_{\mathrm{bo}}^* - \frac{T_{\mathrm{b}}}{T_{\mathrm{b}} - T_{\mathrm{o}}}\left[\frac{1}{2}\left(v_{\max}^2 + fr_{\mathrm{m}}v_{\max}\right) + R_{\mathrm{d}}T_{\mathrm{o}}\ln\left(p_{\mathrm{m}}/p_{\mathrm{e}}\right) - \frac{1}{4}f^2 r_{\mathrm{o}}^2\right],$$

$$(8.A5)$$

where r_{m} is the radius of maximum winds, r_{o} is the outer limit of the vortex, and p_{m} and p_{e} are the surface pressures at the radius of maximum winds and in the unperturbed environment, respectively. Note that boundary-layer quasi-equilibrium, applied to the storm environment, gives $h_{\mathrm{b}+}^* = h_{\mathrm{bo}}$, the boundary-layer moist static energy. Also note that gradient wind balance may be used to find $p_{\mathrm{m}}/p_{\mathrm{e}}$, given $v_{\max}$. This is where the outer wind profile *does* influence the maximum wind speed, albeit weakly. Here we simplify matters by using an empirical relationship, $R_{\mathrm{d}}T_s\ln(p_{\mathrm{m}}/p_{\mathrm{e}}) \cong -bv_{\max}^2$, where b is an empirical constant. Using this, neglecting fr_{m} in comparison to $v_{\max}$ and the difference between T_{b} and the surface temperature T_s, and substituting (8.A5) into (8.A1) gives

$$v_{\max}^2 \cong \frac{C_k}{C_D}\left[\frac{\frac{T_s - T_{\mathrm{o}}}{T_{\mathrm{o}}}(h_s^* - h_{\mathrm{bo}}) - \frac{1}{4}\frac{T_s}{T_{\mathrm{o}}}f^2 r_{\mathrm{o}}^2}{1 - \frac{C_k}{C_D}\left(\frac{1}{2}\frac{T_s}{T_{\mathrm{o}}} - b\right)}\right]. \qquad (8.A6)$$

Note that there is also a pressure dependence of h_s^*, which we have not accounted for in (8.A6). The lower pressure at the radius of maximum winds will increase h_s^* over its ambient value, thus increasing the wind speed over that estimated using the ambient value of the saturation moist static energy of the sea surface. Also note that steeper wind profiles yield smaller values of b and thus greater maximum winds. We estimate a typical value of b by using the idealized wind profile given by (8.45) to evaluate the radial integral of the right-hand side of (8.A3) and then comparing the result with the right-hand side of (8.A4) using $R_{\mathrm{d}}T_s\ln(p_{\mathrm{m}}/p_{\mathrm{e}}) \cong -bv_{\max}^2$. This gives a value of b very close to 1, making the denominator of (8.A6) slightly larger than 1 under typical conditions. Finally, as pointed out by Emanuel (1986), the last term in the numerator shows that the maximum wind speed decreases with storm size, though the effect is not large unless the outer radius becomes quite big, of order 1000 km.

References

Alamaro, M., Emanuel, K. and McGillis, W. (2004). Experimental investigation of air–sea transfer of momentum and enthalpy at high wind speed. *J. Flued Mech.*, submitted.

Andreas, E. L. and Emanuel, K. A. (1999). Effects of sea spray on tropical cyclone intensity. Preprint, *23rd Conf. on Hurricanes and Tropical Meteorology, Dallas, TX*, Boston: Amer. Meteor. Soc.

Bender, M. A., Ginis, I. and Kurihara, Y. Y. (1993). Numerical simulations of tropical cyclone–ocean interaction with a high resolution coupled model. *J. Geophys. Res.*, **98**, 23245–23263.

Bister, M. and Emanuel, K. A. (1998). Dissipative heating and hurricane intensity. *Meteor. Atmos. Physics*, **50**, 233–240.

Byers, H. R. (1944). *General Meteorology*, New York: McGraw-Hill.

Chang, S. W. and Anthes, R. A. (1979). The mutual response of the tropical cyclone and the ocean. *J. Phys. Ocean*, **9**, 128–135.

Charney, J. G. and Eliassen, A. (1964). On the growth of the hurricane depression. *J. Atmos. Sci.*, **21**, 68–75.

Emanuel, K., Rotunno, R. and Lilly, D. K. (1985). An air–sea interaction theory for tropical cyclones. Preprint, *16th Conf. on Hurricanes and Tropical Meteorology, Houston, TX*, Boston: Amer. Meteor. Soc.

Emanuel, K. A. (1986). An air–sea interaction theory for tropical cyclones. Part I. *J. Atmos. Sci.*, **42**, 1062–1071.

(1995a). The behavior of a simple hurricane model using a convective scheme based on subcloud-layer entropy equilibrium. *J. Atmos. Sci.*, **52**, 3959–3968.

(1995b). Sensitivity of tropical cyclones to surface exchange coefficients and a revised steady-state model incorporating eye dynamics. *J. Atmos. Sci.*, **52**, 3969–3976.

(1997). Some aspects of hurricane inner-core dynamics and energetics. *J. Atmos. Sci.*, **54**, 1014–1026.

(1999). Thermodynamic control of hurricane intensity. *Nature*, **401**, 665–669.

(2000). A statistical analysis of tropical cyclone intensity. *Mon. Wea. Rev.*, **128**, 1139–1152.

Gallacher, P. C., Rotunno, R. and Emanuel, K. A. (1989). Tropical cyclogenesis in a coupled ocean–atmosphere model. Preprint, *18th Conf. on Hurricanes and Tropical Meteorology*, Boston: Amer. Meteor. Soc.

Hawkins, H. F. and Imbembo, S. M. (1976). The structure of a small, intense hurricane – Inez 1966. *Mon. Wea. Rev.*, **104**, 418–442.

Holland, G. J. (1980). Analytic model of the wind and pressure profiles in hurricanes. *Mon. Wea. Rev.*, **108**, 1212–1218.

(1997). The maximum potential intensity of tropical cyclones. *J. Atmos. Sci.*, **54**, 2519–2541.

Khain, A. and Ginis, I. (1991). The mutual response of a moving tropical cyclone and the ocean. *Beitr. Phys. Atmosph.*, **64**, 125–141.

Kleinschmidt, E., Jr. (1951). Gundlagen einer Theorie des tropischen Zyklonen. *Archiv fur Meteorologie, Geophysik und Bioklimatologie, Serie A*, **4**, 53–72.

Large, W. G. and Pond, S. (1982). Sensible and latent heat flux measurements over the ocean. *J. Phys. Ocean*, **12**, 464–482.

Lilly, D. K. and Emanuel, K. (1985). A steady-state hurricane model. Preprint, *16th Conf. on Hurricanes and Tropical Meteorology, Houston, TX*, Boston: Amer. Meteor. Soc.

Malkus, J. S. (1958). On the structure of the mature hurricane eye. *J. Meteor.*, **15**, 337–349.

Malkus, J. S. and Riehl, H. (1960). On the dynamics and energy transformations in steady-state hurricanes. *Tellus*, **12**, 1–20.

Miller, B. I. (1958). On the maximum intensity of hurricanes. *J. Meteor.*, **15**, 184–195.

Ooyama, K. (1969). Numerical simulation of the life-cycle of tropical cyclones. *J. Atmos. Sci.*, **26**, 3–40.

Powell, M. D. (1990a). Boundary layer structure and dynamics in outer hurricane rainbands. Part I: Mesoscale rainfall and kinematic structure. *Mon. Wea. Rev.*, **118**, 891–917.

(1990b). Boundary layer structure and dynamics in outer hurricane rainbands. Part II: Downdraft modification and mixed layer recovery. *Mon. Wea. Rev.*, **118**, 918–938.

Powell, M. D., Vickery, P. J. and Reinhold, T. A. (2003). Reduced drag coefficients for high wind speeds in tropical cyclones. *Nature*, **422**, 279–283.

Raymond, D. J. (1995). Regulation of moist convection over the west Pacific warm pool. *J. Atmos. Sci.*, **52**, 3945–3959.

Riehl, H. (1950). A model for hurricane formation. *J. Appl. Phys.*, **21**, 917–925.

(1954). *Tropical Meteorology*, New York: McGraw-Hill.

(1963). Some relations between wind and thermal structure of steady state hurricanes. *J. Atmos. Sci.*, **20**, 276–287.

Riehl, H. and Malkus, J. S. (1961). Some aspects of Hurricane Daisy, 1958. *Tellus*, **13**, 181–213.

Rosenthal, S. L. (1971). The response of a tropical cyclone model to variations in boundary layer parameters, initial conditions, lateral boundary conditions and domain size. *Mon. Wea. Rev.*, **99**, 767–777.

(1978). Numerical simulation of tropical cyclone development with latent heat release by resolvable scales. I: Model description and preliminary results. *J. Atmos. Sci.*, **35**, 258–271.

Rotunno, R. and Emanuel, K. A. (1987). An air–sea interaction theory for tropical cyclones. Part II. *J. Atmos. Sci.*, **44**, 542–561.

Schade, L. R. (1997). A physical interpretation of SST-feedback. Preprint, *22nd Conf. on Hurricanes and Tropical Meteorology*, Boston: Amer. Meteor. Soc.

Schade, L. R. and Emanuel, K. A. (1999). The ocean's effect on the intensity of tropical cyclones: Results from a simple coupled atmosphere–ocean model. *J. Atmos. Sci.*, **56**, 642–651.

Shay, L. K. (1999). Upper ocean response to tropical cyclones. RSMAS Tech. Note 99–003, Miami: Rosentiel School for Marine and Atmospheric Sciences.

Shutts, G. J. (1981). Hurricane structure and the zero potential vorticity approximation. *Mon. Wea. Rev.*, **109**, 324–329.

Sutyrin, G. G. and Khain, A. P. (1984). Effect of the ocean–atmosphere interaction on the intensity of a moving tropical cyclone. *Izvestiya, Atmospheric and Oceanic Physica*, **20**, 697–703.

9

Mountain meteorology and regional climates

Ronald B. Smith

Department of Geology and Geophysics, Yale University, New Haven, USA

9.1 Introduction

The subject of mountain meteorology advanced quickly in the twentieth century. These advances can be organized into four broad themes: (A) local mountain climates and human adaptation; (B) airflow dynamics, the dynamic effect of terrain on winds; (C) the thermal effect of hills and slopes on local circulations; and (D) the influence of major mountain ranges on global circulations and climate. For the most part, researchers in these four fields have worked independently with their own tools, paradigms, meetings, and journals. This is especially true for theme (A), including subjects that are generally found in geography and in the agricultural, soil and social sciences (Peattie, 1936; Price, 1981). By contrast, themes (B), (C), and (D) lie clearly in the realm of atmospheric science. Ranked by the number of papers published, theme (B) has been the most active, perhaps because simple mathematical problems are most easily formulated in this area. This ease of formulation has brought together meteorologists and applied mathematicians to develop elegant theories of stationary mountain waves in two and three dimensions.

As in many areas of science, research in mountain meteorology has benefited from the interaction between different approaches. This synergy between different methods was evident in the 1970s and later, championed by Doug Lilly among others. Lilly used research aircraft to observe the spectacular, and still prototypical, 11 January 1972 severe downslope windstorm (Lilly and Zipser, 1972). He also used aircraft to gather statistics and understand the larger significance of wave momentum flux and turbulence (e.g., Lilly, 1972; Lilly and Kennedy, 1973; Lilly and Lester, 1974). Taking advantage of increasing computer speeds, he sought to understand nonlinear mountain wave behavior using numerical models (Klemp and Lilly, 1975, 1978; Lilly and Klemp, 1979). This mix of theoretical, observational and numerical studies established a standard that continues to influence the field of mountain airflow dynamics.

While researchers can be justly proud of the remarkable advances they have made in airflow dynamics, they face more difficult challenges ahead. First, the linkages between airflow dynamics, climatology and the other traditional areas of mountain meteorology (i.e., themes (A), (C) and (D) above) need to be strengthened. Second, a number of new issues have arisen in areas of natural resources, natural hazards, paleoclimate, and geophysics that deserve attention. These new linkages and issues include:

- Water resources and flood forecasting in mountain catchments.
- Air pollution in basins and on mountain slopes.
- Mountain waves, rotors and clear-air turbulence.
- Mountain waves influence on stratosphere circulation and mixing.
- Mountain glaciers and the growth of continental ice-sheets.
- Orographic convection and its role on chemical transport.
- Orographic air mass transformation, rain shadows and deserts.
- Inter-ocean moisture transport and conveyor-belt circulations.
- Orographic triggering of baroclinic and convective storms.
- Climate of the world's largest mountain ranges and basins (e.g., Tibet, Andes, Rockies, Tauros-Zagros, Africa Rift, etc.).
- Global paleoclimate and the role of tectonic mountain building.
- Paleo lapse rate and snow line as an ancient thermometer.
- Mountain erosion and tectonic uplift.
- Paleo-altitude studies using fossil and geochemical indicators of ancient geography.

With these future challenges in mind, we will review the research tools of airflow dynamics research and a few of its conceptual advances. The reader can also consult reviews by Queney *et al.* (1960), Atkinson (1981), Reiter (1982), Lilly (1983), Smith (1979, 1989a, 2001), Blumen (1990), Barry (1992), Baines (1995), Wurtele (1996), Smith *et al.* (1997b), and Whiteman (2000).

9.2 Research tools of mountain airflow dynamics

Advances in mountain airflow dynamics have required the development of new tools: observational, numerical and theoretical. These are reviewed below.

9.2.1 Observational techniques

The observation of orographically disturbed flows is challenging due to the inaccessibility of mountain regions and the complexity of the terrain-induced flow patterns. Since the 1930s, instrumented aircraft have played a central role in this work, but their limitations have become more and more evident. Foremost is the fact that a single aircraft can only be at one place at one time. This limitation was illustrated in an attempt by Lilly and Zipser (1972) to observe the Boulder windstorm. Orographic flows are often unsteady, and the attempt to measure complete spatial structure competes with full temporal monitoring. As more instrumented aircraft become available in the international research community, multiple aircraft missions have become more common. The use of multiple aircraft, however, as in the recent Mesoscale Alpine Program (MAP, Bougeault *et al.*, 2001), only partly overcomes the time–space sampling problem.

As orographically generated disturbances are associated with particular terrain features, the precise measurement of aircraft position has always been a critical requirement for successful aircraft observations. The development of the inertial navigation systems in the 1950s and 1960s brought new capabilities to aircraft mapping of mountain waves, but not until Global Positioning System (GPS) technology became available in the late 1990s, were position measurements in these studies sufficiently accurate. Today, smaller less expensive research platforms equipped with GPS have sufficient positional accuracy to contribute to mountain wave research.

The development of onboard instrumentation for airflow research has developed steadily. The use of gust probes for turbulence and wave analysis has had a very significant impact on the field. Fast temperature, humidity, and tracer measurement have also played a role.

Current advances are mostly in the area of remote sensing. Only remote sensing can provide the volumes of data needed to trace spatial and temporal changes in flow fields. Important new remote-sensing instruments include airborne radar, lidar and dropsondes, and surface-based wind profilers. Downward-looking airborne lidar and dropsondes have helped to solve the persistent problem of how to sample near mountain peaks. Such areas are unsafe for aircraft surveying.

Over the last three decades these tools have been used in small studies and occasionally in large joint field projects in the Alps (ALPEX and MAP), the Pyrenees (PYREX), the southern Alps of New Zealand (SALPEX), the Wasatch Range (IPEX), and the Cascades (IMPROVE). Large projects not only benefit from the coordinated use of new observing technologies, but also from the cooperation between scientists from different organizations.

Satellites are increasing their impact on mountain meteorology. Space-based microwave radar such as the Tropical Rainfall Measuring Mission (TRMM) can map out regions of orographic precipitation. Space-based limb-scanners can detect mountain waves reaching the stratosphere. Rapid-scan geostationary satellites can follow diurnal orographic convection. Surface properties such as temperature and vegetative cover in mountain areas can be mapped with environmental satellites. Scatterometers can detect orographic influence on wind patterns over the sea.

While these observational tools are quite powerful, most can only be deployed in small regions for brief periods of time. Except for satellites, these techniques have done little to improve our ability to continuously monitor weather and climate in the world's mountainous areas. This weakness in the global climate monitoring system is not widely appreciated. According to a common view, it is the world's oceans that are considered to be the problematic "data void" regions. But recent advances in satellite remote sensing have had by far their biggest impact on the oceanic regions. Satellite-borne scatterometers can deduce global-scale patterns of surface winds from ocean wave properties. No equivalent observations are possible over land. Satellite-derived cloud-vector winds give an idea of winds at various altitudes over the ocean. Over mountainous areas, stationary lenticular clouds confuse cloud-tracking algorithms. Moreover, satellite-borne infrared sounders work best over the oceans, where surface properties are more uniform.

Our observing capability in mountainous regions is not nearly so favorable. Generally, mountain regions have fewer weather and climate stations than flat terrain, and those existing stations are less representative of regional conditions. The common practice of locating climate stations in valleys has the potential to bias our climate estimates. Surface radar stations are usually unable to monitor precipitation in mountains due to beam blockage. To make matters worse, high-resolution terrain datasets are unavailable for most of the world. The well-known dataset GTopo30 is extremely useful, but its quality is rather poor in Africa and Asia. New terrain datasets, such as those from the Shuttle Radar Topographic Mission (SRTM) are just becoming freely available.

We conclude that the mountainous regions of the world are, and will remain, the true "data void" regions on our planet.

9.2.2 Numerical models

Since 1975, numerical models have played an increasing role in research in mountain airflow dynamics. One of the first major contributions by the numerical approach was the use of 2D hydrostatic models to investigate the nonlinear effects on the severe downslope wind storms (Klemp and Lilly, 1975, 1978; Clark and Peltier, 1977; Peltier and Clark, 1979). Clark and Peltier showed that wave breaking triggers a dramatic reorganization of the wave field, in which the lower tropospheric flow decouples from the flow aloft, approaching a spilling "water-over-a-dam" geometry (Clark and Peltier, 1984; Smith, 1985).

Since that time numerous advances in modeling technologies have occurred, see Chapter 6 by Klemp and Skamarock. Such improvements include: nonhydrostatic models, higher spatial resolution, nesting, interchangeable parameterizations, and advanced visualizations. Several problems still remain however. Turbulence and cloud physics parameterizations are complex, slow, and uncertain. Terrain-following coordinates occasionally cause problems in weak flows when flow cannot climb over the mountain. Numerical diffusion and pressure gradient errors along coordinate surfaces may degrade model accuracy. Parameterization algorithms are also more difficult to apply in a curvilinear coordinate system. Additional work on coordinate systems for mesoscale models may be required.

While some progress has been made on the rigorous testing and inter-comparison of numerical models, further work remains. It has become routine to use linear wave theory and wave drag formulae as a reference for numerical models (see Section 9.6). For finite-amplitude disturbances, model inter-comparisons can be useful (Doyle *et al.*, 2000). To identify errors in diffusion or pressure gradient, "no-flow" cases are studied. In spite of these positive examples, the modeling community has yet to accept an extensive protocol for model testing and inter-comparison. Ideally, new models should be tested on a suite of quantitative exercises with known answers.

9.2.3 Theoretical and conceptual tools

Investigators engaged in theoretical work on airflow dynamics have their own set of special tools. The same concepts and mathematical tricks appear over and over again in the mountain wave dynamics literature. Some of the most widely used tools are: the Boussinesq equations that remove the effects of compressibility and flow divergence; linearization for simplifying the governing equations and for identifying nonlinear effects; the hydrostatic approximation for simplifying wave propagation and obtaining closed-form solutions; Long's equation for treating steady, finite-amplitude, 2D stratified flows; hydraulic theory for reducing the dimensionality of the problem; Fourier transforms for solving differential equations; causality and Galilean invariance; dimensional analysis; group velocity and the Eliassen–Palm flux laws; and vorticity and Bernoulli conservation laws for identifying the effects of dissipation. Students entering the field of mountain meteorology need a sound understanding of these concepts.

One can identify a number of important theoretical problems in airflow dynamics that have received attention over the last 50 years. While these problems are each rather narrow in scope, together they form a broad foundation for our field. Many of these issues were controversial when first studied, but most have been settled by now. To give the reader a taste of the field, a few problems are listed below. These problems are fully discussed in the reviews listed above, and in the primary literature.

- How do upper boundary conditions influence mountain wave patterns and drag? (*The radiation condition aloft requires that wave energy propagates upward and that wave patterns tilt upstream, resulting in upstream deceleration and downstream wave drag.*)
- What is the mechanism of severe downslope winds? (*Turbulence and nonlinear gravity wave resonance allows a continuously stratified fluid to behave much like a single hydraulic layer.*)
- How does Coriolis force modify gravity waves? (*It limits horizontal divergences while reducing and altering the phase of vertical motions.*)
- What is the mechanism of gravity wave breaking? (*Both overturning and shear instability can generate turbulence that will dissipate the wave.*)
- What is the reason for upstream blocking and deflection? (*High pressure, caused by tilted wave field aloft, decelerates and deflects the incoming flow.*)
- Why does airflow accelerate through mountain gaps? (*A cross-mountain pressure gradient caused by tilted wave field aloft accelerates the gap flow.*)
- How do mountains produce jets and wakes? (*Lee-side Bernoulli loss and potential vorticity gradients are caused by friction or wave breaking.*)
- What is the role of mountain drag and wave momentum flux on the general circulation? (*Weakens the jet stream and induces meridional overturning in the stratosphere.*)
- What causes lee cyclogenesis? (*Several theories have been proposed. In one, potential vorticity generation provides a trigger for baroclinic development.*)
- Is there a mountain anticyclone induced by vortex shortening in the real atmosphere? (*This looks doubtful, in spite of the theoretical prediction. Reasons are unclear. Diurnal heating and cooling may obscure the vortex effect.*)
- What is the relative importance of forced ascent and elevated heating in orographic disturbances? (*Heating dominates in broad weak flows with strong insolation. Forced ascent dominates in strong flows with weak solar heating.*)

Some of these problems illustrate the pure reductionist approach to mountain meteorology. With extreme simplification, certain questions can be cleanly posed and convincingly answered. Other problems do not allow such simplicity, as competing mechanisms may be involved. A new intellectual approach to mountain meteorology is emerging. Instead of focusing entirely on specific narrow physical or fluid dynamic process, different mechanisms must be quantitatively compared and assessed. As the field matures, new problems often include competing physical mechanisms. Dominant processes may differ from location to location, season to season, or event to event. Questions of dominance are just as subtle and challenging as the earlier work on pure mechanisms. They require more extensive monitoring or even a climatological approach. As progress is made on questions of dominance, mountain meteorology will be able to make a greater contribution to weather prediction and regional climate studies. Specific discussions of how to approach complex problems of regional climate dynamics are given by Reiter (1982), McGregor (1997), Giorgi and Mearns (1999), and Leung *et al.* (2003).

In the rest of this chapter, we consider four subject areas that go beyond the analysis of pure single mechanisms. These are subjects for which competing mechanisms can be identified. A discussion of these topics will illustrate how pure results can be synthesized to attack more complex problems.

9.3 Flow splitting and gravity wave breaking

An important prediction of mountain wave theory is the onset of flow splitting and gravity wave breaking. Flow splitting is defined as the horizontal splitting of the incoming flow so that it passes around rather than over a mountain peak. Wave breaking is usually associated with the vertical overturning of potential temperature surfaces. While geometrically quite different, these two phenomena share a common attribute; they must be preceded by flow deceleration and incipient stagnation. Streamline splitting requires that the low-level flow be decelerated to a stagnation point. At a stagnation point, two wind vector directions can co-exist; a geometry essential for streamline splitting. Gravity wave breaking, in a uniform background state, usually begins by the steepening of the potential temperature surfaces, possible only in decelerated flow. Wave steepening leads to overturning and turbulence.

In the ideal formulation (e.g., neglecting the Coriolis force), the parameters that enter this problem are: upstream wind speed (U), upstream buoyancy frequency (N), mountain width scale (a) and mountain height (h). Work on the stagnation problem has mostly been focused on the hydrostatic limit where the non-dimensional parameter Na/U is large. In this case, the nonlinearity parameter Nh/U plays a dominant role, along with parameters describing the mountain planform shape, such as horizontal aspect ratio (Fig. 9.1). We refer to $\hat{h} = Nh/U$ as the non-dimensional mountain height. The mountain width plays no role in the hydrostatic limit, so intuitive ideas about the importance of mountain steepness must be discarded.

The mechanism of flow deceleration is the same for both flow splitting and wave breaking. In stably stratified air, a positive density anomaly is created by ascent. According to the hydrostatic law, areas of high pressure will exist at the base of these dense fluid anomalies. According to Bernoulli's law for steady incompressible flow,

$$B = p + (1/2)\rho U^2 + \rho g z = const, \tag{9.1}$$

a constant value of Bernoulli function (B) requires that pressure (P) and wind speed behave oppositely. As parcels approach a high-pressure region, their speeds decrease due to the adverse pressure gradient. The height term ($\rho g z$, where ρ is air density, g the acceleration due to gravity and z is altitude) in (9.1), once thought to be dominant, plays only a small role (Sheppard, 1956; Smith, 1989b, 1990).

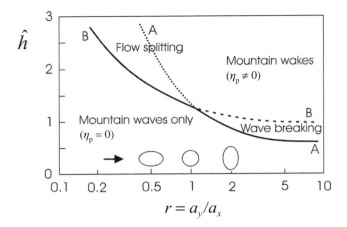

Figure 9.1. Regime diagram for flow splitting and wave breaking. Abscissa is the horizontal aspect ratio of the hill. Ordinate is the non-dimensional mountain height. Line A indicates the onset of stagnation aloft leading to wave breaking. Line B indicates the onset of low-level flow stagnation leading to flow splitting. Both curves are derived from linear hydrostatic mountain wave theory. Above the curves, flows are nonlinear and dissipative with the generation of potential vorticity, η_p.

As the non-dimensional mountain height ($\hat{h}$) increases, the strength of the high-pressure regions increases at two special locations in the flow; on the windward mountain slope (point B) and at a point directly above the hill at an altitude of approximately $z = (3\pi/2)U/N$ (point A). The relative magnitude of these two potential deceleration points determines whether flow splitting or gravity wave breaking occurs first (Smith, 1989b; Stein, 1992; Baines and Smith, 1993; Smith and Grønås, 1993; Ólafsson and Bougeault, 1996). Estimates for the onset of splitting and breaking based on linear theory are given in Fig. 9.1 for a range of hill aspect ratios.

For a long ridge, or in strictly two-dimensional flow, the deceleration at point A is stronger than at point B. Thus wave breaking occurs before low-level blocking, starting approximately when $\hat{h} = 0.85$. For an isolated hill with circular contours, the two points (A and B) are similar in their deceleration potential. Splitting and wave breaking begin approximately when $\hat{h} = 1.3$. In 3D flow, the lateral dispersion of waves aloft weakens the density anomalies, so a larger hill is required to stagnate the flow. For hills aligned with the flow, the more rapid dispersion of the wave field aloft prevents the generation of a deep positive density anomaly. Flow stagnation requires a mountain height far greater than for airflow across a long ridge.

Once flow splitting or wave breaking begins, the flow pattern restructures itself quite dramatically. The lee-side flow region takes on a complex vortical structure about which linear non-dissipative theory can say very little. Vortical wakes in stratified flow have been investigated in the laboratory (Brighton, 1978; Snyder *et al.*, 1985; Gheusi *et al.*, 2000) and with numerical simulation (e.g., Rotunno and Smolarkiewicz, 1991; Miranda and James, 1992).

One important difference between pure gravity wave flow and flow with either flow splitting or wave breaking, is the existence of potential vorticity (PV), which we define as

$$\eta_p = (1/\rho)\boldsymbol{\xi} \cdot \nabla\theta, \tag{9.2}$$

where $\boldsymbol{\xi}$ is the vorticity vector and θ is the potential temperature. Material changes in η_p are governed by

$$\frac{d\eta_p}{dt} = \frac{1}{\rho}(\nabla \times \mathbf{F}) \cdot \nabla\theta + \frac{1}{\rho}\boldsymbol{\xi} \cdot \nabla\dot{H}, \tag{9.3}$$

where $\mathbf{F}$ and $\dot{H}$ are frictional force and internal heating rate, respectively. In pure gravity wave flow, strong vorticity generation occurs due to baroclinic effects (i.e., pressure gradient torques), but these vorticity vectors are tangential to the potential temperature surfaces so that $\eta_p = 0$ in (9.2). When wave breaking or flow splitting occur, boundary-layer effects or internal turbulence produce friction and heat fluxes that act in (9.3) to generate potential vorticity. After air parcels leave the violent dissipative η_p-generating regions, η_p tends to be conserved once again, advecting downwind into the wake region. These plumes of potential vorticity have been called "PV-banners." It is useful to define the "wake" as a region of potential vorticity, to distinguish it from lee-side disturbances that may contain only lee waves (Smith, 1989c). Recent analyses of η_p-generation mechanisms are given by Schär and Durran (1997), Rotunno *et al.* (1999), Epifanio and Durran (2002), and Schneider *et al.* (2003). Hints of vertical vorticity in non-dissipative, weakly nonlinear solutions appear unrelated to real wakes with potential vorticity (Epifanio and Durran, 2002), reinforcing the utility of defining wakes in terms of potential vorticity. Helpful relationships between the Bernoulli constant (9.1) and potential vorticity (9.2) are discussed by Schär and Smith (1993), and Schär (1993).

We conclude that mountain airflow situations can be classified into two categories.

- With strong winds, modest static stability and low mountains, air parcels easily climb over the terrain. Conserved quantities like potential temperature, specific humidity, potential vorticity, and tracer concentrations are flushed and equalized between lowland and highland sites. Mountain wave energy propagates vertically, even reaching the stratosphere. No wake will be present.
- With weak winds, strong stability and higher mountains, low-level airflow will stagnate or run parallel to terrain contours. This stagnant air fills the valleys and upslope regions.

Winds aloft may still hit the exposed mountain peaks, causing gravity waves aloft. Low-level wave breaking may occur. A wake may form downstream, as documented near trade-wind islands such as Hawaii (Smith and Grubišić, 1993) and St. Vincent (Smith *et al.*, 1997a), and higher-latitude islands such as the Aleutians (Pan and Smith, 1999).

This relatively simple picture for flow splitting and wave breaking can be modified when the ambient atmospheric profile has vertical structure or a turbulent boundary layer, or if latent heat or Coriolis force play a role. For example, strong shear or a shallow stable layer aloft may promote wave breaking by a Kelvin–Helmholtz mechanism without requiring deceleration and overturning.

9.4 Lapse rates on mountain slopes

A common observation in mountainous terrain is the decrease of temperature with altitude. Temperature lapse rates along mountain slopes vary widely, but are usually negative. Occasionally, positive lapse rates are seen, almost always connected with the intersection of an elevated atmospheric inversion with the sloping terrain. Examples of positive lapse rate are found in the Los Angeles basin in California and the western slope of Andes in Peru. The cold California and Humboldt currents cool the air from below while descending subtropical air aloft forms the inversion. A third example is central Utah, where radiative heat loss in the winter generates a cold pool of air lying in the Salt Lake basin. A further example is on the big island of Hawaii, where the volcanic peaks penetrate up through the tradewind inversion.

The extensive literature on mountain lapse rates suggests that a value of -5 ± 1 K km^{-1} is a reasonable worldwide average (e.g., McCutchan, 1983; Barry, 1992). Thus, a 3 km mountain would experience a temperature 15 K lower than the surrounding plains. In mid-latitudes, this temperature difference will push the uplands into a colder, less-hospitable climate zone. In the tropics however, the cooler mountain conditions provide welcome relief from extreme heat and humidity.

The mountain lapse rate depends on several competing processes. Consider the following limiting cases (see Fig. 9.2).

- With low terrain in a windy environment, the air is forced to rise along mountain slopes. If this lifting occurs quickly enough, heat supplied by or lost to the earth's surface can be neglected and parcels will conserve potential temperature. The result, in a dry atmosphere, would be an adiabatic lapse rate of $\gamma = -g/c_p = -9.8$ K km^{-1}. In moist saturated ascent, the magnitude of the cooling rate is reduced by the latent heat of condensation.
- With steep high isolated terrain and weak flow, the airflow would split and flow around, exposing the hill to temperature and other properties of the undisturbed free atmosphere. In this case, the mountain lapse rate might equal the lapse rate of the free atmosphere

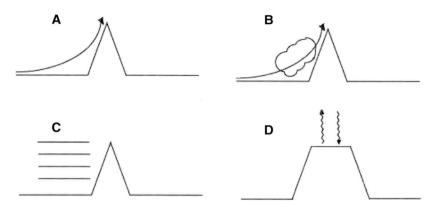

Figure 9.2. Schematic of mountain lapse-rate processes. (A) Dry ascent causing a dry adiabatic lapse rate; (B) moist ascent causing a moist adiabatic lapse rate; (C) stagnant air or airflow splitting allowing the free-atmosphere lapse rate to impact mountain slope temperatures; and (D) isolated elevated surface, local energy budgets control the surface temperature.

(see Section 9.3). The average free-atmosphere lapse rate is -6.5 K km^{-1}. If the flow is too weak, however, advection of heat from the surrounding free atmosphere will be overcome by local surface radiative effects.

- With broad gentle terrain and weak flow, the influences of forced ascent or the free-atmosphere advection are small compared to the local radiative heating and cooling effects of the mountain surface. Such a mountain climate is isolated from both the sea-level climate and the distant free atmosphere. Surface temperature will be controlled by local energy budget factors.

Estimates of the surface temperature in radiative balance could be made from the greenhouse formula

$$T = \left[\frac{S(1 - \alpha_s)(1 + \tau_0)}{4\sigma} \right]^{1/4} = T_{\text{eff}}(1 + \tau_0)^{1/4}, \qquad (9.4)$$

where S is the solar constant, α_s is the surface albedo, τ_0 is the optical depth for longwave radiation and σ is the Stefan–Boltzmann constant (Goody and Walker, 1972). Also, $T_{\text{eff}} = [S(1 - \alpha_s)/4\sigma]^{1/4}$ is the effective temperature for earth (~ 254 K). An inverse calculation for optical depth, using the actual average earth temperature of $T = 288$ K, gives $\tau = (T/T_{\text{eff}})^4 - 1 = 0.65$. Now if the optical depth is proportional to the total atmospheric mass above a point, i.e.,

$$\tau(z) = \tau_0 \exp(-z/H), \qquad (9.5)$$

where $H = RT/g \approx 8.4$ km is the density scale height, we obtain

$$T(z) = T_{\text{eff}}[1 + \tau_0 \exp(-z/H)]^{1/4}. \qquad (9.6)$$

Table 9.1. *Mountain lapse-rate mechanisms.*

Case	Lapse rate (K cm^{-1})
Dry ascent	-9.8
Moist ascent	-3 to -8
Free atmosphere	-6.5, but variable
Isolated greenhouse	-4.8

Differentiating (9.6) and setting $z = 0$ gives a lapse rate for plateaus in the lower atmosphere:

$$\gamma = -\frac{T_{\text{eff}}\tau_0}{4H} \approx -4.8\text{K km}^{-1}. \tag{9.7}$$

This "greenhouse lapse rate" arises from the fact that mountain plateaus lie above some fraction of the atmospheric greenhouse gases. This lapse rate decreases exponentially aloft. A more sophisticated computation of this effect is given by Molnar and Emanuel (1999). Note that decreased surface albedo in (9.4), associated with snow or barren ground, and variations in evaporative cooling (not included) will influence energy budgets on high terrain. We summarize these competing lapse-rate mechanisms in Table 9.1.

To test lapse-rate ideas in real complex terrain, we look at the Sierra Nevada range in California. Detailed mapping of surface properties and temperature are available in clear-sky satellite images such as those from the new MODIS (MODerate resolution Imaging Spectroadiometer) instrument in the Terra and Aqua satellites. An example from 30 October 2002 is shown in Plate X. We use MODIS channel 11 in the infrared window to estimate surface temperature, assuming that the emissivity is equal to unity everywhere. The image was taken at approximately 10:30 local time, with the sun in the southeast quadrant. Temperature data combined with terrain data from GTopo30 allows the influence of elevation on temperature to be studied. This method requires cloud-free skies. Plates X and XI describe the San Joaquin valley in California and the western slopes of the Sierras. Plates XII and XIII describe the eastern slopes and inter-mountain plateau.

An important caveat is that satellite-derived brightness temperatures represent a "skin" temperature rather than a low-level air temperature. During the night, the skin temperature drops below the 2-meter air temperature. During a cloud-free day, as in Plates X to XIII, the skin temperature may exceed the 2-meter air temperature by several degrees. Still, when considering large differences between high and low terrain, brightness temperature gives a useful estimate.

Plates X and XII are scattergrams in which pixel temperature is plotted against altitude. Selected zones defined by height and temperature are identified on the accompanying maps (Plates XI and XIII, respectively). Generally, the west-facing slopes cool rather quickly with altitude, at a rate approaching the adiabatic lapse rate. The negative lapse rate weakens above 2 km.

The highest region (A) is the coldest, with a temperature near the freezing point, consistent with the existence of snow fields near Mt. Whitney. Region B comprises the upper western slopes. Region C, with the clearest and strongest lapse rate, is located in the smooth western foothills. Region D, with an anomalous positive lapse rate, probably cools westward due to maritime influence of the Pacific Ocean.

The eastern slopes show a different pattern. The high eastern slopes (regions E and F) are 5–10 K warmer than the corresponding altitudes on the western slopes (A and B). Regions G and H are two broad plateaus of the inter-mountain region. Region G in the southeast quadrant ranges from 2000 to 2500 m above sea level with temperature higher than 285 K. Region H covers the expansive area in the northeast, with altitudes from 1100 to 1800 m. Temperatures in H exceed 288 K, equal to those in the San Joaquin Valley 1.5 km below (C, D).

The application of the concepts A and B illustrated in Fig. 9.2 is inappropriate in this case. High mountains, weak winds and lack of clouds make it unlikely that air is passing over the range, equilibrating the surface temperatures to adiabatic lapse-rate values. Concept C is also unlikely. Weak winds and the elongated shape of the Sierras suggest that a free air mass cannot efficiently ventilate the slopes. In any case, different air masses probably dominate the eastern and western slopes. These domains are isolated from each other. The atmosphere is more likely accommodating to the surface energy budget than vice versa.

The idea of local surface heat budget (D in Fig. 9.2) is probably most helpful for interpreting our scene, but the simple greenhouse model (9.7) is inappropriate. Strong solar heating is balanced by significant evaporation in the dense irrigated cropland (C), keeping the skin temperature cool. By contrast, on the eastern slopes and plateaus (G, H), lack of rainfall and stored ground-water nearly eliminate evaporative cooling. Upland temperatures rise to high values in spite of the negative free-troposphere lapse rate, increased albedo and reduced greenhouse forcing.

We conclude that mountain lapse rates in our scene are dominated by local surface energy budgets, especially latent heat fluxes. The role of the Sierras is still profound, however. The mountain barrier prevents precipitation from reaching the interior plateaus and it separates contrasting air masses to the east and west. Note that the wet–dry contrast across the Sierras arises from processes not present on the day the MODIS image was taken. Precipitation in the region occurs during cloudy winter days with strong westerly flows (see Section 9.5). On such days, temperature patterns may be dominated by concept B in Fig. 9.2. On subsequent weeks and months, clear-day weak-wind temperature patterns are dominated by evaporation of stored soil moisture.

Clearly, no universal mountain slope lapse rate can be found. Temperatures on mountain slopes respond to many factors. In some cases altitude effects may be overwhelmed by local heat budget factors such as evaporation or surface albedo.

9.5 Orographic precipitation and air mass transformation

A problem of central importance to regional climatology is that of orographic precipitation and air mass transformation. The problem is quite complex as it involves both airflow dynamics and cloud physics. The impact of new knowledge in this area is quite high, however, as it relates to water resources, flood prediction and the role of mountains in the formation of deserts and inter-ocean water transport. Research activity has accelerated recently, with mid-latitude field projects in the southern Alps (Sinclair *et al.*, 1997), the Cascades (Colle and Mass, 1998) the Sierras (Pandey *et al.*, 1999), coastal California (Neiman *et al.*, 2002), the Wasatch Range (Schultz *et al.*, 2002), the European Alps (Medina and Houze, 2003). Some of the questions addressed by these projects are:

- How deep does orographic lifting extend?
- Does the depth of lifting control the amount of precipitation?
- Does latent heat influence the airflow dynamics?
- What fraction of the condensed water reaches the ground as precipitation?
- How well can this fraction be estimated?
- What process controls this fraction?
- How much precipitation spills over onto the lee slopes?
- What factors control spill-over?
- Does lee-side dryness arise from moisture exhaustion or local descent?
- How do mountain scales enter the problem?
- Do small-scale terrain features generate small-scale precipitation patterns?
- Are the cloud physics processes linear, i.e., will twice the lifting and condensed water generate twice the amount of precipitation?
- How important are environmental parameters such as temperature and aerosol content? Can existing mesoscale models capture the essential elements of orographic precipitation?
- Can even simpler models capture these elements?

In the text below we address a few of these questions.

To address the problem of precipitation efficiency, we review some findings from the Mesoscale Alpine Program (MAP) project in 1999. In one well-studied case (20 September 1999) interpolations from several different numerical models were used to estimate water-vapor fluxes, condensation and precipitation rates. During this event, a trough over Spain brought strong moist southerly flow against the Italian Alps. From the low-level humidity field (Fig. 9.3), it is clear that the Alps are drying out the southerly flow. The specific humidity is reduced north of the Alps.

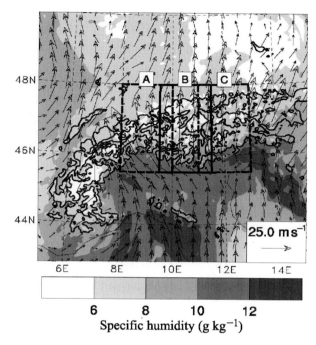

Figure 9.3. Plan view of surface wind (m s^{-1}) and specific humidity (g kg^{-1}) on 20 September 1999. Fields are given by output from COAMPS, see text and Table 9.2. (From Smith *et al.*, 2003.)

It has been shown by several authors that detailed water budgets are difficult to construct using only directly observed wind and humidity fields. To best utilize these measurements, one needs to assimilate them into a numerical model using nudging or other techniques. Models bring the advantages of mass and moisture conservation and accurate advection through time and space. In MAP, four models were used to estimate water-vapor fluxes: COAMPS, MC2 and the ECMWF forecast and post-event analysis. The pattern of water-vapor flux approaching the Alps, crossing the 45.5° latitude line, is shown in Fig. 9.4 for two models. The flux patterns from COAMPS and MC2 are similar. In Fig. 9.4, the Alpine crestline is also shown, to indicate how much lifting must occur for the air stream to cross the range. Time-varying fields from these models were used to compute the inflow and outflow fluxes in the control volumes (A, B, C) shown in Fig. 9.3.

According to Table 9.2, the four models agree fairly well on the flux of water vapor approaching the Alps. Over the 24-hour period, between 42×10^{11} and 56×10^{11} kilograms of water entered the control volume A. The models differ more in the amount of precipitation. The COAMPS model precipitated 19×10^{11} kg, while MC2 precipitated only 12×10^{11} kg. The difference is probably caused by

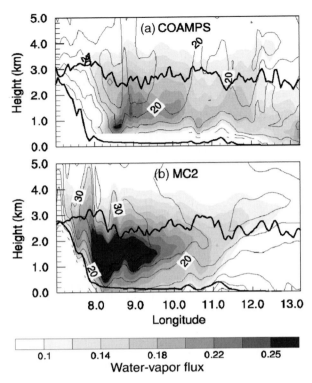

Figure 9.4. Inflow cross-section along 45.5° N. (a) COAMPS, (b) MC2. Water-vapor flux is shaded. The wind speed is contoured 5 m s^{-1} intervals. The Alpine crestline is projected onto this section to indicate how high the air must rise to cross the Alps. (From Smith *et al.*, 2003.)

their different cloud physics parameterizations. At least in this event, MC2 was less efficient at converting cloud water to precipitation. In row (e), we compute the precipitation rate P from the well-known upslope model:

$$P = \rho q_0 \mathbf{U} \cdot \nabla h(x, y), \qquad (9.8)$$

assuming that forced ascent lifts saturated moist-neutral air and that all condensed water falls immediately to the ground (Smith, 1979). In (9.8), q_0 is the specific humidity at the surface and $h(x, y)$ is the mountain height field. Only positive (i.e., upslope) values of (9.8) are retained. The wind and humidity output from COAMPS and MC2 is used, along with a high-resolution 1 km terrain. Precipitation computed in this way, about 100×10^{11} kg, exceeds the actual precipitation by a factor of five. Even more striking is that the precipitation from (9.8) is twice the incoming water-vapor flux; a physical impossibility. The large values of precipitation from (9.8) imply that the fine-scale terrain is repeatedly lifting the same

Table 9.2. *Alpine water-vapor fluxes, precipitation and drying ratios for 20 September 1999, derived from four models. Models and data are for box A in Fig 9.7.*

Row	Quantity	COAMPS	MC2	ECMWF Forecast	ECMWF Analysis
a	WV influx	56	53	42	44
b	WV outflux	33	35	33	29
c	Model precipitation	19	12	17	
d	Actual precipitation	19 (19)	19 (19)	19 (19)	19 (19)
e	Upslope precipitation	95	108		
f	DR (c/a)	34	23	40	
g	DR (d/a)	34	36	45	43

Note: Water-vapor (WV) flux and precipitation values are accumulations over 24 hours, in units 10^{11} kg. The drying ratio (*DR*) is given in percent. The ECMWF Analysis does not have precipitation. Actual precipitation values are independent estimates from the Federal University of Technology, Zurich (Frei and Haeller, private communication), and the University of Vienna (in parentheses). (Data are taken from Smith *et al.*, 2003.)

air parcels, always causing rain in the model. This error is scale dependent. The higher the terrain resolution used, the more lifting events occur and the worse the over-prediction becomes. The upslope model (9.8) goes wrong for three reasons. First, the net air mass drying was neglected; second, the forced ascent was assumed to reach all the moist layers and, third, the condensed water in small orographic clouds was assumed to precipitate rather than being allowed to evaporate on the lee side.

A measure of the air mass drying is the drying ratio

$$DR = \text{Total precipitation/Incoming water vapor flux.} \qquad (9.9)$$

The drying ratio computed from the observed precipitation and mesoscale model fluxes is about 35% (Table 9.2). While this ratio is substantial, it does not seem large enough to fully explain the gross overestimation of precipitation by the upslope model. We therefore conclude that the lack of realism in (9.8) is due mostly to the neglect of airflow dynamics and cloud physics. New attempts to understand these processes have been made by Colle (2004), Jiang and Smith (2003), and Smith and Barstad (2004).

One important constraint on orographic precipitation is the inability of condensed water to fall quickly out of the cloud. A measure of this constraint is the precipitation efficiency: the ratio of precipitation to condensation rate. One potentially useful theory of precipitation efficiency involves the identification of a cloud physics time scale (τ) for hydrometeor generation and fallout. In a simple box model, the precipitation efficiency (e_P) can be written in terms of this time scale, that is

$$e_P = \frac{1}{(1 + \tau U/a)},\qquad(9.10)$$

where U is the wind speed and a is the mountain half-width (Jiang and Smith, 2003). When the wind speed is high and the mountain narrow, the time taken for an air parcel to cross the ridge (a/U) is less than the cloud physics time scale. Condensed water will be carried quickly to the lee slopes where it will evaporate in descending air. The value of e_P from (9.10) is very small in this case. Conversely, for wide hills, slow winds and fast conversion, (9.10) predicts that a large fraction of the condensed water will fall to the ground as precipitation. Further work is required to evaluate this type of model for precipitation efficiency.

9.6 Gravity waves and wave drag over complex terrain

The importance of mountain wave drag and the propagation and deposition of momentum by waves has been appreciated for three decades. While some controversy remains, there is good evidence that these waves have an influence on the large-scale circulation of the earth atmosphere (Lilly, 1972; Palmer *et al.*, 1986). Most general circulation models (GCMs) include a parameterization of wave drag, derived from simple linear models of mountain wave generation. In addition to this application, analytically derived linear drag laws are often used to check numerical models, or used as reference values to examine nonlinear effects. Drag laws are sensitive indicators of errors due to numerical dispersion, dissipation and boundary reflection.

As an aid to the development of drag parameterizations and the use of drag laws as a model-evaluation tool, we have calculated values of appropriate drag coefficients. Values for five simple mountain ridges are presented in Table 9.3. These were

Table 9.3. *Area coefficient, C_A and linear hydrostatic drag coefficient, C_D for ridges.*

Mountain shape	C_A	C_D (exact)	C_D (value)	Reference
Witch	π	$\pi/4$	0.7854	Queney (1947, 1948)
3/2 power bell	2	**	$\sim$0.824	***
Gaussian	$\sqrt{\pi}$	1	1.0	***
Exponential	2	$2/\pi$	0.6366	***
Triangle	1	$4\ln(2)/\pi$	0.8825	***

Note: ** no analytical solution is known; *** derived for the present paper.

Table 9.4. *Volume coefficient, C_V and linear hydrostatic drag coefficient, C_D for circular hills.*

Mountain shape	C_V	C_D (exact)	C_D (value)	Reference
Witch	∞*	**	~0.968	*
3/2 power bell	2π	$\pi/4$	0.7854	Smith (1988)
Gaussian	π	$(2\pi)^{3/2}/16$	0.9843	Smith and Grønås (1993)
Exponential	2π	$\pi^2/16$	0.6168	***
Cone	$\pi/3$	**	~0.555	***

Note: * not convergent; ** no analytical solution is known; *** the present paper or an unknown source.

derived using the following shape relationships, where h_m is the maximum shape (mountain) height:
the Witch of Agnesi,

$$h(\hat{x}) = h_m/(1 + \hat{x}^2), \tag{9.11}$$

the 3/2 power bell-shape,

$$h(\hat{x}) = h_m/(1 + \hat{x}^2)^{3/2}, \tag{9.12}$$

the Gaussian,

$$h(\hat{x}) = h_m \exp(-\hat{x}^2), \tag{9.13}$$

the exponential,

$$h(\hat{x}) = h_m \exp(-|\hat{x}|), \tag{9.14}$$

and the triangle,

$$h(\hat{x}) = h_m(1 - |\hat{x}|) \quad \text{for } |\hat{x}| < 1, h(\hat{x}) = 0 \quad \text{for } |\hat{x}| \geq 1. \tag{9.15}$$

For these shapes, the quantity $\hat{x} = x/a$ is a non-dimensional distance from the ridge crest, where actual distance x is normalized by a, the characteristic half-width of the hill. The definition of a as a half-width does not imply that $h(x = a) = h_m/2$, as this is valid only for (9.11). The title "Witch of Agnesi" for (9.11) arises from an improper translation into English of the term "turning" curve used by Maria Agnesi (1718–99).

In Table 9.4, we construct an analogous list of isolated circular hills for which (9.11)–(9.15) are still valid, but with $\hat{x}$ replaced by a non-dimensional radius $\hat{r} = r/a$. For ridges, the cross-sectional area is given by $A = C_A ha$ and the drag per unit length is

$$D_L = C_D \rho U N h^2.$$ (9.16)

For circular isolated hills, the hill volume is $V = C_V h a^2$ and the drag is

$$D = C_D \rho U N a h^2.$$ (9.17)

Values in Table 9.3 have recently been re-derived and/or verified by the author and compared against numerical Fast Fourier Transform computations.

Other linear wave drag laws are given by Phillips (1984) for elliptical hills, and by Smith (1986) and Grubišić and Smolakiewicz (1997) for shearing flow. The influence of hill shape and atmospheric structure was discussed by Lilly and Klemp (1979), Blumen and Hartsough (1985), and Leutbecher (2001). The roles of friction and Coriolis force were analyzed by Smith (1979), and by Ólafsson and Bougeault (1997). All of these studies considered only smooth terrain of simple shape. Welch *et al.* (2001) considered periodic terrain.

One persistent question has been whether simple linear drag formulae continue to apply to high and/or complex terrain. For high smooth terrain, a sudden transition to a "severe wind" or "high drag" state has been described by Clark and Peltier (1977, 1984), Lilly and Klemp (1980) and Smith (1985), see the review by Smith (1989a). Drag values in this state exceed the prediction of (9.17) by a factor of two or more. The best atmospheric example of this state is still the observation of the 1972 Boulder windstorm by Lilly and Zipser (1972). Bauer *et al.* (2000) and Epifanio and Durran (2001) computed the effect of nonlinearity on drag from elliptical hills.

The analysis of wave drag from terrain that is both high and complex, was undertaken in the Pyrenees (Beau and Bougeault, 1998), and in the Alps in the recent MAP project (Bougeault *et al.*, 2001; Smith *et al.*, 2002). In the MAP project, research aircraft flew coordinated patterns over the Alps, measuring mountain waves, while aircraft-deployed dropsondes and downward-looking lidar observed the flow structures near and below mountain top. One well-studied event that occurred on 2 November 1999 in the vicinity of Mt. Blanc (Fig. 9.5, Smith *et al.*, 2002). The wave pattern found on that day is illustrated in Figs. 9.6 and 9.7. Attempts to reproduce that wave pattern with linear theory were successful only when the concept of reference base altitude was introduced. According to this concept, the air is assumed to be stagnant below z_{ref}. The synoptic-scale flow impacts only the terrain higher than z_{ref}, producing gravity waves (Fig. 9.8). Obviously, the higher z_{ref} becomes, the smaller are the effective mountains and the smaller is the wave drag (Welch *et al.*, 2001)

The only mountain shape in Table 9.4 that allows the influence of a stagnant layer to be easily analyzed is the cone. A cone truncated at z_{ref} retains the perfect shape

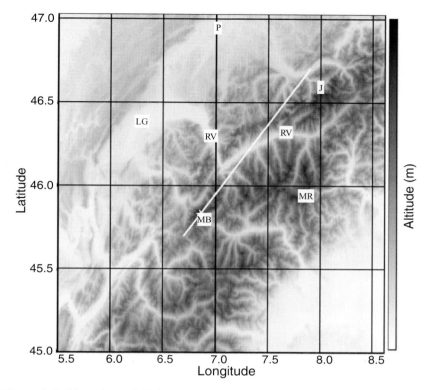

Figure 9.5. Plan view of Alpine terrain near Mt. Blanc (MB). Line indicates the wind direction and the track of the research aircraft on 2 November 1999. Southwesterly flow over Mt. Blanc generates vertically propagating mountain waves. LG is Lake Geneva. (From Smith *et al.*, 2002.)

of a cone. Using the drag and volume formulae (9.17), the drag can be expressed in terms of the exposed volume

$$D = C_D \rho U N (3/\pi)(h_m/a) V, \tag{9.18}$$

where the ratio h_m/a is the mountain slope. If the typical terrain slope is $h_m/a = 0.1$, $C_D = 0.555$, $\rho = 1$ kg m^{-3}, $U = 15$m s^{-1}, $N = 0.01$ s^{-1}, (9.18) reduces to $D = 0.0079V$ newton(N) in SI units. Thus, as low-level blocking reduces the exposed mountain volume (V), the drag is reduced proportionately.

The stagnant air below z_{ref} has another less obvious influence on the wave drag. On the day of the Mt. Blanc observations, a jet stream and reduced static stability in the upper troposphere produced "evanescent" conditions for mountain waves. According to wave theory, partial downward wave reflection will occur under these conditions. In most cases of jet stream wave reflection, trapped lee waves will form and after a few bounces between the earth's surface and the jet stream, waves will

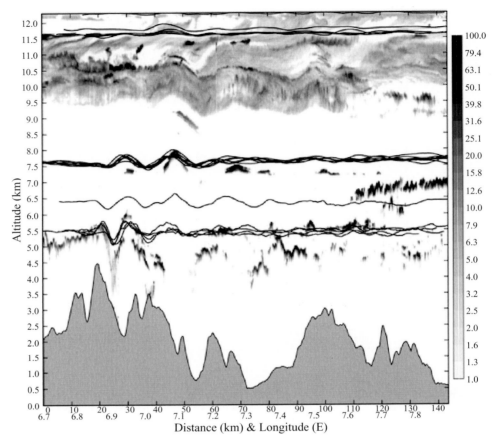

Figure 9.6. Lidar cross-section through the Mt. Blanc wave field. Shaded fields show the back-scatter intensity of laser light from clouds, roughly proportional to cloud particle density. Solid black lines show the vertical displacement of air parcels as they move from left to right in the diagram, computed from vertical velocity measured on the three aircraft: Falcon, C-130, and Electra. Note that the lenticular cloud at $z = 7.5$ km corresponds to a region of uplift detected by the C-130. The Alpine terrain along the flight track is shown at the bottom of the figure. (From Smith *et al.*, 2002.)

leak into the stratosphere. In this case, however, the low-level stagnant air absorbed the downward reflected wave energy on the first bounce. The result is a further reduction in the wave momentum flux reaching the stratosphere. To represent this effect, we introduce a reflection coefficient q at the lower boundary of the wave domain. The quantitative effect of these z_{ref} processes is shown in Table 9.5. As momentum flux is independent of altitude in steady flow, only one value is given for each condition.

Figure 9.7. Photograph of the clouds in the Mt. Blanc wave field, taken from the cockpit of the C-130. Note the three levels of cloud: cirro-stratus, lenticular, and cumulus. Note also the correspondence with Fig. 9.10. (From Smith *et al.*, 2002.)

The results in Table 9.5 support the strong dependence of wave drag on exposed volume, as least as predicted by linear theory. The ratio of drag (momentum flux) to volume in the table (for $q = 0.9$) is about 0.003 N m^{-3}, approximately half the predicted value from (9.18). The difference may arise from a different environment, nonhydrostatic effects, inaccurate slope, mountain interference or nonconical shapes.

Table 9.5. *Mt. Blanc regional momentum flux (M) on 2 November 1999 from a linear model. (From Smith et al., 2002.)*

z_{ref} (m)	Volume (km^3)	M; $q = 0.9$ (10^8N)	M; $q = 0$ (10^8N)
1500	6481	212	133
2500	988	24.4	16.7
3500	25	0.58	0.28

Note: The parameter q is a reflection coefficient (see text and Fig. 9.8).

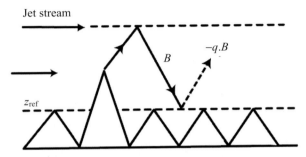

Figure 9.8. Schematic of wave reflections caused by the jet stream, and wave absorption caused by a low-level stagnant layer. A and B are the amplitudes of the up- and down-going waves; q is the reflection coefficient at the lower boundary. (From Smith *et al.*, 2002.)

The drag predictions of linear theories and nonlinear models can be tested in two ways. A dedicated campaign of aircraft surveys for this purpose was described by Lilly and Kennedy (1973), and Lilly (1982). A challenging problem with this approach is achieving statistical significance and understanding the controlling environmental factors. The PYREX and MAP projects used a different approach. They validated numerical models using aircraft data, then used repeated model runs to study the wave mechanisms and to evaluate wave drag parameterizations.

9.7 Conclusions

In this chapter, we have introduced four subjects related to the influence of mountains on regional climates: flow splitting and wave breaking; mountain lapse rates; orographic precipitation; and wave drag. It is clear from the discussions that substantial progress has been made in understanding these aspects of mountain meteorology using the reductionist methods of physics; i.e., the formulation and analysis of simplified problems capturing some aspects of the real world. While these methods will continue to be fruitful, we must also tackle complex problems with competing

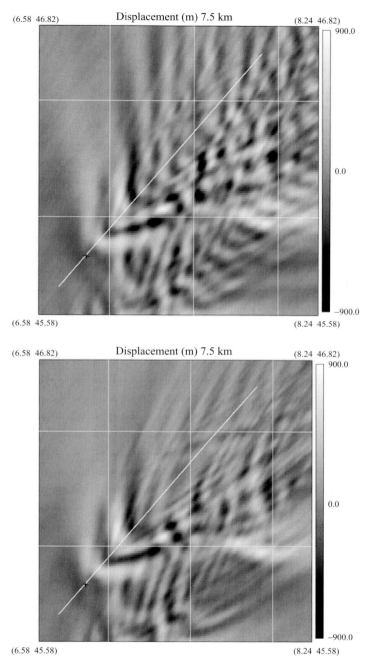

Figure 9.9. Top view of gravity waves generated in the Mt. Blanc region, according to a linear model. Upper panel: reflective lower boundary. Lower panel: absorbing lower boundary. In the figure, the gray scale represents the vertical displacement of the air parcel at each point: white means an upward displacement of 900 m while black means a downward displacement of 900 m. The thin white line shows the wind direction (SW) and track taken by the research aircraft ."X" marks the location of Mt. Blanc. (From Smith *et al.*, 2002.)

mechanisms. Only in this way will our research be relevant to regional climate analyses and other emerging disciplines.

Acknowledgements

The author had the good fortune to work in Doug Lilly's group at NCAR during the summer of 1972, and to continue this association informally during the period 1973 to 1974 while appointed as a visiting assistant professor at the Astro-Geophysics Department of the University of Colorado. During this time Doug and Joe Klemp were working on problems of mountain waves, wave drag, and severe downslope winds.

Part of this material was developed from a review talk at the *10th AMS Conference for Mesoscale Meteorology* in Portland, Oregon. The invitation from the Conference organizers is appreciated. Larry Bonneau provided the Sierra Nevada satellite analysis. Advice and assistance from Idar Barstad, Jason Evans, Steve Sherwood, and Qingfang Jiang is appreciated. Sigrid Smith assisted with the references. The author's research in mountain meteorology is mostly supported by the National Science Foundation, ATM-0112354 and by a grant for Middle East climate analysis by NASA; NAG5–9316.

References

Atkinson, B. W. (1981). *Meso-scale Atmospheric Circulations*. New York: Academic Press.

Baines, P. G. and Smith, R. B. (1993). Upstream stagnation points in stratified flow past obstacles. *Dyn. Atmos. Ocean*, **18**, 105–113.

Baines, P. G. (1995). *Topographic Effects in Stratified Flows*, Cambridge, UK: Cambridge University Press.

Barry, R. G. (1992). *Mountain Weather and Climate*, 2nd edn, New York: Routledge.

Bauer, M. H., Mayr, G. J., Vergeiner, I. and Pichler, H. (2000). Strongly nonlinear flow over and around a three-dimensional mountain as a function of the horizontal aspect ratio. *J. Atmos. Sci.*, **57**, 3971–3991.

Beau, I. and Bougeault, P. (1998). Assessment of a gravity-wave drag parametrization with PYREX data. *Quart. J. Roy. Meteor. Soc.*, **124A**, 1443–1464.

Blumen, W. and Hartsough, C. S. (1985). Reflection of hydrostatic gravity waves in a stratified shear flow. Part II: Application to downslope surface windstorms. *J. Atmos. Sci.*, **42**, 2320–2331.

Blumen, W. ed. (1990). *Atmospheric Processes Over Complex Terrain*. Meteorological Monographs, **23/45** Boston: Amer. Meteor. Soc.

Bougeault, P., Binder, P., Buzzi, A., *et al.* (2001). The MAP special observing period. *Bull. Amer. Meteor. Soc.*, **82**, 433–462.

Brighton, P. W. M. (1978). Strongly stratified flow past 3-dimensional obstacles. *Quart. J. Roy. Meteor. Soc.*, **104**, 289–307.

Clark, T. L. and Peltier, W. R. (1977). On the evolution and stability of finite amplitude mountain waves. *J. Atmos. Sci.*, **34**, 1715–1730.

Clark, T. L. and Peltier, W. R. (1984). Critical level reflection and the resonant growth of non-linear mountain waves. *J. Atmos. Sci.*, **41**, 3122–3134.

Colle, B. A. and Mass, C. F. (1998). Windstorms along the western side of the Washington Cascade Mountains, Part II: Characteristics of past events and three-dimensional idealized simulations. *Mon. Wea. Rev.*, **126**, 53–71.

Colle, B. A. (2004). Sensitivity of orographic precipitation to changing ambient conditions and terrain geometries: An idealized modeling perspective. *J. Atmos. Sci.*, **61**, 588–606.

Doyle, J. D., Durran, D. R., Chen, C., *et al.* (2000). An intercomparison of model-predicted wave breaking for the 11 January 1972 Boulder windstorm. *Mon. Wea. Rev.*, **128**, 901–914.

Epifanio, C. C. and Durran, D. R. (2001). Three-dimensional effects in high-drag-state flows over long ridges. *J. Atmos. Sci.*, **58**, 1051–1065.

(2002). Lee-vortex formation in free-slip stratified flow over ridges. Part I: Comparison of weakly nonlinear inviscid theory and fully nonlinear viscous simulations. *J. Atmos. Sci.*, **59**, 1153–1165.

Gheusi, F., Stein, J. and Eiff, O. S. (2000). A numerical study of three-dimensional orographic gravity-wave breaking observed in a hydraulic tank. *J. Fluid Mech.*, **410**, 67–99.

Giorgi, F. and Mearns, L. O. (1999). Introduction to special section: Regional climate modelling revisited. *J. Geophys. Res.*, **104**, 6335–6352.

Goody, R. M. and Walker, J. C. G. (1972). *Atmospheres. Foundations of Earth Science Series*, A. L. McAlester, ed., Englewood Cliffs, NJ: Prentice-Hall.

Grubišić, V. and Smolarkiewicz, P. (1997). The effect of critical levels on 3D orographic flows: Linear regime. *J. Atmos. Sci.*, **54**, 1943–1960.

Jiang, Q. and Smith, R. B. (2003). Cloud timescales and orographic precipitation. *J. Atmos. Sci.*, **60**, 1543–1559.

Klemp, J. B. and Lilly, D. K. (1975). The dynamics of wave-induced downslope winds. *J. Atmos. Sci.*, **32**, 320–339.

(1978). Numerical simulation of hydrostatic mountain waves. *J. Atmos. Sci.*, **35**, 78–107.

Leung, L. R., Mearns, L. O., Giorgi, F. and Wilby, R. L. (2003). Regional climate research. *Bull. Amer. Meteor. Soc.*, **84**, 89–95.

Leutbecher, M. (2001). Surface pressure drag for hydrostatic two-layer flow over axisymmetric mountains. *J. Atmos. Sci.*, **58**, 797–807.

Lilly, D. K. (1972). Wave momentum flux – a GARP problem. *Bull. Amer. Meteor. Soc.*, **53**, 17–23.

Lilly, D. K. and Zipser, E. J. (1972). The Front Range windstorm of 11 January 1972 – a meteorological narrative. *Weatherwise*, **25**, 56–63.

Lilly, D. K. and Kennedy, P. J. (1973). Observations of a stationary mountain wave and its associated momentum flux and energy dissipation. *J. Atmos. Sci.*, **30**, 1135–1152.

Lilly, D. K. and Lester, P. F. (1974). Waves and turbulence in the stratosphere. *J. Atmos. Sci.*, **31**, 800–812.

Lilly, D. K. and Klemp, J. B. (1979). The effects of terrain shape on nonlinear hydrostatic mountain waves. *J. Fluid Mech.*, **95**, 241–251.

(1980). Comments on "The evolution and stability of finite-amplitude mountain waves. Part II: Surface wave drag and severe downslope windstorms" [by Peltier and Clark, 1979]. *J. Atmos. Sci.*, **37**, 2119–2121.

Lilly, D. K. (1982). Aircraft measurements of wave momentum flux over the Colorado Rocky. *Quart. J. Roy. Meteor. Soc.*, **108**, 625–642.

(1983). Linear theory of internal gravity waves and mountain waves. In *Mesoscale Meteorology – Theories, Observations, and Models*, D. K. Lilly and T. Gal-Chen, eds., Dordrecht: D. Reidel, 13–24.

McCutchan, M. H. (1983). Comparing temperature and humidity on a mountain slope and in the free air nearby. *Mon. Wea. Rev.*, **111**, 836–845.

McGregor, J. L. (1997). Regional climate modelling. *Meteor. Atmos. Phys.*, **63**, 105–117.

Medina, S. and Houze, R. A., Jr. (2003). Air motions and precipitation growth in Alpine storms. *Quart. J. Meteor. Soc.*, **129**, 345–371.

Miranda, P. M. and James, I. N. (1992). Non-linear three-dimensional effects on gravity wave drag: Splitting flow and breaking waves. *Quart. J. Roy. Meteor. Soc.*, **118**, 1057–1081.

Molnar, P. and Emanuel, K. A. (1999). Temperature profiles in radiative–convective equilibrium above surfaces at different heights. *J. Geophys. Res. – Atmos.*, **104**, 24265–24271.

Neiman, P. J., Ralph, F. M., White, A. B., *et al.* (2002). The statistical relationship between upslope flow and rainfall in the California's coastal mountains: Observations during CALJET. *Mon. Wea. Rev.*, **130**, 1468–1492.

Ólafsson, H. and Bougeault, P. (1996). Nonlinear flow past an elliptic mountain ridge. *J. Atmos. Sci.*, **53**, 2465–2489.

(1997) The effect of rotation and surface friction or orographic drag. *J. Atmos. Sci.*, **54**, 193–210.

Palmer, T. N., Shutts, G. J. and Swinbank, R. (1986). Alleviation of a systematic westerly bias in general circulation and numerical weather prediction models through an orographic gravity wave drag parameterization. *Quart. J. Roy. Meteor. Soc.*, **112**, 1001–1039.

Pan, F. and Smith, R. B. (1999). Gap winds and wakes: SAR observations and numerical simulations. *J. Atmos. Sci.*, **56**, 905–923.

Pandey, G. R., Cayan, D. R. and Georgakakos, K. P. (1999). Precipitation structure in the Sierra Nevada of California during winter. *J. Geophys. Res.*, **104**, 12019–12030.

Peattie, R. (1936). *Mountain Geography*, Cambridge, MA: Harvard University Press.

Peltier, W. R. and Clark, T. L. (1979). Evolution and stability of finite-amplitude mountain waves. Part II: Surface-wave drag and severe downslope windstorms. *J. Atmos. Sci.*, **36** (8), 1498–1529.

Phillips, D. S. (1984). Analytical surface pressure and drag for linear hydrostatic flow over three-dimensional elliptical mountains. *J. Atmos. Sci.*, **41**, 1073–1084.

Price, L. (1981). *Mountains and Man*, Berkeley, CA: University of California Press.

Queney, P. (1947). *Theory of Perturbations in Stratified Currents with Applications to Airflow over Mountain Barriers*. Misc. Report No. 23, Dept. of Meteorology, University of Chicago.

(1948). The problem of airflow over mountains: A summary of theoretical studies. *Bull. Amer. Meteor. Soc.*, **29**, 16–26.

Queney, P., Corby, G., *et al.* (1960). *The Airflow over Mountains*. Tech. Note No. 34, World Meterological Organization.

Reiter, E. R. (1982). Where we are and where we are going in mountain meteorology. *Bull. Amer. Meteor. Soc.*, **63**, 1114–1122.

Rotunno, R. and Smolarkiewicz, P. K. (1991). Further results on lee vorticies in low-froude-number flow. *J. Atmos. Sci.*, **48**, 2204–2211.

Rotunno, R., Grubišić, V. and Smolarkiewicz, P.K. (1999). Vorticity and potential vorticity in mountain wakes. *J. Atmos. Sci.*, **56**, 2796–2810.

Schär, C. (1993). A generalization of Bernoulli's Theorem. *J. Atmos. Sci.*, **50**, 1437–1443.

Schär C. and Smith, R. B. (1993). Shallow-water flow past isolated topography. Part I: Vorticity production and wake formation. *J. Atmos. Sci.*, **50**, 1373–1400.

Schär, C. and Durran, D. R. (1997). Vortex formation and vortex shedding in continuously stratified flows past isolated topography. *J. Atmos. Sci.*, **54**, 534–554.

Schneider, T., Held, I. M. and Garner, S. T. (2003). Boundary effects in potential vorticity dynamics. *J. Atmos. Sci.*, **50**, 1024–1040.

Schultz, D. M., Steenburgh, W. J., Trapp, R. J., *et al.* (2002). Understanding Utah winter storms – The Intermountain Precipitation Experiment. *Bull. Amer. Meteor. Soc.*, **83**, 189–210.

Sheppard, P. A. (1956). Airflow over mountains. *Quart. J. Roy. Meteor. Soc.*, **82**, 528–529.

Sinclair, M. R., Wratt, D. S., Henderson, R. D. and Gray, W. R. (1997). Factors affecting the distribution and spillover of precipitation in the Southern Alps of New Zealand – a case study. *J. Appl. Meteor.*, **36**, 428–442.

Smith, R. B. (1979). The influence of mountains on the atmosphere. In *Advances in Geophysics*, **21**, B. Saltzman, ed., New York: Academic Press, 87–230.

   (1985). On severe downslope winds. *J. Atmos. Sci.*, **42**, 2597–2603.

   (1986). Further development of a theory of lee cyclogenesis. *J. Atmos. Sci.*, **43**, 1582–1602.

   (1988). Linear-theory of stratified flow past an isolated mountain in isosteric coordinates. *J. Atmos. Sci.*, **45**, (24), 3889–3896.

   (1989a). Hydrostatic airflow over mountains. In *Advances in Geophysics*, vol. 31, B. Saltzman, ed., New York: Academic Press, 1–41.

   (1989b). Mountain-induced stagnation points in hydrostatic flow. *Tellus*, **41A**, 270–274.

   (1989c). Comment on "Low Froude number flow past three dimensional obstacles. Part I: Baroclinically generated lee vorticies". *J. Atmos. Sci.*, **46**, 3611–3613.

   (1990). Why can't stably stratified air rise over high ground? In *Atmospheric Processes over Complex Terrain*, W. Blumen, ed., Meteorological Monographs, **23**, 105–107, Boston: Amer. Meteor. Soc.

Smith, R. B. and Grønås, S. (1993). The 3-D mountain airflow bifurcation and the onset of flow splitting. *Tellus*, **45A**, 28–43.

Smith, R. B. and Grubišić, V. (1993). Aerial observations of Hawaii's wake. *J. Atmos. Sci.*, **50**, 3728–3750.

Smith, R. B., Gleason, A. C., Gluhosky, P. A., and Grubišić, V. (1997a). The wake of St. Vincent. *J. Atmos. Sci.*, **54**, 606–623.

Smith, R. B., Paegle, J., Clark, T., *et al.* (1997b). Local and remote effects of mountains on weather: Research needs and opportunities. *Bull. Amer. Meteor. Soc.*, **78**, 877–892.

   (2001). Stratified flow over topography. In *Environmental Stratified Flows*, R. Grimshaw, ed., *Topics in Environmental Fluid Mechanics*, EFMS 3, Boston: Kluwer, 121–159.

Smith, R. B., Skubis, S., Doyle, J. D., *et al.* (2002). Mountain waves over Mont Blanc: Influence of a stagnant boundary layer. *J. Atmos. Sci.*, **59**, 2073–2092.

Smith, R. B., Jiang, Q., Fearon, M. G., *et al.* (2003). Orographic precipitation and air mass transformation: An Alpine example. *Quart. J. Roy. Meteor. Soc.*, **129**, 433–454.

Smith, R. B. and Barstad, I. (2004). A linear theory of orographic precipitation. *J. Atmos. Sci.*, **61**, 1377–1391.

Snyder, W. H., Thompson, R. S., Eskridge, R. E., *et al.* (1985). The structure of strongly stratified flow over hills: dividing streamline concept. *J. Fluid Mech.*, **152**, 249–288.

Stein, J. (1992). Investigation of the regime diagram of hydrostatic flow over a mountain with a primitive equation model, Part I: Two-dimensional flows. *Mon. Wea. Rev.*, **120**, 2962–2976.

Welch, W. T., Smolarkiewicz, P., Rotunno, R. and Boville, B. A. (2001). The large-scale effects of flow over periodic mesoscale topography. *J. Atmos. Sci.*, **58**, 1477–1492.

Whiteman, C. D. (2000). *Mountain Meteorology: Fundamentals and Applications*, New York: Oxford University Press.

Wurtele, M. G. (1996). Atmospheric lee waves. *Ann. Rev. Fluid Mech.*, **28**, 429–476.

10

Dynamic processes contributing to the mesoscale spectrum of atmospheric motions

Kenneth S. Gage

Aeronomy Laboratory, NOAA, Boulder, USA

10.1 Introduction

This chapter is a review of recent developments in our understanding of the mesoscale spectrum of atmospheric motions. This topic has received considerable attention in the two decades since Doug Lilly's (1983) seminal paper on stratified turbulence. The subject has not been without controversy as atmospheric scientists and fluid dynamicists have debated the relative contributions of turbulent processes and internal waves to the spectrum of atmospheric motions. In this review we focus attention on the lower atmosphere, which is of primary interest to meteorologists.

Several papers that preceded Lilly's work are worth noting. Gage and Jasperson (1979) noted the variability in high-resolution sequential wind observations taken with a novel balloon sounding system. Gage (1979) placed these observations in a turbulence context and attributed much of the variability in these observations to two-dimensional turbulence arguing that the scales were too large to be associated with three-dimensional turbulence. Dewan (1979) examined stratospheric spectra and concluded that while the spectra had many of the features generally associated with turbulence they could also be explained by a spectrum of internal waves. Similar arguments were made by VanZandt (1982) who argued for a universal spectrum of internal waves analogous to the Garrett–Munk spectrum of internal waves in the ocean (Garrett and Munk, 1972).

The importance of an improved understanding of mesoscale variability has recently become evident as increasing importance is attached to the assimilation of diverse atmospheric data into numerical models. The current situation is summarized by Daley (1997) who argues that model forecasts depend critically on the assimilation of data with a specified error covariance. The error covariance has three components: measurement error and model error are quantifiable by observationalists and modelers, respectively, but the variability of the fields being measured is often unknown and must be estimated.

Sources of mesoscale variability have only recently come into focus. The advent of radar wind profiling has contributed substantially to our ability to observe rapidly changing wind fields and begin to resolve the space–time variability of mesoscale fields of motion. Profilers have also enabled resolution of short-period internal waves and inertia-gravity waves as well as continuous measurement of vertical motions (Gage, 1990; Gage and Gossard, 2003). Aircraft observations have also contributed substantially to our ability to resolve mesoscale atmospheric motion fields. Mesoscale spectra from aircraft motion sensors have provided a wealth of information on the spectrum of mesoscale motions (Nastrom *et al.*, 1984; Nastrom and Gage, 1985; Cho *et al.*, 1999a, b).

The aim of this review is to synthesize some of the most important developments in understanding the dynamics of mesoscale atmospheric variability since the publication of Lilly's work. The advances reported here have been made by many groups working in different disciplines.

The review begins with an examination of highly resolved samples of wind variability. Examples of horizontal and vertical velocities in Section 10.2 are shown to illustrate the variability intrinsic to these fields of motion even in the absence of extreme weather events. The observed spectrum of mesoscale wind variability is examined in Section 10.3, where spectra of horizontal and vertical velocities are considered separately. Dynamic processes that contribute to the observed mesoscale spectrum of motions are discussed in Section 10.4, followed by a consideration of some spectral models that have been developed for internal waves (Section 10.5) and stratified turbulence (Section 10.6). The consistency of the observed and modeled spectra is examined in Section 10.7. Contributions of topography and convection to meteorological variability on mesoscales are considered in Section 10.9. The review concludes with a summary (Section 10.9) of some recent developments and some thoughts about the direction of future research.

10.2 Space–time variability of mesoscale meteorological fields

As mentioned in Section 10.1, mesoscale variability of meteorological fields is assuming greater importance owing to the fact that the success of numerical forecasts requires a specification of the variability of the fields being simulated. Perhaps the best-studied meteorological field is wind. Wind variability research has a long history and has benefited from remote sensing provided by radar profilers. Even before the advent of wind profilers, experiments were conducted to examine the variability of winds seen in sequential wind soundings from tracking rising balloons.

Early results in wind variability research are summarized by Elsaesser (1969). Temporal wind variability is typically expressed in terms of $\sigma_\tau^{(V)}$ defined in the equation:

$$\sigma_\tau^{(V)} \equiv \{\overline{[V(t) - V(t + \tau)]^2}\}^{1/2}, \tag{10.1}$$

where V is wind velocity, t is time, and τ is the time lag. In early work $\sigma_\tau^{(V)}$ has been expressed as a power law so that

$$\sigma_\tau^{(V)} = a + b\tau^p, \tag{10.2}$$

and p is close to 1/3 although values closer to 1/2 have been reported (Elsaesser, 1969). Note that

$$\sigma_\tau^{(V)} \equiv [D(\tau)]^{1/2} = 2\overline{(V')^2}[1 - R(\tau)], \tag{10.3}$$

where D is the temporal structure function, $\overline{V}$ is the mean wind velocity, $V' = V - \overline{V}$ and $R(\tau)$ is the Eulerian autocorrelation function defined by

$$R(\tau) \equiv \frac{\overline{V'(t + \tau)V'(t)}}{\overline{(V')^2}}. \tag{10.4}$$

The expressions (10.3) and (10.4) are familiar from turbulence theory and help place wind variability research within the context of turbulence (Elsaesser, 1969; Gage and Jasperson, 1979; Jasperson, 1982). While Elsaesser made the connection with turbulence theory and argued that the $\tau^{1/3}$ dependence is consistent with Kolmogorov turbulence theory, Gage (1979) argued that Kolmogorov 3D turbulence theory could not possibly be germane to the mesoscale and that the existence of a $\tau^{1/3}$ dependence suggests two-dimensional turbulent processes may be relevant to the atmospheric mesoscale.

Gage and Jasperson (1979) and Jasperson (1982) examined the mesoscale variability of lower tropospheric winds under fair weather conditions utilizing a novel balloon sounding system known as METRAC (Gage and Jasperson, 1974). The METRAC system was a self-contained differential Doppler navigational system that permitted the location of a lightweight transmitter within 10–30 cm inside the baseline of the receiving array. With this novel system it was possible to obtain high-resolution wind soundings with sufficient precision (with error from all sources less than about 1 m s^{-1}) to determine mesoscale variability at very short spatial and temporal scales (Gage and Jasperson, 1979; Jasperson, 1982).

The wind field below 5 km is shown in Fig. 10.1. It reveals the temporal variability of the lower tropospheric winds seen on 31 March 1976 at St. Cloud, Minnesota. The top panel shows the variability of the zonal wind and the bottom panel shows the variability of the meridional wind. Both panels contain a composite of wind

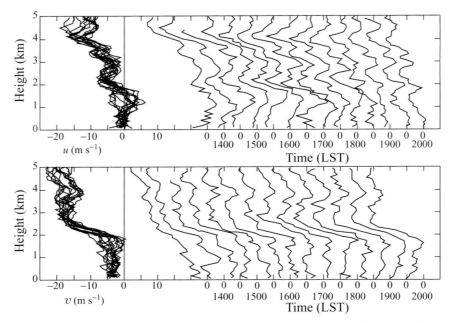

Figure 10.1. Sequential vertical profiles of (*top panel*) the zonal, *u*, and (*bottom panel*) the meridional, *v*, wind components (in m s^{-1}) from high-resolution soundings taken at St. Cloud, MN, on 31 March 1976. (After Gage and Jasperson, 1979.)

profiles on the left side, which show clearly the variability of the winds during the seven-hour field campaign. Altogether fourteen flights spaced thirty minutes apart were made during this brief campaign.

The statistics of the variability observed in this campaign are shown in Fig. 10.2. In this figure the variability was first determined for 100 m intervals and averaged for 500 m height increments. The results for 500 m height intervals are contained in Fig. 9 of Gage and Jasperson (1979). Combining the statistics for all heights leads to the results shown in Fig. 10.2. For this campaign both components of the wind followed a 1/3 power law.

In a subsequent series of field experiments, Jasperson (1982) examined the space and time variability of the lower tropospheric winds at St. Cloud. In his larger sample of data Jasperson found the wind variability fit the 1/3 power law under anticyclonic conditions but that under cyclonic conditions, when the transverse wind was changing with time, the transverse component of winds more closely fit a 1/2 power law even though the longitudinal component still fit the 1/3 power law. Jasperson attributed the 1/2 power law to the influence of a changing mean wind.

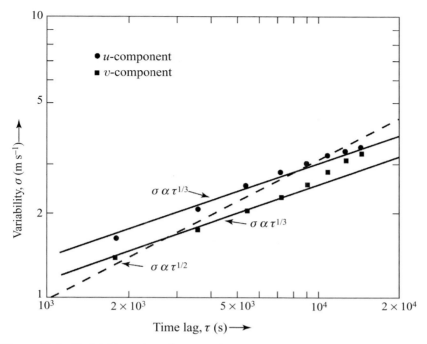

Figure 10.2. Variability σ (m s^{-1}) of zonal, u, and meridional, v, wind components averaged over fifty 100 m intervals. (After Gage and Jasperson, 1979.)

Doppler-radar profilers also provide a wealth of high-resolution wind information since they observe continuously over a range of altitudes (Gage and Balsley, 1978; Balsley and Gage, 1982). Early observations of jet-stream winds observed by the Sunset radar near Boulder, Colorado, were analyzed by Gage and Clark (1978). These authors showed that the temporal variability of winds also followed a 1/3 power-law relationship for the component of the wind, as can be seen in Fig. 10.3.

Since the work reported above was completed, it has become increasingly evident that the representativeness of observed winds or any observable is an important issue when assimilating observations into models. Kitchen (1989) has considered the representativeness of radiosonde soundings of wind, temperature, and humidity. Basically, the issue is whether individual soundings can possibly replicate the large-scale wind field simulated in a numerical model. From the modeler's perspective a measure of the wind that is representative of the larger scale is of far more value than a measurement that is influenced by local effects. This issue is of greater impact in models that do not resolve mesoscale features. For such models the high-resolution details introduce noise into the model even though the observed features are real. From this perspective, these high-resolution features represent errors to the large-scale fields and are generally regarded as observational errors by modelers.

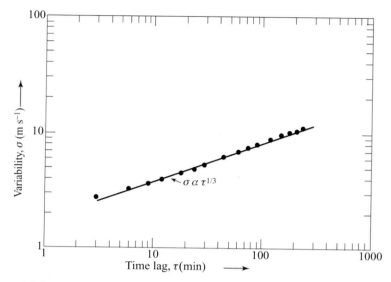

Figure 10.3. Variability of wind observed by the Sunset Radar during a jet-stream passage. (After Gage and Clark, 1978.)

To grasp the magnitude of these representativeness errors it is necessary to have knowledge of the natural variability of the mesoscale fields. Clearly, time-averaged or volume-averaged quantities are of greater value to the large-scale model than are point measurements.

Kitchen (1989) examined representativeness errors within the context of the UK radiosonde network. His analysis includes special radiosonde observations during campaigns designed to document mesoscale variability at time and space scales that lie within the synoptic-scale sampling routinely provided by operational radiosonde networks. Kitchen's analysis also considers temperature and humidity in addition to wind fields. The rms difference in a measured quantity q is related to the temporal variability and spatial variability and measurement error ϵ_q by the equations

$$(\Delta q)^2 = \left(\sigma_\tau^{(q)}\right)^2 + 2\epsilon_q^2, \tag{10.5}$$

or

$$(\Delta q)^2 = \left(\sigma_x^{(q)}\right)^2 + 2\epsilon_q^2. \tag{10.6}$$

Equations (10.5) and (10.6) apply to temporal variability at a fixed location, and spatial variability at a fixed time, respectively. More generally, the rms difference will depend on both time and space variability so that

$$(\Delta q)^2 = \left(\sigma_\tau^{(q)}\right)^2 + \left(\sigma_x^{(q)}\right)^2 + 2\epsilon_q^2 + F_{\tau,x}\left(\sigma_\tau^{(q)}, \sigma_x^{(q)}\right), \tag{10.7}$$

where $F_{\tau,x}$ represents a function of both temporal and spatial variability that vanishes when either time lag or spatial difference becomes very small, and ϵ_q refers to the measurement error. In practice $(\Delta q)^2$ is typically dominated by the variability of the q field except at the smallest spatial differences and time lags. Indeed, one way to estimate the measurement error is to take the limit of vanishingly small τ and x in (10.7). In common practice temporal and spatial variability are examined separately.

The analysis by Kitchen (1989) extends the earlier work of Gage and Jasperson (1979) and Jasperson (1982) in several respects. For the most part the time scales considered by Kitchen are synoptic scales varying from 6–60 hr while space scales range from roughly 100–1000 km. An exception is the special soundings taken from Larkhill and Beaufort Park. Jasperson (1982) considers temporal variability primarily in the range 30–360 min and at fixed separations of 20 m, 4.415 km and 20.91 km. While the experiments of Jasperson were primarily in fair weather, the observations analyzed by Kitchen cover all weather conditions. The results of Kitchen's analysis yield a power-law dependence of 0.3–0.6 for temporal variability of the vector wind over the range 6–24 hr and 0.3–0.8 for spatial variability in the range 200–500 km. Kitchen also finds a power-law dependence of 0.4–0.5 for temporal variability of temperature and a power-law dependence of 0.4–0.6 for spatial variability of temperature. In comparing the results of Kitchen and Jasperson it is worth noting that Kitchen's results extend to higher altitudes as well as considerably larger time and space scales. Kitchen comments that while the power laws for spatial and temporal variability are not equal they are close enough that over limited ranges the Taylor transformation may be a useful approximation.[1] Kitchen also gives some limited information on the variability of relative humidity. He finds that the variability is dominated by gradients associated with stable layers that do not correlate well on synoptic scales. Historically, relative humidity has been prone to measurement errors making it difficult to determine variability statistics. Observations taken in the recent International H_2O Project (IHOP) campaign should help quantify the variability of humidity fields (Weckwerth *et al.*, 2003).

Direct vertical wind measurement is not routinely made except by radar profilers. Figure 10.4 contains a composite profile of vertical velocities measured on the vertical beam of a 50 MHz profiler operated at Liberal, Kansas, during the Pre-STORM experiment. This figure illustrates the fact that local vertical velocity variability is very large compared to the small values associated with mean synoptic-scale motions. While it is possible to resolve coherent vertical motions associated with convection and gravity waves, vertical velocities are typically dominated by an incoherent spectrum of gravity waves having an rms value on the order of a few tens

---

[1] A Taylor transformation relates spatial to temporal scales by means of a mean advection velocity and enables temporal variability to be converted to spatial variability.

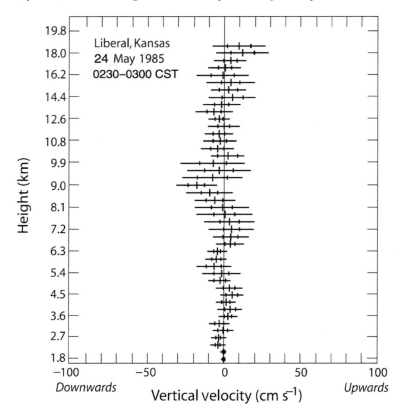

Figure 10.4. Vertical velocity observed at Liberal, Kansas during Pre-STORM. (After Gage, 1990.)

of centimeters per second. It has been found that substantial averaging is required of vertical velocity data to obtain a value representative of atmospheric mean motions. Unlike (horizontal) wind and temperature, a single sounding of vertical velocity is not representative of large scales of motion. Vertical velocity can be averaged to reduce the geophysical noise under certain circumstances (Nastrom *et al.*, 1985, 1990a; Gage *et al.*, 1991). This topic will be explored further in Section 10.3 where observations of the mesoscale spectrum of atmospheric motions are reviewed.

10.3 Observed spectra of mesoscale variability

10.3.1 Frequency spectra of horizontal and vertical velocities

Radar wind profilers provide a nearly continuous measurement of atmospheric velocities in the same volume of the atmosphere. These data are ideal for the determination of the frequency spectrum of wind throughout the range of altitudes

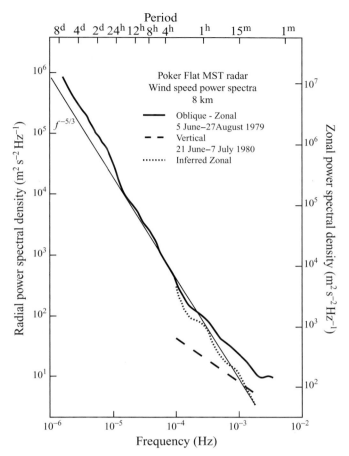

Figure 10.5. Power spectrum of zonal wind observed at Poker Flat, Alaska. (After Balsley and Carter, 1982.)

accessible to wind-profiler measurement. The first spectral analysis of profiler-measured wind reproduced in Fig. 10.5 was published by Balsley and Carter (1982) using observations from the Poker Flat MST (Mesosphere–Stratosphere–Troposphere) radar in Alaska. The remarkable result of the Poker Flat frequency spectra was how faithfully they followed a $-5/3$ power law.

Profilers can also measure vertical motions and have been utilized by many authors to study gravity waves; see, for example, Gage and Gossard (2003) and references therein. The spectrum of vertical velocities, which are seen in the vertical beam of wind profilers, has a nearly universal shape as demonstrated by Ecklund and colleagues (Ecklund *et al.*, 1985, 1986). Reproduced in

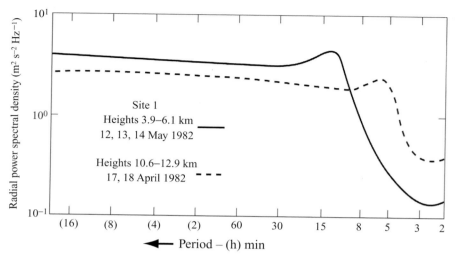

Figure 10.6. ALPEX vertical velocity spectra seen in quiet conditions. (After Ecklund *et al.*, 1985.)

Fig. 10.6 is the spectrum of vertical motions observed in weak wind conditions during the ALPEX (ALPine Experiment) campaign in southern France (Ecklund *et al.*, 1985). The observed spectra illustrate a systematic change in shape and amplitude between the troposphere and stratosphere consistent with the concept of a nearly universal spectrum of internal waves similar to the Garrett–Munk spectrum in the ocean (Garrett and Munk, 1972, 1975). Figure 10.7 shows the contrast of active and quiet spectra seen at ALPEX. As can be seen in the figure, these spectra are changed dramatically under disturbed conditions at least at sites in the vicinity of complex terrain. In order to understand the nature of the vertical velocity spectra and their relationship to underlying terrain the Flatland radar was constructed in central Illinois in 1988. Vertical velocity spectra observed at Flatland are discussed in Section 10.8.

10.3.2 Wavenumber spectra of horizontal velocities

The ability to resolve the frequency spectrum of atmospheric motions has led to a great improvement in our knowledge of mesoscale variability of the atmosphere and the contributions of internal waves to that variability. However, there is uncertainty about the nature of the dynamic processes that contribute to the observed spectra. In order to gain further insight into these dynamic processes it is important to examine the wavenumber spectrum. In the mid 1980s several studies (Lilly and Petersen, 1983; Nastrom *et al.*, 1984; Nastrom and Gage, 1985; Gage and Nastrom, 1986)

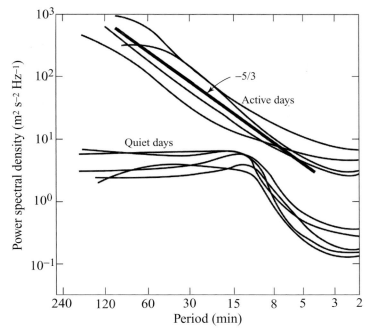

Figure 10.7. Spectra of vertical velocities observed in southern France during the ALPEX field campaign in quiet and active periods. (After Ecklund *et al.*, 1985.)

were made of aircraft spectra and an effort was made to place the aircraft and profiler spectra in a common framework (Gage and Nastrom, 1985).

Spectra of horizontal velocities measured with an on-board inertial navigation system on commercial aircraft were analyzed by Nastrom and colleagues (Nastrom *et al.*, 1984; Nastrom and Gage, 1985). These data had been collected during the NASA Global Atmospheric Sampling Program (GASP) and provide a reasonable climatology of the wavenumber spectra of horizontal velocities and temperature over the North American continent. Figure 10.8 contains the wavenumber spectra of horizontal velocity deduced from several hundred flights. These spectra show definite power-law dependence very close to $-5/3$ for both components of the horizontal velocity. Note that the $-5/3$ power law extends to scales of about 700 km and that at larger scales there is a transition to a power-law dependence of -3.

10.3.3 Vertical wavenumber spectra

Vertical wavenumber spectra have been reported by many authors (Smith *et al.*, 1987; Fritts *et al.*, 1988; Tsuda *et al.*, 1989; VanZandt and Fritts, 1989). Example

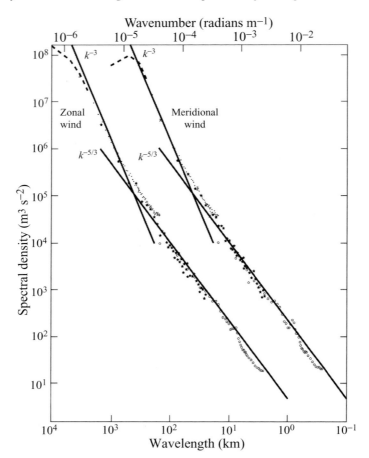

Figure 10.8. Mesoscale spectra of horizontal velocities deduced from aircraft velocity measurements. Note that the meridional wind spectrum has been shifted one decade to the right for clarity of presentation. (After Nastrom *et al.*, 1984.)

spectra (Fritts *et al.*, 1988) from the MU (Middle and Upper Atmosphere) radar in Japan are shown in Fig. 10.9(a, b). Viewed as energy density $F(k_z)$ these spectra fall off as k_z^{-3}, where k_z is the vertical wavenumber. Figure 10.9(c, d) show the same spectra plotted in energy-content form, $k_z F(k_z)$, and clearly show dominant vertical scales in the range 2–3 km. In the energy-content form equal areas make equal contributions to the energy spectrum. The vertical wavenumber spectra are usually interpreted in the framework of internal waves although, as Lilly (1983) pointed out, vertical variability is an important part of stratified turbulence and so a vertical wavenumber spectrum is also expected within the context of stratified turbulence.

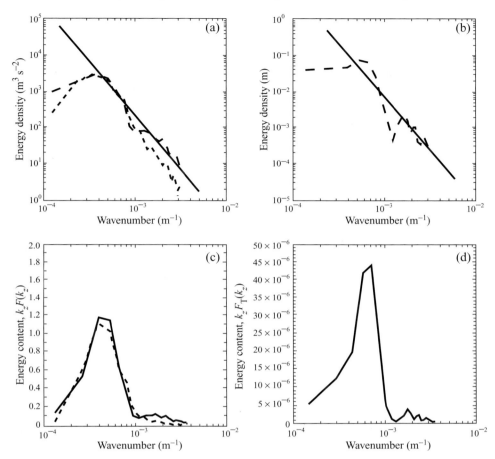

Figure 10.9. Power spectral densities of (a) eastward (solid line) and north-ward (short-dash line) radial velocity, (b) normalized temperature between 13 and 20.5 km for 24–25 October 1986. Long-dash lines show saturated PSDs. The corresponding area-preserving spectra are of (c) radial velocity and (d) normal-ized temperature. Observations are from the MU radar observatory near Shigaraki, Japan (After Fritts *et al.*, 1988.)

10.4 Dynamic processes contributing to mesoscale variability

In this section, some of the dynamic processes that contribute to mesoscale variabil-ity are considered separately. These processes include internal waves and "turbu-lence." Short-period internal gravity waves and low-frequency inertia-gravity waves are important contributors to vertical velocity and horizontal velocity, respectively. In addition, a class of motions often referred to as the vortical mode appears to be important. Mesoscale turbulent processes referred to as stratified turbulence or quasi-two-dimensional turbulence are thought to make important contributions to the mesoscale variability of horizontal winds.

Internal gravity waves represent oscillations of air parcels in the gravitational force field that arise from vertical displacements which can be caused by turbulence, convection, or flow over complex terrain, etc. These waves can be found almost everywhere except the convective boundary layer where unstable thermal stratification precludes their existence. For a more complete description of waves in the atmosphere see Gossard and Hooke (1975).

Much progress has been made in the past decade in recognizing various modes of atmospheric waves and identifying their sources. It has been known for some time that the wave motions in the lower atmosphere have a profound influence on the dynamics of the middle and upper atmosphere (e.g., Hodges, 1967; Lindzen, 1981; Fritts and Alexander, 2003). Indeed, as waves propagate into the middle and upper atmosphere their amplitude increases as atmospheric density decreases. This fact is responsible for the dominant role that waves play in the dynamics of the middle and upper atmosphere. Also, as waves propagate to higher altitudes the waves of short vertical wavelength become unstable and break. The breaking of these waves produces a stress on the winds of the middle and upper atmosphere.

10.4.1 Short-period internal gravity waves

Ecklund *et al.* (1982) reported observations of waves in vertical motions in the lee of the Rocky Mountains in Colorado using wind-profiler observations. These authors showed a clear relationship between the magnitude of vertical velocity fluctuations and the strength of winds over the nearby mountains. Indeed, when the winds blew from the east where the terrain is relatively flat, much of the vertical wind variability ceased. The Flatland radar was constructed in Illinois to contrast the vertical velocity variability over flat terrain with what was observed in Colorado. VanZandt *et al.* (1991) report results of an analysis of vertical velocity variability at Flatland Atmospheric Observatory and show that the variability is substantially less in Illinois than in Colorado. Furthermore, there appears to be a background spectrum of vertical motions seen at Flatland that is reasonably consistent with the idea of a universal spectrum of internal waves (VanZandt, 1982) as will be discussed in Section 10.5.

Several studies have been made using profilers to identify the dynamics of internal gravity waves in the lower atmosphere. Three spaced profilers were operated in southern France during the Alpine Experiment (ALPEX, Ecklund *et al.*, 1985). The profilers were located on the southern coast of France in relatively flat terrain downwind of the Alps. During mistral winds, the magnitude of the vertical wind variability was clearly seen to be related to the strength of winds and their direction relative to the mountains. The existence of multiple profiler sites, separated by

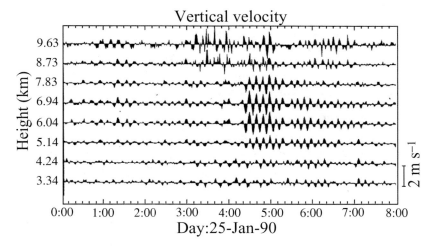

Figure 10.10. Short-period quasi-sinusoidal internal gravity waves observed 25 January 1990 using the 50 MHz profiler at Piura, near the Andes in northern Peru. (After Liziola and Balsley, 1998.)

5–6 km, provided an opportunity to explore the propagation characteristics of the waves. Data collected in the ALPEX campaign were carefully analyzed by Carter *et al.* (1989). Only a few cases of monochromatic waves could be identified at each of the sites and cross-correlated to yield wave parameters. Typical horizontal wavelengths were 10–20 km and horizontal phase speeds were in the range 4–15 m s^{-1}.

Before the advent of profilers, detailed case studies of wave events that revealed their vertical structure were rare. Ralph *et al.* (1993) analyzed a mesoscale ducted gravity wave observed during FRONTS 84 in southwestern France. Vertical velocities observed by a 50 MHz profiler and surface-pressure fluctuations observed by a network of surface pressure sensors revealed a trapped 90 min-period wave on 19 June 1984 that was found to have an approximate 76 km wavelength. The wave was trapped within a duct formed between the ground and a stable lower tropospheric layer bounded above by an unstable or near-neutral layer. In addition there was a critical layer (where the mean wind speed was equal to the ground-relative phase velocity) present near the top of the duct.

Liziola and Balsley (1998) have investigated the occurrence of quasi-horizontally propagating waves in the troposphere using the Piura radar, which is part of the NOAA/CU Trans-Pacific Profiler Network (TPPN). Figure 10.10 contains an example of multiheight time series of vertical velocities observed at Piura on 25 January 1990. In this example the waves have peak amplitude near 1.5 m s^{-1} and period close to 10 min. Note that there is very little or no phase change with altitude which is typical of the waves seen in the profiler observations. It is very likely that these are trapped waves that are propagating horizontally. Liziola and Balsley (1998) constructed a closely spaced antenna array with spacing of a few hundred meters and were able to determine phase velocities of propagating waves up to a few meters per second. Horizontal wavelengths were found to be in the range 1–3 km. They also investigated the climatology of the waves and found that they were most common during August when easterlies (from the Andes) were strongest.

The observations of vertical motions using wind profilers have provided much new information on the dynamics of wave motions in the atmosphere. The relatively frequent occurrence of quasi-periodic disturbances at locations near significant topography and the relative lack of such quasi-periodic disturbances at places like Christmas Island and at the Flatland Atmospheric observatory (Nastrom *et al.*, 1990b) in central Illinois implicates orography as the cause of these disturbances. These facts suggest the development of trapped non-stationary waves associated with wind flowing over mountains and other complex terrain.

10.4.2 Inertia-gravity waves

In addition to the short-period internal gravity waves discussed above, longer-period inertia-gravity waves have been observed at many locations. The inertia-gravity waves are long-period waves with an intrinsic period somewhat shorter than the local inertial period. They are most clearly identified in hodographs of wind soundings, which reveal a characteristic turning of the horizontal wind with height (Cadet and Teitelbaum, 1979; Gill, 1982; Sato, 1989, 1993, 1994). An example of an inertia-gravity wave observed by the MU radar in Japan during the passage of Typhoon Kelly in October 1987 is reproduced in Fig. 10.11 from Sato (1993). The wave

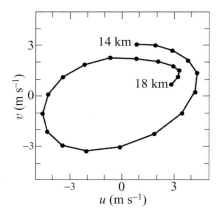

Figure 10.11. Inertia-gravity wave observed by the MU radar near Shigaraki, Japan. (After Sato, 1993.)

parameters found were: a period of 20 h, a vertical wavelength of 2.7 km and a horizontal wavelength of 300 km. Many studies have shown that inertia-gravity waves are commonly observed in wind-profiler soundings at many locations (Cornish and Larsen, 1989; Thomas *et al.*, 1992, 1999; Sato, 1994; Riggin *et al.*, 1995; Yamanaka *et al.*, 1996; Sato *et al.*, 1997). Fritts and Luo (1992) in a theoretical study have examined the excitation of inertia-gravity waves by the geostrophic adjustment process. Their work builds on earlier studies by many authors and concludes that the geostrophic adjustment process is a likely source for low-frequency gravity waves.

10.4.3 Vortical modes and stratified turbulence

Fluid motions can be partitioned into divergent and non-divergent vorticity-bearing flows. Lilly (1983) following Riley *et al.* (1981) demonstrated that the equations of motions governing mesoscale motions could be separated in the low-Froude-number limit into a set of equations governing internal waves and a set of equations governing stratified turbulence. According to this separation only the vortical mode carries potential vorticity. Riley and Lelong (2000) have reviewed what they refer to as "potential vorticity modes" and their significance in geophysical fluid dynamics. Müller *et al.* (1988) review the evidence for the existence of the vortical mode in the ocean to explain the observed variability within the range of frequencies that lie between the Coriolis frequency f and the Brunt–Väisälä frequency N. Earlier work by Müller *et al.* (1978) cited the difficulties of attributing the entire observed variability in the ocean to the spectrum of internal waves.

10.5 Spectral models for mesoscale atmospheric variability based on internal waves

Internal atmospheric waves can arise from many sources. Of course, they are most easily identified as monochromatic waves and other large-amplitude wave disturbances. However, waves are ubiquitous and as they interact they lose their identity with their sources and create a background spectrum of waves that can be nearly universal in character (Staquet and Sommeria, 2002). The situation in the atmosphere is similar in some respects to what is found in the ocean except that the speed of ocean currents is much smaller than atmospheric winds and the ocean has a definite upper boundary that reflects waves. VanZandt (1982) pointed out that a spectrum of waves similar to the Garrett–Munk spectrum found in the ocean should be present in the atmosphere. In this section we review the empirical evidence supporting the existence of a nearly universal spectrum of waves in the atmosphere. We focus attention here on the spectrum of vertical motions.

The temperature lapse rate is the most important factor governing small-scale waves in the atmosphere. For example, a parcel of air displaced upward for whatever reason will tend to oscillate with the Brunt–Väisälä (buoyancy) frequency N defined by $N^2 = (g/\theta)(d\theta/dz)$ where θ is potential temperature and g is acceleration due to gravity. A buoyancy wave or internal gravity wave typically has periods between the Brunt–Väisälä period and the local inertial period defined by $2\pi/2\Omega \sin \varphi$; where Ω is the angular velocity of the earth in radians per second and ϕ is latitude. A typical period of buoyancy waves in the free troposphere is 15 min and the inertial period at 45 ° latitude is close to 17 hr. Here the term internal gravity waves is used to refer to those waves which may have an intrinsic frequency ω anywhere in the range $N \geq \omega \geq 2\Omega \sin \varphi$. Short-period internal gravity waves have intrinsic frequencies close to the Brunt–Väisälä frequency while low-frequency internal gravity waves possess quasi-horizontal motions and have intrinsic frequencies closer to the inertial frequency. These waves are usually referred to as inertia-gravity waves.

The simple physical description of air motion within internal waves refers to a frame of reference moving with a background wind velocity U. The intrinsic frequency ω is related to the observed frequency ω_o seen by a ground-based observer by

$$\omega = \omega_o - Uk, \tag{10.8}$$

where k is a characteristic wavenumber, so that the frequency of an internal wave seen by a ground-based radar or pressure sensor differs from the intrinsic frequency by an amount Uk that is due to Doppler shifting. While Doppler shifting occurs in a uniform wind, a more complex form of Doppler shifting occurs when waves propagate in a height-dependent mean wind (Bretherton, 1969). Under these circumstances waves propagating vertically conserve their phase speed c and their horizontal wavelength relative to the ground. The observed frequency $\omega_o = kc$ does not vary as the wave propagates but the intrinsic frequency changes in accord with (10.8). The intrinsic frequency ω is related to other wave parameters through the gravity wave dispersion relation:

$$\omega^2 = \frac{N^2 k_{x,y}^2 + f^2 k_z^2}{k_{x,y}^2 + k_z^2} \tag{10.9}$$

where f is the inertial frequency ($= 2\Omega \sin \varphi$), $k_{x,y}$ is horizontal wavenumber and k_z is vertical wavenumber.

In accordance with (10.9), changes in ω must be accompanied by changes in k_z. For example, if the intrinsic frequency of a wave increases as it propagates through a wind shear it will become trapped if $\omega = N$ and the vertical wavelength λ_z becomes infinite. If the wave encounters a critical level where $c = U$ then $\omega \rightarrow 0$

and λ_z becomes very small consistent with critical-layer absorption. Internal waves do not propagate through critical layers or regions of the atmosphere where their intrinsic frequency would be greater than N.

As pointed out in Section 10.3, fluctuations of velocity measured by Doppler-radar profilers show evidence of a broad spectrum of incoherent motions. This suggests that one interpretation of the velocity fluctuations seen on Doppler radars is that they are comprised of a broad spectrum of internal waves similar to what is found in the ocean.

The spectrum of internal waves is generally thought to dominate the spectrum of ocean current variability over periods ranging from the Brunt–Väisälä period to the inertial period; see Olbers (1983) for a review. The Garrett–Munk spectrum is essentially an empirical model of internal wave spectra that provides a framework for the synthesis of diverse ocean spectra. These spectra include dropped spectra (k_z spectra), towed spectra (k_x spectra), and moored spectra (ω spectra).

The Garrett–Munk spectrum is assumed to be comprised of linear internal waves and Doppler-shifting effects are neglected. Internal waves must satisfy the dispersion relation (10.9) and the polarization relation (Fofonoff, 1969; Eriksen, 1978) that can be expressed in the form

$$E_\mathrm{h}(\omega) = \left(\frac{N^2 - \omega^2}{\omega^2} \right) \left(\frac{\omega^2 + f^2}{\omega^2 - f^2} \right) E_\mathrm{v}(\omega), \qquad (10.10)$$

where $E_\mathrm{h}(\omega)$ is the frequency spectrum of horizontal motions and $E_\mathrm{V}(\omega)$ is the frequency spectrum of vertical motions.

VanZandt (1982) examined atmospheric spectra of $E_\mathrm{h}(\omega)$, $E_\mathrm{h}(k_x)$ and $E_\mathrm{h}(k_z)$ and concluded that it was possible to define a model spectrum in the atmosphere analogous to the Garrett–Munk spectrum. In order to fit the model spectrum to the atmosphere, VanZandt chose $\omega^{-5/3}$ for the frequency dependence of the horizontal velocity spectrum whereas the ocean's spectrum is generally regarded as following ω^{-2}.

The VanZandt model spectrum represents an early attempt to fit atmospheric spectra to an internal wave spectral model. The frequency spectrum of vertical motions $E_\mathrm{v}(\omega)$ was not considered and Doppler-shifting effects were not taken into account. As pointed out by Gage (1990), the $E_\mathrm{v}(\omega)$ is implicitly determined in the VanZandt model through (10.10). Thus the $E_\mathrm{v}(\omega)$ spectrum provides an important test of consistency for the VanZandt model and any internal wave model designed to fit atmospheric spectra, as will be shown in Section 10.7.

With a few notable exceptions, the horizontal wind velocities in the atmosphere are several orders of magnitude larger than the vertical velocities. While typical instantaneous vertical motions can be as large as 1 m s^{-1} in the free troposphere, vertical velocity standard deviations are typically at most a few tens of cm s^{-1}

(Ecklund *et al.*, 1986; VanZandt *et al.*, 1991; Williams *et al.*, 2000) and long-term mean vertical motions averaged over many hours rarely exceed a few cm s^{-1}.

As shown in Section 10.3, the spectrum of vertical motions has been described by Ecklund *et al.* (1985, 1986) based on observations with vertically directed wind profilers in a number of locations. The characteristic shape of the vertical velocity spectrum is fairly flat with a peak near the Brunt–Väisälä period. Indeed, Ecklund *et al.* (1986) showed that the amplitude of the spectrum is less in the stratosphere than it is in the troposphere, as expected for internal waves (cf Fig. 10.6). The spectrum of vertical motions as observed by profilers appears to have a fairly universal shape under undisturbed conditions and far from sources, very similar to what is found in the ocean for the Garrett–Munk spectrum.

Over the past two decades considerable progress has been made in identifying wave motions in atmospheric observations. Much of this work has been facilitated with the advent of wind-profiling radars that have permitted the direct measurement of vertical velocities simultaneously over a large range of altitudes. The efforts of VanZandt (1982) and Scheffler and Liu (1985) have established a model spectrum for the atmosphere analogous to the spectrum of internal waves in the ocean commonly referred to as the Garrett–Munk spectrum. However, in the initial stages of formulating the atmospheric models the vertical velocity spectral amplitude and shape were largely unknown. Yet the vertical velocity spectral shape and amplitude are implicitly determined by the internal wave spectral models. It is therefore a critical test for the internal wave spectral models to see how consistent they are with the frequency spectrum of vertical motions. Another important issue is Doppler shifting which is more important in the atmosphere than in the ocean. These topics will be revisited in Section 10.7.

10.6 Stratified turbulence and mesoscale variability

It is well known in fluid mechanics that, as turbulence decays in a stably stratified environment, propagating waves are radiated and quasi-horizontal motions possessing vorticity are left behind (Müller *et al.*, 1978; Lilly, 1983; Hopfinger, 1987; Riley and Lelong, 2000). The vorticity-possessing quasi-horizontal motions, often referred to as the vortical mode, have been reproduced in the laboratory and by Direct Numerical Simulation (DNS). This topic has been reviewed recently by Riley and Lelong (2000) who point out that these motions take place under conditions of low Froude number ($Fr = u_{rms}/NL$) where u_{rms} is a rms turbulence velocity, L is a length scale of the energy-containing motions and N is the buoyancy frequency introduced in Section 10.5. Riley and Lelong refer to this component of motion as PV modes since they possess potential vorticity and the propagating waves do not.

Rotation is known to have an important influence on these flows and is characterized by the Rossby number ($Ro = u_{rms}/fL$) where f is one-half the system rotation rate. Rossby numbers ≥ 1 are relevant to stratified turbulence that can be regarded as a regime of strong stability and weak rotation.

In the atmosphere this class of motions is thought to contribute substantially to the mesoscale spectrum of atmospheric motions (Gage, 1979; Lilly, 1983; Gage and Nastrom, 1986). Lilly referred to these motions as stratified turbulence although the same class of motions is often referred to as quasi-two-dimensional turbulence because the vertical motion is suppressed in strongly stable environments and because the motion fields have considerable vertical structure. Their frequency spectrum occupies the same spectral range as internal wave motions which makes it difficult to differentiate stratified turbulence from a spectrum of waves (e.g., Dewan, 1979; VanZandt, 1982). Riley *et al.* (1981), Lilly (1983), and Riley and Lelong (2000) give scaling arguments for two sets of equations of motion governing the wave component and the vortical mode, respectively.

Section 10.5 developed some of the background for the spectrum of internal waves in the atmosphere. In this section emphasis will be placed on stratified turbulence as a source of much of the fine structure observed in the atmosphere. It is important to recognize, however, that low-frequency inertia-gravity waves can also produce vertical structure. Stratified turbulence has been investigated in the laboratory and by DNS. Laboratory studies are necessarily restricted in their size and cannot possibly replicate the range of scales and parameter space covered in the oceanic or atmospheric mesoscale. Nevertheless, considerable insight has been obtained concerning stratified turbulence from laboratory experiments. Several laboratory experiments have simulated some of the fundamental dynamics of stratified turbulence (Itsweire *et al.*, 1986; Itsweire and Helland, 1989; Maxworthy, 1990; Narimousa *et al.*, 1991; Yap and van Atta, 1993). Narimousa *et al.* (1991), and Yap and van Atta (1993) concentrate on the spectra and energy transfers and show that in the laboratory it is possible to simulate the inverse cascade of stratified turbulence with $k^{-5/3}$ spectra at scales larger than the scale of energy input and steeper spectral slopes at smaller scales. Hopfinger (1987), and Riley and Lelong (2000) concentrate on issues surrounding the collapse of three-dimensional turbulence in strongly stratified flows and its relationship to the residual motions that are left behind in wakes after waves are radiated away. Laboratory experiments demonstrate that a region of strongly stratified turbulence will decay into a field of quasi-two-dimensional horizontal motions that possess vertical vorticity with a characteristic collapse of vertical motions, and an increase in the horizontal dimension of the layer similar to the dynamics of wake collapse described in Hopfinger (1987).

In addition to the laboratory experiments, considerable progress has been made in the DNS of stratified flows. These studies help to interpret laboratory experiments, understand the energy transfers within the stratified flows and the structures that develop as these flows evolve. One long-standing issue is the difficulty of simulating high-Reynolds-number flows, which makes it difficult to simulate many atmospheric flows with high resolution.

Numerical studies to date have focused on decaying turbulence in stratified flows and forced turbulence. Métais and Herring (1989) analyzed decaying turbulence in a strongly stratified flow and found the same general features reported in laboratory studies by Itsweire *et al.* (1986). Bartello (1995) has examined the interaction of PV and wave modes in decaying turbulence and found them to be closely coupled. Herring and Métais (1989) examined the dynamics of forced, non-rotating, strongly stratified flows with varying amounts of vertical variability. The existence of the vertical variability, which reflects the tendency of layered structure to form in stratified turbulence, was found to inhibit the $k^{-5/3}$ inverse cascade to larger scales of motion. Lelong and Riley (1991) consider wave/vortical mode interactions in strongly stratified flows. Riley and Lelong (2000) point out that the inverse cascade is a result of PV/PV mode interactions and, in the presence of vertical variability, wave/PV mode interactions dominate the dynamics and result in the cascading of PV mode energy to smaller scales. Métais *et al.* (1994) considered both two- and three-dimensional forcing with varying degrees of rotation. The results showed that in the presence of rotation the inverse cascade was more likely to develop. Vallis *et al.* (1997) and Lilly *et al.* (1998) used low-resolution meteorological models to show that a $k^{-5/3}$ regime develops when convectively driven turbulence is present at the mesoscale with or without rotation. Bartello (2000a) suggests that the model atmosphere behaves more two-dimensionally in a low-resolution model because of the lack of vertical variability.

Many of the laboratory experiments cited above have been designed primarily to simulate oceanographic structure and dynamics while the numerical studies have been motivated also by the need to improve the understanding of the dynamics of waves and turbulence in the atmosphere. Observational studies addressing the issues discussed in this section have been relatively sparse perhaps owing to the difficulties of unraveling the issues involved. Nevertheless, several studies have been reported that are summarized next.

10.7 Comparison of observed mesoscale spectra with model spectra

The concept advanced by Lilly (1983) of stratified turbulence placed the earlier work of Gage (1979) on a more solid theoretical footing incorporating the idea developed by Riley *et al.* (1981) of a separation of the governing dynamic equations for waves and turbulence into a set governing waves and a set governing stratified turbulence. Fundamental to the concept of stratified turbulence was the idea that a

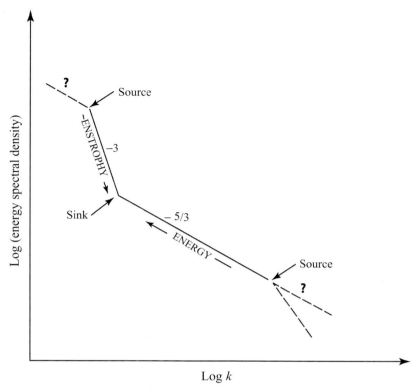

Figure 10.12. Schematic of quasi-two-dimensional or stratified turbulence. (After Larsen *et al.*, 1982.)

small-scale source of turbulence could lead to a reverse cascade filling the mesoscale with quasi-two-dimensional turbulence. This concept has much appeal since the free atmosphere is known to be stably stratified under most conditions and the diffusivity of the atmosphere is highly anisotropic. Indeed, layered structure is commonly observed even on fairly small vertical scales, see e.g., Gage and Green (1978), Gossard *et al.* (1985), Salathé and Smith (1992), and Dalaudier *et al.* (1994). A schematic of the inverse cascade is reproduced in Fig. 10.12.

Gage and Nastrom (1985) transformed the radar-derived horizontal kinetic-energy frequency spectra to wavenumber spectra using a Taylor transformation and combined the transformed radar spectra with energy spectra from other sources in a format first used by Lilly and Petersen (1983) as reproduced in Fig. 10.13.

Masmoudi and Weil (1988) analyzed sodar observations collected from four sites during the MESOGERS 84 campaign (Gers is a region in southwest France and MESO refers to the mesoscale nature of the campaign). The analysis tested the Taylor hypothesis over the domain of the network of stations and examined

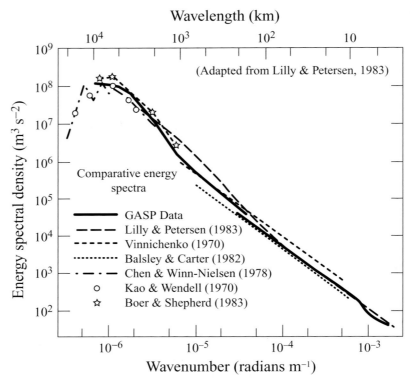

Figure 10.13. Synthesis of horizontal kinetic energy spectra deduced from different sources. (Adapted from Lilly and Petersen, 1983.)

structure functions of horizontal wind speed. The spectra of the wind speed followed a $k^{-5/3}$ spectral slope and otherwise appeared consistent with stratified turbulence. Högström *et al.* (1999) have analyzed the mesoscale velocities over the Baltic Sea also finding consistency with stratified turbulence. Cho *et al.* (1999a) reported horizontal wavenumber spectra of winds, temperature and trace gases measured by aircraft during the Pacific Exploratory Mission (PEM). The spectra determined from the PEM flights were similar to those reported by Nastrom and Gage (1985) for the Global Atmospheric Sampling Program (GASP). A detailed analysis of the PEM spectra was reported by Cho *et al.* (1999b). They tentatively concluded that the observed spectra were likely from several sources including quasi-two-dimensional turbulence. Cho and Lindborg (2001) and Lindborg and Cho (2001) analyzed data collected from the European MOZAIC (Measurement of Ozone and water vapor by Airbus in service) program and, using a third-order velocity structure function (Lindborg, 1999), could find no support for an inverse cascade even though the analysis yielded spectra with slopes similar to those reported by Nastrom and Gage

(1985). However, Cho *et al.* (2001) reported an analysis of aircraft observations that does give support to a reverse cascade. Clearly, more research is needed to sort out these issues.

In the remainder of this section observed spectra are examined for their internal consistency with models. In particular, the observed frequency spectra of horizontal and vertical velocity will be compared to the internal wave spectral model without Doppler shifting. Then the influence of Doppler shifting will be considered by comparing observations with theoretical models.

10.7.1 Observed vs model spectral amplitudes

While models are constructed to fit observations they can be tested by examining their internal consistency and by checking implications of the models. In this section we presume the spectrum of vertical velocities observed under quiet conditions represents an unambiguous spectrum of internal waves and consider the consistency of the horizontal velocity spectrum of internal waves associated with this spectrum compared to the observed spectra.

The gravity wave spectral model advanced by VanZandt (1982, 1985) and Scheffler and Liu (1985) provides a way to predict the horizontal velocity spectrum due to waves when the vertical wave spectrum is known. Gage (1990) compared the observed horizontal velocity spectra with the horizontal velocity spectrum due to waves and argued that there was more energy in the observed spectrum than was consistent with the gravity wave model fit to the observed vertical velocity spectrum. More recently Högström *et al.* (1999) also examined the frequency spectrum of vertical velocity deduced from aircraft observations and compared it to the horizontal velocity spectrum. They concluded that the vertical velocity spectrum was due to waves and that the horizontal velocity spectrum was due to quasi-two-dimensional turbulence.

The VanZandt spectral model was normalized to fit the observed horizontal velocity spectrum in frequency, horizontal wavenumber and vertical wavenumber space. Since it was not normalized to fit the vertical velocity spectrum, comparison of the vertical velocity spectrum with the model spectrum is a critical test for the model. Figure 10.14 contains a comparison of the tropospheric vertical velocity spectrum presented in Fig. 10.6 with the horizontal velocity spectra of Vinnichenko (1970) and Balsley and Carter (1982). The calculated horizontal velocity spectrum was obtained using the polarization relations that relate the energy in the horizontal velocity spectrum with the energy in the vertical velocity spectrum (Fofonoff, 1969; Eriksen, 1978; Olbers, 1983). Note that the horizontal velocity spectrum contains much more energy than the vertical velocity spectrum. A similar result was found by Högström *et al.* (1999).

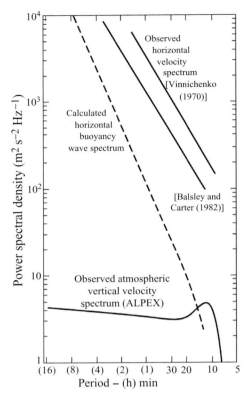

Figure 10.14. Model internal wave horizontal frequency spectrum without Doppler shifting (dash line) compared with two observed horizontal frequency spectra (upper solid lines). (After Gage and Nastrom, 1985.)

10.7.2 Doppler-shifting effects

Gravity wave spectral models considered so far do not account for the influence of Doppler shifting, which is more important in the atmosphere than in the ocean. Next the influence of Doppler shifting on a spectrum of internal waves is considered. Then the consistency of spectral models that account for Doppler shifting with observations made at Platteville, Colorado, is examined.

Fritts and VanZandt (1987) and Scheffler and Liu (1986) have carefully considered the effect of Doppler shifting on a model internal wave spectrum. According to Fritts and VanZandt (1987), the influence of a mean wind on the spectrum of internal waves can be parameterized by

$$\beta = \overline{U} m_* / N \qquad (10.11)$$

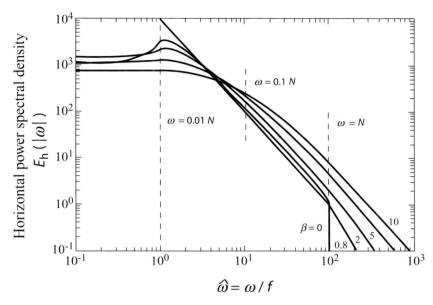

Figure 10.15. Doppler-shifting effect on a model spectrum of internal waves. (After Fritts and VanZandt, 1987.)

where $\overline{U}$ is the mean wind, m_* is a characteristic vertical wavenumber and N is the Brunt–Väisälä frequency. The parameter β provides a convenient measure of the mean wind scaled to the phase velocity as shown by Fritts and VanZandt (1987). Doppler shifting only affects the gravity wave spectrum in the direction of the mean velocity.

The results of the Fritts and VanZandt (1987) analysis on the horizontal velocity spectrum are summarized in Fig. 10.15, which shows that qualitatively the influence of the mean wind is to increase the spectral amplitude at the high-frequency end of the spectrum and to decrease the spectral amplitude at the low-frequency end of the spectrum. The net effect of the Doppler shifting is to increase the slope of the frequency spectrum above the −2 value of the non-Doppler-shifted spectrum. While this is in the sense needed to fit the observed horizontal frequency spectrum it is important to examine quantitatively the magnitude of this effect compared to observations.

Observations of zonal velocity spectra from Platteville stratified by wind speed have been examined by Gage and Nastrom (1988) and used for a quantitative comparison with the Doppler-shifting model spectrum. To complete the comparison a simple turbulence model was used by Gage and Nastrom to provide a quantitative measure of Doppler-shifting influence on a spectrum of turbulent motions.

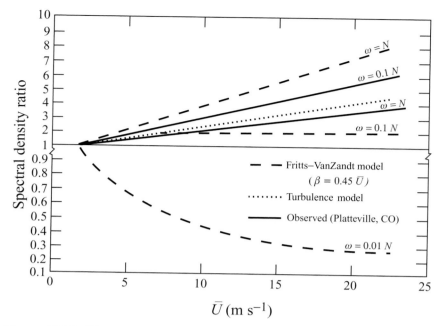

Figure 10.16. Observed dependence (solid lines) of zonal velocity spectrum on mean wind speed compared with predictions from wave (dash line) and turbulence (dotted line) models. (After Gage and Nastrom, 1988.)

The turbulence model was based on a horizontal wavenumber spectrum that approximates the observed GASP spectra (Nastrom and Gage, 1985). The Taylor transformation was used to determine the frequency spectrum of turbulence seen by a fixed observer when the specified wavenumber spectrum was advected past at a specified velocity (Gage and Nastrom, 1988).

The results for the turbulence model and the wave model are compared with observations in Fig. 10.16. In this figure, the changes of spectral density ratio with mean wind speed are plotted for several frequencies. The frequencies for comparison are the Brunt–Väisälä frequency N, and frequencies equal to $0.1N$ and $0.01N$. While the observed frequency spectra did not extend to $0.01N$, the model results are shown for these frequencies. To determine the change in spectral density ratio for the wave model it is necessary to assign a value to the characteristic wavenumber m_*. For this purpose a value of $m_* = 0.75 \times 10^{-3}$ m^{-1} has been used which is consistent with values anticipated by Fritts and VanZandt (1987). For the summertime troposphere pertinent to the observed Platteville spectra a value of $N = 1.67 \times 10^{-3}$ s^{-1} has been used, implying $\beta = 0.45\overline{U}$.

There are six curves in Fig. 10.16. Each curve gives the values of the spectral density ratio as a function of wind speed. Curves with positive (negative) slope have increasing (decreasing) spectral density ratio with increasing wind speed. Two curves are plotted for the observed change corresponding to frequencies of N and $0.1N$. These curves show a modest *increase* in slope with *decreasing* frequency. The turbulence model result does not depend on frequency so only a single curve is plotted. Three curves are plotted for the wave model for frequencies N, $0.1N$ and $0.01N$, respectively. The slopes for the three curves *decrease* markedly with *decreasing* frequency. This is the opposite sense to the change noted in the observed spectra.

Eckermann (1990) introduced the idea that the polarization of internal waves could be used to discriminate between waves and turbulence. The internal gravity waves are elliptically polarized with the major axis aligned with the direction of propagation. In the case of aircraft measurements (Cho *et al.*, 1999b) the velocity variances parallel to the aircraft heading would be associated with maximally Doppler-shifted waves, whereas velocity variances transverse to the aircraft heading would be caused by waves suffering no Doppler shift. Consequently, if $R_{pt} \equiv \overline{(u')^2_{\parallel}}/\overline{(u')^2_{\perp}}$, the ratio of variances between parallel and transverse directions, $R_{pt} > 1$ is expected for waves and $R_{pt} < 1$ is expected for stratified turbulence and vortical modes. Cho *et al.* (1999b) found that $R_{pt} < 1$ for PEM westward flights at latitudes greater than $15°$ whereas $R_{pt} > 1$ for latitudes smaller than $15°$.

Cho *et al.* (1999b) also considered application of the Stokes parameter analysis of Vincent and Fritts (1987) to the PEM westward spectra. Waves should have definite phase relations between u and v components. Velocity and potential temperature (θ) fluctuations should have phase differences close to ± 90 degrees. Only occasionally were Cho *et al.* (1999b) able to find evidence of definite phase relations between u and v and θ fluctuations. These authors concluded that their observations support the dominance of rotational modes in the mesoscale spectrum with the exception of near-equatorial latitudes.

In this section the consistency of horizontal and vertical velocity frequency spectra have been examined within the framework of a non-Doppler-shifted model of internal wave spectra analogous to the Garrett–Munk spectrum in the ocean. While the vertical velocity frequency spectra bear a striking resemblance to the model wave spectra, the horizontal velocity frequency spectra appear to be too energetic to also be due to waves. Even accounting for Doppler shifting does not explain the differences seen between observed and model internal wave spectra. Furthermore, critical tests designed to differentiate between waves and turbulence have thus far been inconclusive and do not support the idea that the entire spectrum of mesoscale variability is due to waves. Thus, it would appear to be reasonable to conclude that the vertical velocity spectra are due to waves but that the horizontal velocity frequency spectra are influenced or even dominated by stratified turbulence. The idea of co-existing spectra of waves and turbulence is consistent with the ideas contained in Lilly (1983). Clearly, much more work will be required to establish

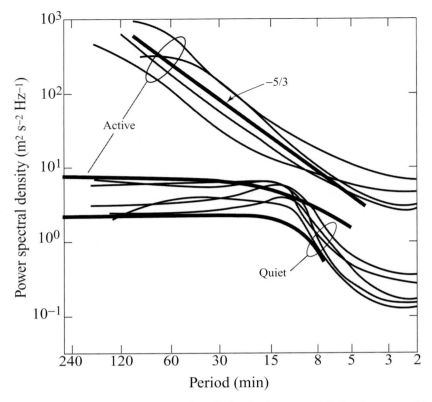

Figure 10.17. Frequency spectra of vertical velocity seen at Flatland compared to those observed during ALPEX (see Fig. 10.7). The Flatland spectra are contained within the dark solid lines that delineate the quiet spectra seen during ALPEX. (After VanZandt *et al.*, 1991.)

quantitatively the relative contributions of wave and turbulent processes to the mesoscale spectra of atmospheric motions and to mesoscale variability.

10.8 The role of topography and convection as a source of mesoscale variability

In order to investigate the influence of complex terrain on vertical motions, the Flatland radar was constructed in central Illinois. At Flatland, the vertical velocity spectrum was found to be very similar to that found in other locations of complex terrain under undisturbed conditions of low winds or winds blowing over flat terrain. As reported by VanZandt *et al.* (1991) and shown in Fig. 10.17, a flat spectrum is seen at Flatland Atmospheric Observatory (FAO) under all wind conditions although the magnitude of the spectrum does vary with wind speed. VanZandt and colleagues were able to demonstrate, using a spectral model, that the magnitude variations seen in the Flatland spectra are explicable assuming they are indeed a spectrum of waves. This success in modeling the dependence of the spectrum on wind speed by incorporating Doppler-shifting effects considerably strengthens the interpretation of the vertical velocity spectrum as a wave spectrum. The behavior of the vertical velocity spectrum under disturbed conditions near mountains is considered next.

Also shown in Fig. 10.17 is the spectrum under disturbed conditions of strong winds blowing over complex terrain. Under these circumstances the spectrum is substantially modified with a greatly enhanced magnitude and a slope approaching $-5/3$. Worthington and Thomas (1998) have shown that similar spectra are observed during mountain wave events at Aberystwyth, UK. Several authors have considered possible explanations for the occurrence of the $-5/3$ spectral slope in the frequency spectra during disturbed mountain wave conditions. Gage and Nastrom (1990) hypothesized that quasi-horizontal rotational motions along tilted isentropic surfaces might be responsible for these frequency spectra owing to the fact that the vertical velocity frequency spectra resembled the horizontal velocity spectra observed at the same time albeit with reduced amplitude. However, Worthington and Thomas argued that at Aberystwyth the frequency spectra are actually vertical motions of the mountain waves and not due to quasi-horizontal motions created by the tilting of isentropic surfaces. Instead, they show that a simple stochastic model with a pattern of mountain waves that moves stochastically with respect to the ground produces a spectrum that resembles the observed frequency spectrum.

The role of topography in the generation of gravity waves has been documented by several studies in addition to the work reviewed in the previous section. Nastrom *et al.* (1987) and Jasperson *et al.* (1990) examined the relationship of the variance of mesoscale winds measured by commercial aircraft during GASP with the underlying surface. The variance was found to be highly correlated with the roughness of the topography. Variance in the troposphere was also related to wind speed. The variance of wind and temperature on scales of 4–80 km was found to be as much as six times greater over mountains than over oceans. At larger scales variance was also greater over mountains compared to oceans but by a smaller amount.

Nastrom and Fritts (1992) extended the earlier work with two case studies aimed at an improved understanding of the linkage between wind variance observed in the GASP flights, underlying terrain and meteorological conditions. They found that, while in most instances variances were greater over mountains than over plains and oceans, in some cases the variances were not enhanced by topography. Instances of small variances over the mountains were clearly related to regimes of weak wind over the local terrain and high static stability in the lower troposphere that would suppress convection. These results can be compared with the wind-profiler results from Platteville (Ecklund *et al.*, 1982) and Flatland results of Nastrom *et al.* (1990b).

For Platteville the vertical velocity variance was directly related to wind speed at a pressure of 500 mbar, and at Flatland the vertical velocity variance was most closely related to stability in the lower troposphere independent of wind speed.

In the aircraft data the variability was due either to waves or to stratified turbulence (Nastrom *et al.*, 1987), so a less ambiguous determination of the influence of topography on the generation of gravity waves would be examination of vertical motions. This has been done at many locations using profilers as reviewed in the previous section. Nastrom *et al.* (1990b) investigated sources of gravity wave activity seen in the vertical velocities at Flatland. Generally, the magnitude of vertical velocity variance was found to be less than at Platteville in the lee of the Colorado Rockies. Observations from Flatland have provided an opportunity to investigate the contributions of other sources to gravity wave generation over very flat terrain. The occurrence of large vertical velocity variance in Flatland observations is related to specific events such as frontal passages or thunderstorms. Statistical studies showed variances systematically larger under cloudy skies than under clear skies suggesting a link with convection. Variances under these disturbed conditions were typically 50% greater than under clear undisturbed conditions where the variance was typically in the range 100–300 $cm^2 \ s^{-2}$.

Fritts and Nastrom (1992) used the GASP dataset to investigate the relationship of enhanced wind and temperature variability to fronts, convection and jet streams. In case studies they were able to relate enhanced variability with one or more sources and found enhancements as large as one to two orders of magnitude extending to scales of 64 km or more. Horizontal velocity variances in apparently source-free regions are typically near 0.1–0.4 $m^2 \ s^{-2}$ and temperature variance is less than 0.4 K^2.

Near-surface outflow from deep convective storm systems often creates a solitary wave that can be observed by radar and lidar (Fulton *et al.*, 1990; Doviak *et al.*, 1991; Koch *et al.*, 1991). Solitary waves are long nonlinear waves that are trapped below an inversion and propagate intact for long distances (Christie, 1989). Trapping mechanisms for the low-level internal gravity waves were discussed by Crook (1988) who pointed out that while trapping requires an inversion it is aided by opposing winds. The theory for solitary waves in the atmosphere has been reviewed by Rottman and Einaudi (1993). They developed the theory for two classes of solitary waves. The first class of solitary waves is confined to the lowest few kilometers of the troposphere possessing horizontal scales of a few kilometers and phase speeds of the order of 10 $m \ s^{-1}$. The second class of solitary waves occupies the entire troposphere and has horizontal scales on the order of 100 km and phase speeds on the order of 25–100 $m \ s^{-1}$. The solitary waves may play an important role in the propagation of squall lines (Carbone *et al.*, 1990; Koch *et al.*, 1993; Trexler and Koch, 2000) with important implications for mesoscale predictability (Koch and O'Handley, 1997).

Another class of waves, known as convection waves, develops over convective boundary layers in clear weather. While internal gravity waves are precluded from propagating within the hydrostatically unstable atmosphere, convection waves often develop in the stable free troposphere above the convective boundary layer. These waves have been seen in aircraft observations (Kuettner *et al.*, 1987; Hauf, 1993) and are observed by profilers as reported by Gage *et al.* (1989). Their observations were made in Liberal, Kansas, on 29 June 1985, on a clear day, during Pre-STORM. Liberal, Kansas, is located on very flat terrain far from any mountains. The observed waves appear to fill the entire free troposphere and possess little or no phase progression with altitude suggesting that these are trapped modes. Clark *et al.* (1986) have simulated convection waves in a model. Convection waves appear to develop when the convective boundary layer possesses a corrugated top and winds flow over this surface under favorable synoptic conditions.

10.9 Concluding remarks

In this chapter, the spectrum of mesoscale atmospheric variability and its dynamic causes have been reviewed. Special emphasis has been placed here on the developments surrounding the seminal work of Lilly (1983) introducing the concept of stratified turbulence with an implicit inverse cascade from a small-scale source of turbulence. Stratified turbulence accounts for the rotational modes of energy that occupy the atmospheric mesoscale horizontal velocity, temperature and related fields. Internal wave spectral models analogous to the Garrett–Munk spectrum in the ocean have also been considered in some detail.

Since Lilly published his paper on stratified turbulence and the mesoscale variability of the ocean there has been considerable research to establish in a quantitative manner the dynamics of stratified turbulence. Laboratory studies have established the essential features of stratified turbulence. Direct numerical simulations have also contributed substantially to the current state of our knowledge. The contributions of stratified turbulence to ocean dynamics are addressed by Müller (1995) and Thorpe (1998). These authors view the contributions of rotational modes of stratified turbulence as important to fully account for ocean variability but there appears to be substantially less energy in stratified turbulence than in the internal wave field in the ocean. In the atmosphere it appears that stratified turbulence plays a much more important role than in the ocean. A likely scenario consistent with observations is that the internal wave field and stratified turbulence co-exist with stratified turbulence dominating the mesoscale variability of horizontal velocity at least in the lower atmosphere. Stratified turbulence may also play an important role in the maintenance of persistent turbulent layers in stable environments (Riley and deBruynKops, 2003).

Issues involving atmospheric mesoscale variability and stratified turbulence are important to resolve in the context of numerical modeling of atmospheric circulation. Bartello (2000b) points out that at larger scales where k^{-3} spectrum is found low-resolution wind observations are sufficient to define scalar fields but are not sufficient to do so at smaller scales where the energy spectrum has $k^{-5/3}$ dependence. Palmer (2001) suggests that imperfect parameterization of subgrid-scale processes in large-scale numerical models may be causing problems in predicting large-scale circulation because of the neglect of processes such as stratified turbulence.

Detailed knowledge of atmospheric mesoscale variability is important for several reasons. The mesoscale variability of meteorological fields is often the dominant component of observational error since it determines the representativeness of observations. The spectrum of mesoscale variability also provides a good test for models. Koshyk *et al.* (1999), and Koshyk and Hamilton (2001) have been able to simulate the mesoscale spectrum of variability in the GFDL (Geophysics Fluid Dynamics Laboratory) SKYHI model. Tung and Orlando (2003) have simulated the Nastrom–Gage spectrum in a two-level quasi-geostrophic model. Klemp and Skamarock (see Chapter 6 of this book) have shown that the new Weather Research and Forecasting (WRF) model being developed by NCAR (National Center for Atmospheric Research) is able to replicate the main features of the Nastrom–Gage spectrum especially for high-resolution model runs. Of course, the simulation of the observed spectrum in a numerical model is insufficient to guarantee that the model physics is realistic.

While the basic ideas considered in this review were developed over twenty years ago, many issues remain to be resolved fully, and active research is continuing on many fronts. Much of the ongoing research has focused on DNS although there are still issues about resolving the full range of motions especially at large Reynolds numbers typical of geophysical flows. The topic of mesoscale variability is taking on renewed importance because of the need to specify error fields for assimilation of meteorological data into numerical models. As models become faster and more sophisticated it is likely that the dynamic processes discussed in this review can be studied more completely, yet observational studies will still be needed to fully understand the processes involved.

References

Balsley, B. B. and Carter, D. A. (1982). The spectrum of atmospheric velocity fluctuations at 8 km and 86 km. *Geophys. Res. Lett.*, **9**, 465–468.

Balsley, B. B. and Gage, K. S. (1982). On the use of radars for operational wind profiling. *Bull. Amer. Meteor. Soc.*, **63**, 1009–1018.

Bartello, P. (1995). Geostrophic adjustment and inverse cascades in rotating stratified turbulence. *J. Atmos. Sci.*, **52**, 4410–4428.

(2000a). Potential vorticity, resonance and dissipation in rotating, convective turbulence. *Fluid Mech. Astrophys. Geophys.*, **8**, 309–322.

(2000b). Using low-resolution winds to deduce fine structure in tracers. *Atmos.-Ocean*, **38**, 303–320.

Boer, G. J. and Shepherd, T. G. (1983). Large-scale two-dimensional turbulence in the atmosphere. *J. Atmos. Sci.*, **40**, 164–184.

Bretherton, F. P. (1969). Waves and turbulence in stably stratified fluids. *Radio Sci.*, **4**, 1279–1287.

Cadet, D. and Teitelbaum, H. (1979). Observational evidence of internal inertia-gravity waves in the tropical stratosphere. *J. Atmos. Sci.*, **36**, 892–907.

Carbone, R. E., Conway, J. E., Crook, N. A. and Moncrieff, M. W. (1990). The generation and propagation of a nocturnal squall line, Part 1: Observations and implications for mesoscale predictability. *Mon. Wea. Rev.*, **118**, 26–49.

Carter, D. A., Balsley, B. B., Ecklund, W. L., Gage, K. S., *et al.* (1989). Investigations of internal gravity waves using three vertically directed closely spaced wind profilers. *J. Geophys. Res.*, **94**, 8633–8642.

Chen, T.-C. and Winn-Nielsen, A. (1978). Non-linear cascades of atmospheric energy and enstrophy. *Tellus*, **30**, 313–322.

Cho, J. Y. N., Zhu, Y., Newell, R. E., *et al.* (1999a). Horizontal wavenumber spectra of winds, temperature, and trace gases during the Pacific Exploratory Missions: 1. Climatology. *J. Geophys. Res.*, **104**, 5697–5716.

Cho, J. Y. N., Newell, R. E. and Barrick, J. D. (1999b). Horizontal wavenumber spectra of winds, temperature, and trace gases during the Pacific Exploratory Missions: 2. Gravity waves, quasi-two-dimensional turbulence, and vortical modes. *J. Geophys. Res.*, **104**, 16297–16308.

Cho, J. Y. N. and Lindborg, E. (2001). Horizontal velocity structure functions in the upper troposphere and lower stratosphere, 1. Observations. *J. Geophys. Res.*, **106**, 10223–10232.

Cho, J. Y. N., Anderson, B. E., Barrick, J. D. W. and Thornhill, K. L. (2001). Aircraft observations of boundary layer turbulence: Intermittency and the cascade of energy and passive scalar variance. *J. Geophys. Res.*, **106**, 32469–32479.

Christie, D. R. (1989). Long nonlinear waves in the lower atmosphere. *J. Atmos. Sci.*, **46**, 1462–1490.

Clark, T. L., Hauf, T. and Kuettner, J. P. (1986). Convectively forced internal gravity waves: Results from two-dimensional numerical experiments. *Quart. J. Roy. Meteor. Soc.*, **112**, 899–925.

Cornish, C. R. and Larsen, M. F. (1989). Observations of low-frequency inertia-gravity waves in the lower stratosphere over Arecibo. *J. Atmos. Sci.*, **46**, 2428–2439.

Crook, N. A. (1988). Trapping of low-level internal gravity waves. *J. Atmos. Sci.*, **45**, 1533–1541.

Dalaudier, F., Sidi, C., Crochet, M. and Vernin, J. (1994). Direct evidence of "sheets" in the atmospheric temperature field. *J. Atmos. Sci.*, **51**, 237–248.

Daley, R. (1997). Atmospheric data assimilation. *J. Meteor. Soc. Japan*, **75**, 319–329.

Dewan, E. M. (1979). Stratospheric wave spectra resembling turbulence. *Science*, **204**, 832–835.

Doviak, R. J., Chen, S. S. and Christie, D. R. (1991). A thunderstorm-generated solitary wave observation compared with theory for nonlinear waves in a sheared atmosphere. *J. Atmos. Sci.*, **48**, 87–111.

Eckermann, S. D. (1990). Effects of nonstationarity on spectral analysis of mesoscale motions of the atmosphere. *J. Geophys. Res.*, **95**, 16685–16703.

Ecklund, W. L., Gage, K. S., Balsley, B. B., *et al.* (1982). Vertical wind variability observed by VHF radar in the lee of the Colorado Rockies. *Mon. Wea. Rev.*, **110**, 1451–1457.

Ecklund, W. L., Balsley, B. B., Carter, D. A., *et al.* (1985). Observations of vertical motions in the troposphere and lower stratosphere using three closely-spaced ST radars. *Radio Sci.*, **20**, 1196–1206.

Ecklund, W. L., Gage, K. S., Nastrom, G. D. and Balsley, B. B. (1986). A preliminary climatology of the spectrum of vertical velocity observed by clear-air Doppler radar. *J. Clim. Appl. Meteor.*, **25**, 885–892.

Elsaesser, H. W. (1969). Wind variability as a function of time. *Mon. Wea. Rev.*, **97**, 424–428.

Eriksen, C. C. (1978). Measurements and models of fine structure, internal gravity waves, and wave breaking in the deep ocean. *J. Geophys. Res.*, **83**, 2989–3009.

Fofonoff, N. P. (1969). Spectral characteristics of internal waves in the ocean. *Deep Sea Res.*, **16**, Suppl., 58–71.

Fritts, D. C. and VanZandt, T. E. (1987). The effects of Doppler shifting on the frequency spectra of atmospheric gravity waves. *J. Geophys. Res.*, **92**, 9723–9732.

Fritts, D. C., Tsuda, T., Sato, T., *et al.* (1988). Observational evidence of a saturated gravity wave spectrum in the troposphere and lower stratosphere. *J. Atmos. Sci.*, **45**, 1741–1759.

Fritts, D. C. and Luo, Z. (1992). Gravity wave excitation by geostrophic adjustment of the jet stream. Part I: Two-dimensional forcing. *J. Atmos. Sci.*, **49**, 681–697.

Fritts, D. C. and Nastrom, G. D. (1992). Sources of mesoscale variability of gravity waves. Part II: Frontal, convective, and jet stream excitation. *J. Atmos. Sci.*, **49**, 111–127.

Fritts, D. C. and Alexander, M. J. (2003). Gravity wave dynamics and effects in the middle atmosphere. *Rev. Geophys.*, **41**, 1003, doi:10,1029/2001RG000106.

Fulton, R., Zrnic, D. S. and Doviak, R. J. (1990). Initiation of a solitary wave family in the demise of a nocturnal thunderstorm density current. *J. Atmos. Sci.*, **47**, 319–337.

Gage, K. S. and Jasperson, W. H. (1974). Prototype Metrac™ balloon tracking system yields accurate, high-resolution winds in Minneapolis field test. *Bull. Amer. Meteor. Soc.*, **55**, 1107–1114.

Gage, K. S. and Balsley, B. B. (1978). Doppler radar probing of the clear atmosphere. *Bull. Amer. Meteor. Soc.*, **59**, 1074–1094.

Gage, K. S. and Clark, W. L. (1978). Mesoscale variability of jet stream winds observed by the Sunset VHF Doppler radar. *J. Appl. Meteor.*, **17**, 1412–1416.

Gage, K. S. and Green, J. L. (1978). Evidence for specular reflection from monostatic VHF radar observations of the stratosphere. *Radio Sci.*, **13**, 991–1001.

Gage, K. S. (1979). Evidence for a $k^{-5/3}$ law inertial range in mesoscale two-dimensional turbulence. *J. Atmos. Sci.*, **36**, 1950–1954.

Gage, K. S. and Jasperson, W. H. (1979). Mesoscale wind variability below 5 km as revealed by sequential high-resolution wind soundings. *Mon. Wea. Rev.*, **107**, 77–86.

Gage, K. S. and Nastrom, G. D. (1985). On the spectrum of atmospheric velocity fluctuations seen by MST/ST radar and their interpretation. *Radio Sci.*, **20**, 1339–1347.

  (1986). Theoretical interpretation of atmospheric wavenumber spectra of wind and temperature observed by commercial aircraft during GASP. *J. Atmos. Sci.*, **43**, 729–740.

  (1988). Further discussion of the dynamical processes that contribute to the spectrum of mesoscale atmospheric motions. In *Proc. 8th Symp. on Turbulence and Diffusion, April 25–29, San Diego, CA*, Amer. Meteorol. Soc., 217–220.

Gage, K. S., Ecklund, W. L. and Carter, D. A. (1989). Convection waves observed using a VHF wind-profiling Doppler radar during the Pre-STORM experiment. In *Proc. 24th Conf. on Radar Meteorology, March 27–31, Tallahassee, FL*, Amer. Meteorol. Soc., 705–708.

Gage, K. S. (1990). Radar observations of the free atmosphere: Structure and dynamics. In *Radar in Meteorology: Battan Memorial and 40th Anniversary Radar Meteorology Conference*, D. Atlas, ed., Amer. Meteor. Soc., 534–574.

Gage, K. S. and Nastrom, G. D. (1990). A simple model for the enhanced frequency spectrum of vertical velocity based on tilting of atmospheric layers by lee waves. *Radio Sci.*, **25**, 1049–1056.

Gage, K. S., McAfee, J. R., Carter, D. A., *et al.* (1991). Long-term mean vertical motion over the tropical Pacific: Wind-profiling Doppler radar measurements. *Science*, **254**, 1771–1773.

Gage, K. S. and Gossard, E. E. (2003). Recent developments in observations, modeling and understanding of atmospheric turbulence and waves, *Meteorological Monographs* **30**, Amer. Meteor, Soc., 139–174.

Garrett, C. and Munk, W. (1972). Space-time scales of internal waves. *Geophys. Fluid Dyn.*, **2**, 225–264.

    (1975). Space-time scales of internal waves: A progress report. *J. Geophys. Res.*, **80**, 291–297.

Gill, A. E. (1982). *Atmosphere-Ocean Dynamics*. New York: Academic Press.

Gossard, E. E. and Hooke, W. H. (1975). *Waves in the Atmosphere. Atmospheric Infrasound and Gravity Waves: Their Generation and Propagation*, Amsterdam: Elsevier Scientif.

Gossard, E. E., Gaynor, J. E., Zamora, R. J. and Neff, W. D. (1985). Fine structure of elevated stable layers observed by sounder and *in-situ* tower sensors. *J. Atmos. Sci.*, **42**, 2156–2169.

Hauf, T. (1993). Aircraft observation of convection waves over southern Germany – A case study. *Mon. Wea. Rev.*, **121**, 3282–3290.

Herring, J. R. and Métais, O. (1989). Numerical experiments in forced stably stratified turbulence. *J. Fluid Mech.*, **202**, 97–115.

Hodges, R. R., Jr. (1967). Generation of turbulence in the upper atmosphere by internal gravity waves. *J. Geophys. Res.*, **72**, 3455–3458.

Högström, U., Smedman, A.-S. and Bergström, H. (1999). A case study of two-dimensional stratified turbulence. *J. Atmos. Sci.*, **56**, 959–976.

Hopfinger, E. J. (1987). Turbulence in stratified fluids: a review. *J. Geophys. Res.*, **92**, 5287–5303.

Itsweire, E. C., Helland, K. N. and van Atta, C. W. (1986). The evolution of grid-generated turbulence in a stably stratified fluid. *J. Fluid Mech.*, **162**, 299–338.

Itsweire, E. C. and Helland, K. N. (1989). Spectra and energy transfer in stably stratified turbulence. *J. Fluid Mech.*, **207**, 419–452.

Jasperson, W. H. (1982). Mesoscale time and space wind variability. *J. Appl. Meteor.*, **21**, 831–839.

Jasperson, W. H., Nastrom, G. D. and Fritts, D. C. (1990). Further study of the terrain effects on the mesoscale spectrum of atmospheric motions. *J. Atmos. Sci.*, **47**, 979–987.

Kao, S.-K. and Wendell, L. L. (1970). The kinetic energy of the large-scale atmospheric motion in wavenumber-frequency space: 1. Northern Hemisphere. *J. Atmos. Sci.*, **37**, 359–375.

Kitchen, M. (1989). Representativeness errors for radiosonde observations. *Quart. J. Roy. Meteor. Soc.*, **115**, 673–700.

Koch, S. E., Dorian, P. B., Ferrare, R., *et al.* (1991). Structure of an internal bore and dissipating gravity current as revealed by Raman lidar. *Mon. Wea. Rev.*, **119**, 857–887.

Koch, S. E., Einaudi, F., Doran, P. B., Lang, S. and Heymsfield, G. M. (1993). A mesoscale gravity-wave event observed during CCOPE. Part IV: Stability analysis and Doppler-derived wave vertical structure. *Mon. Wea. Rev.*, **121**, 2483–2510.

Koch, S. E. and O'Handley, C. (1997). Operational forecasting and detection of mesoscale gravity waves. *Weather Forecasting*, **12**, 253–281.

Koshyk, J. N., Hamilton, K. and Mahlman, J. D. (1999). Simulation of $k^{-5/3}$ mesoscale spectral regime in the GFDL SKYHI general circulation model. *Geophys. Res. Lett.*, **26**, 843–846.

Koshyk, J. N. and Hamilton, K. (2001). The horizontal kinetic energy spectrum and spectral budget simulated by a high-resolution troposphere–stratosphere–mesosphere GCM. *J. Atmos. Sci.*, **58**, 329–348.

Kuettner, J. P., Hildebrand, P. A. and Clark, T. L. (1987). Convection waves: Observations of gravity wave systems over convectively active boundary layers. *Quart. J. Roy. Meteor. Soc.*, **113**, 445–467.

Larsen, M. F., Kelley, M. C. and Gage, K. S. (1982). Turbulence spectra in the upper troposphere and lower stratosphere at periods between 2 hours and 40 days. *J. Atmos. Sci.*, **39**, 1035–1041.

Lelong, M. P. and Riley, J. J. (1991). Internal wave–vortical mode interactions in strongly stratified flows. *J. Fluid Mech.*, **232**, 1–19.

Lilly, D. K. (1983). Stratified turbulence and the mesoscale variability of the atmosphere. *J. Atmos. Sci.*, **40**, 749–761.

Lilly, D. K. and Petersen, E. L. (1983). Aircraft measurements of atmospheric kinetic energy spectra. *Tellus*, **35A**, 379–382.

Lilly, D. K., Bassett, G., Droegemeier, K. and Bartello, P. (1998). Stratified turbulence in the atmospheric mesoscales. *Theor. Comp. Fluid Dyn.*, **11**, 139–154.

Lindborg, E. (1999). Can the atmospheric kinetic energy spectrum be explained by two-dimensional turbulence? *J. Fluid Mech.*, **388**, 259–288.

Lindborg, E. and Cho, J. Y. N. (2001). Horizontal velocity structure functions in the upper troposphere and lower stratosphere, 2. Theoretical considerations. *J. Geophys. Res.*, **106**, 10233–10241.

Lindzen, R. S. (1981). Turbulence and stress owing to gravity wave and tidal breakdown. *J. Geophys. Res.*, **86**, 9707–9714.

Liziola, L. E. and Balsley, B. B. (1998). Studies of quasi horizontally propagating gravity waves in the troposphere using the Piura ST wind profiler. *J. Geophys. Res.*, **103**, 8641–8650.

Masmoudi, M. and Weil, A. (1988). Atmospheric mesoscale spectra and structure functions of mean horizontal velocity fluctuations measured with a Doppler sodar network. *J. Appl. Meteor.*, **27**, 864–873.

Maxworthy, T. (1990). The dynamics of two-dimensional turbulence. In *The Physical Oceanography of Sea Straits*, L. J. Pratt, ed., Dordrecht, Holland: Kluwer Academic Publishers, 567–574.

Métais, O. and Herring, J. R. (1989). Numerical simulations of freely evolving turbulence in stably stratified fluids. *J. Fluid Mech.*, **202**, 117–148.

Métais, O., Bartello, P., Garnier, E., Riley, J. J. and Lesieur, M. (1994). Inverse cascade in stably stratified rotating turbulence. *Dyn. Atmos. Oceans*, **23**, 193–203.

Müller, P., Olbers, D. J. and Willebrand, J. (1978). The IWEX spectrum. *J. Geophys. Res.*, **83**, 479–500.

Müller, P., Lien, R.-C. and Williams, R. (1988). Estimates of potential vorticity at small scales in the ocean. *J. Phys. Oceanogr.*, **18**, 401–416.

Müller, P. (1995). Ertel's potential vorticity theorem in physical oceanography. *Rev. Geophys.*, **33**, 67–97.

Narimousa, S., Maxworthy, T. and Spedding, G. R. (1991). Experiments on the structure of forced, quasi-two-dimensional turbulence. *J. Fluid Mech.*, **223**, 113–133.

Nastrom, G. D., Gage, K. S. and Jasperson, W. H. (1984). Kinetic energy spectrum of large- and mesoscale atmospheric processes. *Nature*, **310**, 36–38.

Nastrom, G. D. and Gage, K. S. (1985). A climatology of atmospheric wavenumber spectra of wind and temperature observed by commercial aircraft. *J. Atmos. Sci.*, **42**, 950–960.

Nastrom, G. D., Ecklund, W. L. and Gage, K. S. (1985). Direct measurement of large-scale vertical velocities using clear-air Doppler radars. *Mon. Wea. Rev.*, **113**, 708–718.

Nastrom, G. D., Fritts, D. C. and Gage, K. S. (1987). An investigation of terrain effects on the mesoscale spectrum of atmospheric motions. *J. Atmos. Sci.*, **44**, 3087–3096.

Nastrom, G. D., Gage, K. S. and Ecklund, W. L. (1990a). Uncertainties in estimates of the mean vertical velocity from MST radar observations. *Radio Sci*, **25**, 933–940.

Nastrom, G. D., Peterson, M. R., Green, J. L., Gage, K. S. and VanZandt, T. E. (1990b). Sources of gravity wave activity seen in the vertical velocities observed by the Flatland VHF radar. *J. Appl. Meteor.*, **29**, 783–792.

Nastrom, G. D. and Fritts, D. C. (1992). Sources of mesoscale variability of gravity waves. Part I: Topographic excitation. *J. Atmos. Sci.*, **49**, 101–110.

Olbers, D. J. (1983). Models of the oceanic internal wave field. *Rev. Geophys. Space Phys.*, **21**, 1567–1606.

Palmer, T. N. (2001). A nonlinear dynamical perspective on model error: A proposal for non-local stochastic–dynamic parameterization in weather and climate prediction models. *Quart. J. Roy. Meteor. Soc.*, **127**, 279–304.

Ralph, F. M., Crochet, M. and Venkateswaran, S. V. (1993). Observations of a mesoscale ducted gravity wave. *J. Atmos. Sci.*, **50**, 3277–3291.

Riggin, D., Fritts, D. C., Fawcett, C. D. and Kudeki, E. (1995). Observations of inertia-gravity wave motions in the stratosphere over Jicamarca, Peru. *Geophys. Res. Lett.*, **22**, 3239–3242.

Riley, J. J., Metcalfe, R. W. and Weissman, M. A. (1981). Direct numerical simulations of homogeneous turbulence in density stratified fluids. In *Nonlinear Properties of Internal Waves*, B. J. West, ed., La Jolla Institute, American Institute of Physics Conf. Proc. #76, 79–112.

Riley, J. J. and Lelong, M.-P. (2000). Fluid motions in the presence of strong stable stratification. *Ann. Rev. Fluid Mech.*, **32**, 613–657.

Riley, J. J. and deBruynKops, S. M. (2003). Dynamics of turbulence strongly influenced by buoyancy. *Phys. Fluids*, **15**, 2047–2059.

Rottman, J. W. and Einaudi, F. (1993). Solitary waves in the atmosphere. *J. Atmos. Sci.*, **50**, 2116–2136.

Salathé, E. P., Jr. and Smith, R. B. (1992). In-situ observations of temperature microstructure above and below the tropopause. *J. Atmos. Sci.*, **49**, 2032–2036.

Sato, K. (1989). Inertial gravity wave associated with a synoptic-scale pressure trough observed by the MU radar. *J. Meteor. Soc. Japan*, **67**, 325–334.

(1993). Small-scale wind disturbances observed by the MU radar during the passage of Typhoon Kelly. *J. Atmos. Sci.*, **50**, 518–537.

(1994). A statistical study of the structure, saturation and sources of inertia-gravity waves in the lower stratosphere observed with the MU radar. *J. Atmos. Terr. Phys.*, **56**, 755–774.

Sato, K., O'Sullivan, D. J. and Dunkerton, T. J. (1997). Low-frequency inertia-gravity waves in the stratosphere revealed by three-week continuous observation with the MU radar. *Geophys. Res. Lett.*, **24**, 1739–1742.

Scheffler, A. O. and Liu, C. H. (1985). On observation of gravity wave spectra in the atmosphere by using MST radars. *Radio Sci.*, **20**, 1309–1322.

(1986). The effects of Doppler shift on the gravity wave spectra observed by MST radar. *J. Atmos. Terr. Phys.*, **48**, 1225–1231.

Smith, S. A., Fritts, D. C. and VanZandt, T. E. (1987). Evidence for a saturated spectrum of atmospheric gravity waves. *J. Atmos. Sci.*, **44**, 1404–1410.

Staquet, C. and Sommeria, J. (2002). Internal gravity waves: from instabilities to turbulence. *Ann. Rev. Fluid Mech.*, **34**, 559–593.

Thomas, L., Pritchard, T. and Astin, I. (1992). Inertia-gravity waves in the troposphere and lower stratosphere. *Ann. Geophys.*, **10**, 690–697.

Thomas, L., Worthington, R. M. and McDonald, A. J. (1999). Inertia-gravity waves in the troposphere and lower stratosphere associated with a jet stream exit region. *Ann. Geophys.*, **17**, 115–121.

Thorpe, S. A. (1998). Turbulence in the stratified and rotating world oceans. *Theor. Comp. Fluid Dyn.*, **11**, 171–181.

Trexler, C. M. and Koch, S. E. (2000). The life cycle of a mesoscale gravity wave observed by a network of Doppler wind profilers. *Mon. Wea. Rev.*, **128**, 2423–2466.

Tsuda, T., Inoue, T., Fritts, D. C., *et al.* (1989). MST radar observations of a saturated gravity wave spectrum. *J. Atmos. Sci.*, **46**, 2440–2447.

Tung, K. K. and Orlando, W. W. (2003). The k^{-3} and $k^{-5/3}$ energy spectrum of atmospheric turbulence: quasi-geostrophic two-level model simulation. *J. Atmos. Sci.*, **60**, 824–835.

Vallis, G. K., Shutts, G. J. and Gray, M. E. B. (1997). Balanced mesoscale motion and stratified turbulence forced by convection. *Quart. J. Roy. Meteor. Soc.*, **123**, 1621–1652.

VanZandt, T. E. (1982). A universal spectrum of buoyancy waves in the atmosphere. *Geophys. Res. Lett.*, **9**, 575–578.

(1985). A model for gravity wave spectra observed by Doppler sounding system. *Radio Sci.*, **20**, 1323–1330.

VanZandt, T. E. and Fritts, D. C. (1989). A theory of enhanced saturation of the gravity wave spectrum due to increases in atmospheric stability. *Pure Appl. Geophys.*, **130**, 399–420.

VanZandt, T. E., Nastrom, G. D. and Green, J. L. (1991). Frequency spectra of vertical velocity from Flatland VHF radar data. *J. Geophys. Res.*, **96**, 2845–2855.

Vincent, R. A. and Fritts, D. C. (1987). A climatology of gravity wave motions in the mesopause region at Adelaide, Australia. *J. Atmos. Sci.*, **44**, 748–760.

Vinnichenko, N. K. (1970). The kinetic energy spectrum in the free atmosphere – 1 second to 5 years. *Tellus*, **22**, 158–166.

Weckwerth, T. M., Parsons, D. B., Koch, S. E., *et al.* (2003). An overview of the International H2O Project (IHOP-2002) and some preliminary highlights. *Bull. Amer. Meteor. Soc.*, **85**, 253–277.

Williams, C. R., Ecklund, W. L., Johnston, P. E. and Gage, K. S. (2000). Cluster analysis techniques to separate air motion and hydrometeors in vertical incident profiler observations. *J. Atmos. Oceanic Technol.*, **17**, 949–962.

Worthington, R. M. and Thomas, L. (1998). The frequency spectrum of mountain waves. *Quart. J. Roy. Meteor. Soc.*, **124**, 687–703.

Yamanaka, M. D., Ogino, S., Kondo, S., *et al.* (1996). Inertia-gravity waves and subtropical multiple tropopauses: Vertical wavenumber spectra of wind and temperature observed by the MU radar, radiosondes and operational rawinsonde network. *J. Atmos. Terr. Phys.*, **58**, 785–805.

Yap, C. T. and van Atta, C. W. (1993). Experimental studies of the development of quasi-two-dimensional turbulence in stably stratified fluid. *Dyn. Atmos. Oceans*, **19**, 289–323.

Appendix A

Douglas K. Lilly: positions, awards, and students

Scientific and academic positions

Research Meteorologist, General Circulation Laboratory, US Weather Bureau	1959–1964
Senior Scientist, National Center for Atmospheric Research (NCAR)	1964–1982
Professor of Meteorology, University of Oklahoma	1982–1995
Professor Emeritus, University of Oklahoma	1995–2002
Distinguished Senior Research Scientist, National Severe Storms Laboratory (NSSL)	1997–2002
Adjunct Professor of Physics, University of Nebraska–Kearney	2002–present

Administrative positions

Program Scientist, Turbulent Interactions Program, NCAR	1965–1972
Director, Atmospheric Dynamics Department, NCAR	1973
Project Leader, Small-Scale Analysis and Prediction Project, NCAR	1974–1975
Director, Atmospheric Analysis and Prediction Division, NCAR	1975–1976
Head, Mesoscale Research Section, NCAR	1978–1981
Director, Cooperative Institute for Mesoscale Meteorological Studies (CIMMS), University of Oklahoma and NOAA	1987–1991
Director, Center for Analysis and Prediction of Storms (CAPS), University of Oklahoma	1989–1994

Visiting positions

Affiliate Professor, New Mexico Institute of Mining and Technology	1966–1969
Visiting Fellow, Cooperative Institute for Research in Environmental Sciences, NOAA and University of Colorado	1977
Guest Scientist, Institute of Atmospheric Physics, German Aerospace Agency (DLR)	1980
Affiliate Professor, School of Geophysical Sciences, Georgia Institute of Technology	1980–1982
Guest Lecturer, University of Wyoming	1981
Guest Lecturer, Pennsylvania State University	1981
Lecturer, American Meteorological Society Intensive Short Course on Mesoscale Meteorology and Forecasting	1984
Visiting Professor, Monash University, Melbourne, Australia	1984
Visiting Scientist, Center for Turbulence Research, Stanford University	1988
Houghton Visiting Lecturer, Massachusetts Institute of Technology	1990

Honors and awards

Fellow, American Meteorological Society	1971
Second Half-Century Award, American Meteorological Society	1973
Outstanding Publication Award, NCAR (with J.B. Klemp)	1975
Fellow, Cooperative Institute for Mesoscale Meteorological Studies	1982
Distinguished Lectureship, University of Oklahoma	1985–1989
George Lynn Cross Research Professor, University of Oklahoma	1986
Carl-Gustav Rossby Medal, American Meteorological Society	1986
Symons Memorial Lecturer, Royal Meteorological Society, United Kingdom	1989
Robert Lowry Endowed Chair in Meteorology, University of Oklahoma	1992–1995
Symons Gold Medal, Royal Meteorological Society, United Kingdom	1993
Member, National Academy of Sciences, USA	1999–
Honorary Member of the American Meteorological Society	2001

Students

Doug Lilly supervised eighteen successful degree candidates: six for M.S., and twelve for Ph.D.:

Michael Foster	M.S.	1984
Brian Jewett	M.S.	1986
Jen-Jing Lin	M.S.	1988
Fushen Zhang	M.S.	1990
Limin Zhao	M.S.	1994
Yuan Ho	M.S.	1995
Dawn Wolfsburg Flicker	Ph. D.	1987
Gregory Byrd	Ph. D.	1987
Eugene McCaul Jr.	Ph. D.	1989
Rodger Brown	Ph. D.	1989
Wan-Shu Wu	Ph. D.	1990
Jeanne Schneider	Ph. D.	1991
Juanzhen Sun	Ph. D.	1992
Litao Deng	Ph. D.	1993
Mei Xu	Ph. D.	1995
Yu-Chieng Liou	Ph. D.	1996
Daniel Weber	Ph. D.	1997
Katharine Kanak	Ph. D.	1999

Appendix B

List of publications by Douglas K. Lilly

Peer-reviewed articles (in chronological order)

Lilly, D. K. (1960). On the theory of disturbances in a conditionally unstable atmosphere. *Monthly Weather Review*, **88**, 1–17.

(1961). A proposed staggered-grid system for numerical integration of dynamic equations. *Monthly Weather Review*, **89**, 59–65.

(1962). On the numerical simulation of buoyant convection. *Tellus*, **14**, 148–172.

Turner, J. S. and Lilly, D. K. (1963). The carbonated-water tornado vortex. *Journal of the Atmospheric Sciences*, **20**, 468–471.

Lilly, D. K. (1964). Numerical solutions for the shape-preserving two-dimensional thermal convection element. *Journal of the Atmospheric Sciences*, **21**, 83–98.

(1965). Experimental generation of convectively driven vortices. *Geofísica Internacional*, **5**, 43–48.

(1965). On the computational stability of numerical solutions of time-dependent non-linear geophysical fluid dynamics problems. *Monthly Weather Review*, **93**, 11–26.

Hidy, G. M. and Lilly, D. K. (1965). Solutions to the equations for the kinetics of coagulation. *Journal of Colloid Science*, **20**, 867–874.

Lilly, D. K. (1966). On the speed of surface gravity waves propagating on a moving fluid, Appendix to "Wind action on water standing in a laboratory channel", by G. Hidy and E. J. Plate. *Journal of Fluid Mechanics*, **26**, 651–687.

(1966). On the instability of Ekman boundary flow. *Journal of the Atmospheric Sciences*, **23**, 481–494.

(1968). Models of cloud-topped mixed layers under a strong inversion. *Quarterly Journal of the Royal Meteorological Society*, **94**, 292–309.

Kuettner, J. and Lilly, D. K. (1968). Lee waves in the Colorado Rockies. *Weatherwise*, **21**, 180–197.

Deardorff, J. W., Willis, G. E. and Lilly, D. K. (1969). Laboratory investigation of non-steady penetrative convection. *Journal of Fluid Mechanics*, **35**, 7–31.

Lilly, D. K. (1969). Numerical simulation of two-dimensional turbulence. High-speed computing in fluid dynamics. *The Physics of Fluids* (Supplement II), vol. 12, 240–249.

Vergeiner, I. and Lilly, D. K. (1970). The dynamic structure of lee wave flow as obtained from balloon and airplane observations. *Monthly Weather Review*, **98**, 220–232.

Lilly, D. K. (1971). Comments on "Case studies of a convective plume and a dust devil". *Journal of Applied Meteorology*, **10**, 590–591.

(1971). Comments on "Venus's general circulation is a merry-go-round". *Journal of the Atmospheric Sciences*, **28**, 827–828.

(1971). Numerical simulation of developing and decaying two-dimensional turbulence. *Journal of Fluid Mechanics*, **45**(2), 395–415.

(1971). Observations of mountain-induced turbulence. *Journal of Geophysical Research – Atmospheres*, **76**, 6585–6588.

(1971). Progress in research on atmospheric turbulence. *EOS*, **52**, International Union of Geodesy and Geophysics, 332–341.

Fox, D. G. and Lilly, D. K. (1972). Numerical simulation of turbulent flows. *Reviews of Geophysics and Space Physics*, **10**, 51–72.

Lilly, D. K. (1972). Numerical simulation of two-dimensional turbulence, Part I: Models of statistically steady turbulence. *Geophysical Fluid Dynamics*, **3**, 289–319.

(1972). Numerical simulation of two-dimensional turbulence, Part II: Stability and predictability studies. *Geophysical Fluid Dynamics*, **4**, 1–28.

(1972). Wave momentum flux – a GARP problem. *Bulletin of the American Meteorological Society*, **53**, 17–23.

Lilly, D. K. and Zipser, E. J. (1972). The front range windstorm of January 11, 1972 – a meteorological narrative. *Weatherwise*, **25**, 56–63.

Lilly, D. K. (1973). Book review of "Buoyancy Effects on Fluids" by J.S. Turner. *Bulletin of the American Meteorological Society*, **54**, 854.

(1973). A note on barotropic instability and predictability. *Journal of the Atmospheric Sciences*, **30**, 145–147.

Lilly, D. K. and Kennedy, P. J. (1973). Observations of a stationary mountain wave and its associated momentum flux and energy dissipation. *Journal of the Atmospheric Sciences*, **30**, 1135–1152.

Deardorff, J. W., Willis, G. E. and Lilly, D. K. (1974). Comments on the paper by A. K. Betts "Non-precipitating cumulus convection and its parameterization". *Quarterly Journal of the Royal Meteorological Society*, **100**, 122–123.

Lilly, D. K and Lester, P. F. (1974). Waves and turbulence in the stratosphere. *Journal of the Atmospheric Sciences*, **31**, 800–812.

Lilly, D. K., Waco, D. E. and Adelfang, S. I. (1974). Stratospheric mixing estimated from high-altitude turbulence measurements. *Journal of Applied Meteorology*, **13**, 488–493.

Klemp, J. B. and Lilly, D. K. (1975). The dynamics of wave-induced downslope winds. *Journal of the Atmospheric Sciences*, **32**, 320–339.

Lilly, D. K. (1975). Severe storms and storm systems: scientific background, methods, and critical questions. *Pure and Applied Geophysics*, **113**, 713–734.

Kessler, E., Manton, M. J., Lilly, D. K., Darkow, G. L. and Court, A. (1976). Tornado forum. Letters to *Nature*, **260**, 457–461, responding to a paper by Isaacs, J. D., Stork, J. W., Goldstein, D. B. and Wick, G. L., *Nature*, **253**, 254–255 (1975).

Lilly, D. K. (1978). A severe downslope windstorm and aircraft turbulence event induced by a mountain wave. *Journal of the Atmospheric Sciences*, **35**, 59–77.

Klemp, J. B. and Lilly, D. K. (1978). Numerical simulation of hydrostatic mountain waves. *Journal of the Atmospheric Sciences*, **35**, 78–107.

Lilly, D. K. (1979). The dynamical structure and evolution of thunderstorms and squall lines. *Annual Review of Earth and Planetary Sciences*, **7**, 117–161.

Lilly, D. K. and Klemp, J. B. (1979). The effects of terrain shape on nonlinear hydrostatic mountain waves. *Journal of Fluid Mechanics*, **95** (2), 241–261.

Lilly, D. K. and Schubert, W. H. (1980). The effects of radiative cooling in a cloud-topped mixed layer. *Journal of the Atmospheric Sciences*, **37**, 482–487.

Lilly, D. K. and Klemp, J. B. (1980). Comments on "The evolution and stability of finite-amplitude mountain waves. Part II: Surface wave drag and severe downslope windstorms". *Journal of the Atmospheric Sciences*, **37**, 2119–2121.

Lilly, D. K. (1981). Doppler radar observations of upslope snowstorms. *Bulletin of the American Meteorological Society*, **62**, 571–572.

　(1981). Doppler radar observations of upslope snowstorms. *Bulletin of the American Meteorological Society*, **62**, 940.

　(1981). Wave-permeable lateral boundary conditions for convective cloud and storm simulations. *Journal of the Atmospheric Sciences*, **38**, 1313–1316.

　(1982). The development and maintenance of rotation in convective storms. In *Topics in Atmospheric and Oceanographic Sciences: Intense Atmospheric Vortices*, L. Bengtsson and J. Lighthill, eds., New York, Berlin, and Heidelberg: Springer-Verlag, 149–160.

Lilly, D. K., Nichols, J. M., Chervin, R. M., Kennedy, P. J. and Klemp, J. B. (1982). Aircraft measurements of wave momentum flux over the Colorado Rocky Mountains. *Quarterly Journal of the Royal Meteorological Society*, **108**, 625–642.

Lilly, D. K. and Gal-Chen, T. (1983). *Mesoscale Meteorology – Theories, Observations and Models . . . , (1982), Bonas, France*. Dordrecht: Holland; Boston: published in cooperation with NATO Scientific Affairs Division by D. Reidel Publishing Co.; Hingham, MA: sold and distributed in the USA and Canada by Kluwer Academic Publishers.

Lilly, D. K. (1983). Dynamics of rotating thunderstorms. In *Mesoscale Meteorology – Theories, Observations and Models*, D. K. Lilly and T. Gal-Chen, eds., Dordrecht: D. Reidel, 531–543.

　(1983). Linear theory of internal gravity waves and mountain waves. In *Mesoscale Meteorology – Theories, Observations and Models*, D. K. Lilly and T. Gal-Chen, eds., Dordrecht: D. Reidel, 13–24.

　(1983). Stratified turbulence and the mesoscale variability of the atmosphere. *Journal of the Atmospheric Sciences*, **40**, 749–761.

Lilly, D. K. and Petersen, E. L. (1983). Aircraft measurements of atmospheric kinetic energy spectra. *Tellus*, **35A**, 379–382.

　(1983). NCAR, NOAA, Oklahoma University scientists gear up for field project. *News and Notes, Bulletin of the American Meteorological Society*, **65**, 721.

Lilly, D. K. (1984). Some facets of the predictability problem for atmospheric mesoscales. In *Predictability of Fluid Motions*, G. Holloway and B. J. West, eds., New York: American Institute of Physics, 287–294.

　(1985). Theoretical predictability of small scale motions. In *Turbulence and Predictability in Geophysical Fluid Dynamics and Climate Dynamics*, M. Ghil, ed., North Holland Publishing Company, 281–289.

　(1986). Atmospheric instabilities. In *Mesoscale Meteorology and Forecasting*, P. Ray, ed., Boston, MA: American Meteorological Society, 259–271.

　(1986). On the structure, energetics and propagation of rotating convective storms. Part I: Energy exchange with the mean flow. *Journal of the Atmospheric Sciences*, **43**, 113–125.

(1986). On the structure, energetics and propagation of rotating convective storms. Part II: Helicity and storm stabilization. *Journal of the Atmospheric Sciences*, **43**, 126–140.

(1988). Cirrus outflow dynamics. *Journal of the Atmospheric Sciences*, **45**, 1594–1605.

(1989). Two-dimensional turbulence generated by energy sources at two scales. *Journal of the Atmospheric Sciences*, **46**, 2026–2030.

Rothfusz, L. P. and Lilly, D. K. (1989). Quantitative and theoretical analyses of an experimental helical vortex. *Journal of the Atmospheric Sciences*, **46**, 2265–2279.

Lilly, D. K. (1989). Helicity. In *Lecture Notes on Turbulence*, J. R. Herring and J. C. McWilliams, eds., Singapore: World Scientific Publishing Company, Incorporated, 171–218.

Lilly, D. K. and Jewett, B. F. (1990). Momentum and kinetic energy budgets of simulated supercell thunderstorms. *Journal of the Atmospheric Sciences*, **47**, 707–726.

Lilly, D. K. and Gal-Chen, T. (1990). Can dryline mixing create buoyancy? *Journal of the Atmospheric Sciences*, **47**, 1170–1171.

Parsons, D. B., Small, B. F. and Lilly, D. K. (1990). Mesoscale organization and processes. In *Radar in Meteorology*, D. Atlas, ed., American Meteorological Society, 461–472.

Lilly, D. K. (1990). Numerical prediction of thunderstorms – has its time come? (Symons Memorial Lecture.) *Quarterly Journal of the Royal Meteorological Society*, **116A**, 779–798.

Sun, J., Flicker, D. W. and Lilly, D. K. (1991). Recovery of three-dimensional wind and temperature fields from simulated single-Doppler radar data. *Journal of the Atmospheric Sciences*, **48**, 876–890.

Lilly, D. K. (1991). Thunderstorms and helicity. In *McGraw-Hill 1991 Yearbook of Science and Technology*, McGraw-Hill, 425–427.

(1992). A proposed modification of the Germano subgrid-scale closure method. *Physics of Fluids A*, **4**(3), 633–635.

(1992). Meteorology. In *1992 Science Year*, The World Book Annual Science Supplement, 311–313.

Wu, W.-S., Lilly, D. K. and Kerr, R. M. (1992). Helicity and thermal convection with shear. *Journal of the Atmospheric Sciences*, **49**, 1800–1809.

Lilly, D. K. (1993). Comments on AMS meeting procedures. *Bulletin of the American Meteorological Society*, **74**, 854.

Filyushkin, V. V. and Lilly, D. K. (1993). Application of a 3D delta-Eddington radiative transfer model to calculation of solar heating and photolysis rates in a stratocumulus cloud layer. *Atmospheric Radiation*, **2049**, 56–66.

Scotti, A., Meneveau, C. and Lilly, D. K. (1993). Generalized Smagorinsky model for anisotropic grids. *Physics of Fluids A*, **5**(9), 2306–2308.

Wong, V. C. and Lilly, D. K. (1994). A comparison of two dynamic subgrid closure methods for turbulent thermal convection. *Physics of Fluids A*, **6**(2), 1016–1023.

Kogan, Y. L, Lilly, D. K., Kogan, Z. N. and Filyushkin, V. V. (1994). The effect of CCN regeneration on the evolution of stratocumulus cloud layers. *Atmospheric Research*, **33**, 137–150.

Emanuel, K. A., Raymond, D., Betts, A., . . . , Lilly, D. K., *et al.* (1995). Report of the first prospectus development team of the U. S. Weather Research Program to NOAA and the NSF. *Bulletin of the American Meteorological Society*, **76**, 1194–1208.

Lilly, D. K. (1995). Tzvi Gal-Chen, 1941–1994. *Bulletin of the American Meteorological Society*, **76**, 68–69.

Kogan, Y. L., Khairoutdinov, M. P., Lilly, D. K., Kogan, Z. N. and Filyushkin, V. V. (1995). Modeling of stratocumulus cloud layers in a large eddy simulation model with explicit microphysics. *Journal of the Atmospheric Sciences*, **52**, 2923–2940.

Kogan, Z. N., Lilly, D. K., Kogan Y. L. and Filyushkin, V. V. (1995). Study of the effects of cloud microphysics on solar radiative properties of stratocumulus cloud layers. *Atmospheric Research*, **35**, 157–172.

Lilly, D. K. (1996). Foreword to the Gal-Chen Memorial Issue. *Journal of the Atmospheric Sciences*, **53**, 2561.

(1996). A comparison of incompressible, anelastic and Boussinesq dynamics. *Atmospheric Research*, **40**, 143–151.

Kanak, K. M. and Lilly, D. K. (1996). The linear stability and structure of convection in a mean circular shear. *Journal of the Atmospheric Sciences*, **53**, 2578–2593.

Kogan, Z. N., Kogan, Y. L. and Lilly, D. K. (1996). Evaluation of sulfate aerosols indirect effect in marine stratocumulus clouds using observation-derived cloud climatology. *Geophysical Research Letters*, **23**, 1937–1940.

(1997). Cloud factor and seasonality of the indirect effect of anthropogenic sulfate aerosols. *Journal of Geophysical Research*, **102**, 25927–25939.

Liou, Y.-C. and Lilly, D. K. (1997). Numerical study of the structure and evolution of a heated planetary boundary layer with a jet. *Journal of Geophysical Research*, **102**(D4), 4447–4462.

Liu, Q., Kogan, Y. L., Lilly, D. K. and Khairoutdinov, M. P. (1997). Variational optimization method for calculation of cloud drop growth in an Eulerian drop-size framework. *Journal of the Atmospheric Sciences*, **54**, 2493–2504.

Lilly, D. K. (1997). Introduction to "Computational design for long-term numerical integration of the equations of fluid motion. Two-dimensional incompressible flow. Part I." *Journal of Computational Physics*, **135**, 101–102.

Lilly, D. K., Bassett, G., Droegemeier, K. and Bartello, P. (1998). Stratified turbulence in the atmospheric mesoscales. *Theoretical and Computational Fluid Dynamics*, **11**, 139–153.

Lilly, D. K (1999). Comments on "From the tropics to the poles in forty days". *Bulletin of the American Meteorological Society*, **80**, 904–905.

(1999). Review of *Buoyant Convection in Geophysical Flows*, E. J. Plate, E. E. Fedorovich, D. X. Viegas and J. C. Wyngaard, eds., 1998, Kluwer *Bulletin American Meteorological Society*, **80**, 937–938.

Schneider, J. M. and Lilly, D. K. (1999). An observational and numerical study of a sheared, convective boundary layer. Part I: Phoenix II observations, statistical description and visualization. *Journal of the Atmospheric Sciences*, **56**, 3059–3078.

Lilly, D. K. (2000). Helical buoyant convection. In *Geophysical and Astrophysical Convection*, P. Fox and R. Kerr, eds., Amsterdam: Gordon and Breach, 241–256.

Liu, Q., Kogan, Y. L., Lilly, D. K., *et al.* (2000). Modeling of ship effluent transport and its sensitivity to boundary layer structure. *Journal of the Atmospheric Sciences*, **57**, 2779–2791.

Kanak, K. M., Lilly, D. K. and Snow, J. T. (2000). The formation of vertical vortices in the convective boundary layer. *Quarterly Journal of the Royal Meteorological Society*, **126A**, 2789–2810.

Lilly, D. K. (2001). Comments on "Changes in the onset of Spring in the Western United States". *Bulletin of the American Meteorological Society*, **82**, 2265–2266.

(2002). Entrainment into mixed layers. Part I: Sharp-edged and smoothed tops. *Journal of the Atmospheric Sciences*, **59**, 3340–3352.

(2002). Entrainment into mixed layers. Part II: A new closure. *Journal of the Atmospheric Sciences*, **59**, 3353–3361.

Stevens, B., Lenschow, D. H., Vali, G., . . . , Lilly, D. K., *et al.* (2003). Dynamics and Chemistry of Marine Stratocumulus – DYCOMS-II. *Bulletin of the American Meteorological Society*, **84**, 579–593.

Stevens, B., Lenschow, D. H., Faloona, I., . . . , Lilly, D. K., *et al.* (2003): On entrainment rates in nocturnal marine stratocumulus. *Quarterly Journal of the Royal Meterological Society*, **129**, 3469–3499.

Other articles (in chronological order)

Hidy, G. M. and Lilly, D. K. (1964). Solutions to the equations for the kinetics of coagulation. NCAR Report, Boulder, CO: National Center for Atmospheric Research.

Lilly, D. K. (1966). On the application of the eddy viscosity concept in the inertial sub-range of turbulence. NCAR Manuscript No. 123, Boulder, CO: National Center for Atmospheric Research.

  (1966). The representation of small-scale turbulence in numerical simulation experiments. NCAR Manuscript No. 281, Boulder, CO: National Center for Atmospheric Research.

  (1967). The representation of small-scale turbulence in numerical simulation experiments. In *Proceedings, IBM Scientific Computing Symposium on Environmental Sciences, November 14–16, 1966*. IBM Form No. 320–1951, H. H. Goldstein, ed., Yorktown Heights, New York: Thomas J. Watson Research Center, 195–210.

Lilly, D. K. and Panofsky, H. A. (1967). Summary of progress in research in atmospheric turbulence and diffusion. US National Report 1963–1967. In *Transactions, Fourteenth General Assembly of American Geophysical Union, Vol. 2.*

Simpson, J., Lilly, D. K., Twomey, S. and Kuo, H. L. (1967). Report of working group on convection (GARP). Boulder, CO: National Center for Atmospheric Research.

Lilly, D. K. (1967). Models of cloud layers under a strong inversion. NCAR Manuscript No. 386, Boulder, CO: National Center for Atmospheric Research.

  (1968). Tornado dynamics. NCAR Manuscript No. 69–117. Boulder, CO: National Center for Atmospheric Research.

Lilly, D. K. and Toutenhoofd, W. (1969). The Colorado Lee Wave Program. In *Proceedings, Symposium on Clear Air Turbulence and its Detection, August 14–16, 1968, Seattle WA*, Y. H. Pao and A. Goldburg eds., New York: Plenum Press, 232–245.

Lilly, D. K. (1969). Mountain Wave Equations – Adiabatic, Frictionless Gas Equations. NCAR Report, Boulder, CO: National Center for Atmospheric Research.

  (1969). The simulation of three-dimensional turbulent flow in two dimensions. NCAR Report, Boulder, CO: National Center for Atmospheric Research.

Deardorff, J. W., Willis, G. E. and Lilly, D. K. (1969). Laboratory investigation of non-steady penetrative convection. NCAR Manuscript No. 68–72, Boulder, CO: National Center for Atmospheric Research.

Lilly, D. K. (1970). Lectures in sub-synoptic scale of motion and two-dimensional turbulence. NCAR Manuscript No. 70–162. Boulder, CO: National Center Atmospheric Research.

  (1970). Progress in research on atmospheric turbulence. NCAR Manuscript No. 70–182. Boulder, CO: National Center Atmospheric Research.

  (1970). The numerical simulation of three-dimensional turbulence with two dimensions. *Society of Industrial and Applied Mathematics – American Mathematical Society, Proceedings*, **2**, 41–53.

Lilly, D. K., Pann, Y., Kennedy, P. and Tottenhoofd, W. (1971). Data catalog for the 1970 Colorado Lee Wave Observational Program. NCAR Technical Note NT/STR-72. Boulder, CO: National Center Atmospheric Research.

Lilly, D. K. (1971). Wave momentum flux – A GARP problem. NCAR Manuscript No. 71–186. Boulder, CO: National Center Atmospheric Research.

(1971). Observations of mountain-induced turbulence. In *Fifty-Second Annual Meeting of the American Geophysical Union, April 12–16, Washington, DC*: American Geophysical Union.

Lilly, D. K. (1972). Numerical simulation of turbulent flow. In *Dynamics of the Tropical Atmosphere: Notes from a Colloquium, Summer 1972*. Boulder, CO: National Center for Atmospheric Research, 379–384.

(1972). Measurement of temperature, pressure, humidity and air velocity from aircraft. In *Dynamics of the Tropical Atmosphere: Notes from a Colloquium, Summer 1972*. Boulder, CO: National Center for Atmospheric Research, 364–376.

(1973). Lectures in sub-synoptic scale of motion and two-dimensional turbulence. In *Dynamic Meteorology, Proceedings of the Lannion 1970 Centre National d'Etudes Spatiales (CNES) Summer School of Space Physics*. P. Morel ed., Dordrecht, Holland: Reidel Publishing Company.

(1974). Open SESAME. In *Proceedings of the Opening Meeting at Boulder, Colorado, September 4–6, 1974*. D. K. Lilly ed., Boulder, CO: National Center for Atmospheric Research.

(1974). NCAR interest in the SESAME Program. In *Proceedings of the Opening Meeting at Boulder, Colorado, September 4–6, 1974*. Boulder, CO: National Center for Atmospheric Research, 29–30.

(1974). Some comments on fine mesh modeling. In *Proceedings of the Opening Meeting at Boulder, Colorado, September 4–6, 1974*. Boulder, CO: National Center for Atmospheric Research, 277–280.

Lilly, D. K. and Lenschow, D. H. (1974). Aircraft measurements of the atmospheric mesoscale using an inertial reference system. In *First Symposium on Flow, Its Measurement and Control in Science and Industry*, Rodger B. Dowdell, ed., Instrument Society of America, 369–377.

Barnes, S. L. and Lilly, D. K. (1975). Covariance analysis of severe storm environments. *Preprints, Ninth Conference on Severe Local Storms, October 21–23, 1975. Norman*, OK: American Meteorological Society, 301–306.

Lilly, D. K., Waco, D. E. and Adelfang, S. I. (1975). Stratospheric mixing estimated from high altitude turbulence measurements by using energy budget techniques. In Climate Impact Assessment Program Monograph 1, Final Report DOT-TST-75–51, Springfield, VA: National Technical Information Service, 6:81–6:90.

Chervin, R. M. and Lilly, D. K. (1975). Boundary layer sensitivity studies with NCAR GCM. In *GARP Report No. 8, January 1975, The Global Atmospheric Research Program on Numerical Experimentation, Research Activities in Atmospheric and Oceanic Modeling*, A. Robert, ed., 38–39.

Lilly, D. K. (1976). Severe storms and storm systems: scientific background, methods and critical questions. NCAR Reprint No. 2510. Boulder, CO: National Center for Atmospheric Research.

(1976). Sources of rotation and energy in the tornado. In *Proceedings, Symposium on Tornadoes: Assessment of Knowledge and Implications for Man, June 22–24, 1976, Texas Technology University, Lubbock, TX*, 145–150.

Atlas, D., Shenk, W. E., Barnes, S. L., Golden, J. H. and Lilly, D. K. (1977). SESAME updated. *Preprints, Tenth Conference on Severe Local Storms, October 18–21, 1977, Omaha, NB*, American Meteorological Society.

Lilly, D. K. (1978). On the upper boundary condition of a cloud-topped mixed layer. *Paper for Boundary Layer Workshop, Boulder, CO, August 14–18, 1978.*

Lilly, D. K. and Klemp, J. B. (1978). Downslope mountain windstorms. *Preprints, Third US National Conference on Wind Engineering Research, February 26–March 1, 1978, University of Florida, Gainesville, FL*, I8:1–I8:8.

Lilly, D. K. (1979). Results from Project SESAME. *Preprints, Eleventh Conference on Severe Local Storms, October 2–5, 1979, Kansas City, MO*, American Meteorological Society.

Klemp, J. B. and Lilly, D. K. (1980). Mountain waves and momentum flux. In *Orographic Effects in Planetary Flows*, GARP Publication Series No. 23, Geneva: International Council of Scientific Unions World Meteorological Organization, 116–141.

Lilly, D. K. (1981). Doppler radar observations of upslope snowstorms. In *Second Conference on Mountain Meteorology, November 8–12, 1981, Snowmass, CO*, American Meteorological Society.

(1981). Doppler radar observations of upslope snowstorms. In *Twentieth Conference on Radar Meteorology, November 30–December 3, 1981, Boston, MA*, American Meteorological Society, 638–645.

Rotunno, R. and Lilly, D. K. (1981). A numerical model pertaining to the multiple vortex phenomenon. Final Report prepared for Division of Reactor Safety Research, Office of Nuclear Regulatory Research, US Nuclear Regulatory Commission, Washington, DC. NCR FIN B6238, under Contract No. NRC-04–78247. NUREG/CR-1840, R6, RB.

Lilly, D. K. (1983). Stratified turbulence and the mesoscale variability of the atmosphere. In *Proceedings, Sixth Symposium on Turbulence and Diffusion, March 22–25, 1983, Boston, MA*, American Meteorological Society.

(1983). Helicity as a stabilizing effect on rotating storms. *Preprints, Thirteenth Conference on Severe Local Storms, October 17–20, 1983, Tulsa, OK*, American Meteorological Society, 219–222.

(1984). Atmospheric noise spectrum: waves or stratified turbulence? In *Proceedings, "Aha Huliko" a Hawaiian Winter Workshop on Internal Gravity Waves and Small-Scale Turbulence, University of Hawaii, Manoa, January 17–20, 1984*, Honolulu: Hawaii Institute of Geophysics, 285–294.

Lilly, D. K. and Moeng, C.-H. (1984). A technique for recovering the total velocity field from single Doppler measurements. *Preprints, Twenty-Second Conference on Radar Meteorology, September 10–13, 1984*, American Meteorological Society, 506–508.

Emanuel, K., Rotunno, R. and Lilly, D. K. (1985). An air–sea interaction theory for tropical cyclones. *Preprints, Sixteenth Conference on Hurricanes and Tropical Meteorology, Houston, TX*, Boston: American Meterological Society.

Lilly, D. K. and Emanuel, K. (1985). A steady-state hurricane model. *Preprints, Sixteenth Conference on Hurricanes and Tropical Meteorology, Houston, TX*, Boston: American Meterological Society.

Lilly, D. K., Gal-Chen, T., Lin, J.-J. and Schneider, J. M. (1988). Phoenix II analysis results – mean state, variances, and spectra. *Proceedings, Eighth Symposium on Turbulence and Diffusion, April 25–29, 1988, San Diego, CA*, American Meteorological Society.

Lilly, D. K. and Lin, J.-J. (1988). Kinetic energy balances in and above a heated boundary layer as observed by aircraft and Doppler radar. *Proceedings, Eighth Symposium on Turbulence and Diffusion, April 25–29, San Diego, CA*, American Meteorological Society.

Schneider, J. M. and Lilly, D. K. (1988). Convective boundary layer fields as observed by Doppler radar. *Proceedings, Eighth Symposium on Turbulence and Diffusion, April 25–29, San Diego, CA*, American Meteorological Society.

Lilly, D. K. (1990). Initialization of storm prediction models from single-Doppler radar. *Preprints, Sixteenth Conference on Severe Local Storms, October 22–26, Kananaskis Park, Alberta, Canada.* American Meteorological Society.

Wu, W.-S. and Lilly, D. K. (1990). Kinetic energy budgets of rotational and divergent components of simulated supercell thunderstorms. *Preprints, Sixteenth Conference on Severe Local Storms, October 22–26, Kananaskis Park, Alberta, Canada.* American Meteorological Society, 24–29.

Sun, J., Flicker, D. W. and Lilly, D. K. (1990). Recovery of three-dimensional wind and thermodynamic variables from single-Doppler radar data. *Preprints, Sixteenth Conference on Severe Local Storms, October 22–26, Kananaskis Park, Alberta, Canada.* American Meterological Society, 219–224.

Lilly, D. K. and Mason, P. J. (1990). A numerical simulation of an observed heated and sheared boundary layer with mesoscale forcing. *Preprints, Ninth Symposium on Atmospheric Turbulence and Diffusion, April 30–May 3, Roskilde, Denmark,* 258–261.

Lilly, D. K. and Schneider, J. M. (1990). Dual-Doppler measurement of momentum flux: Results from the Phoenix II study of the convective boundary layer. *Preprints, Ninth Symposium on Atmospheric Turbulence and Diffusion, April 30–May 3, Roskilde, Denmark,* 98–101.

Schneider, J. M. and Lilly, D. K. (1990). The hunt for the big eddy: A visual exploration of the vortical structures in a heated, sheared, planetary boundary layer. *Preprints, Ninth Symposium on Atmospheric Turbulence and Diffusion, April 30–May 3, Roskilde, Denmark.*

Lilly, D. K. (1991). Invited lecture: Toward storm scale numerical weather prediction. *Preprints, Ninth Conference on Numerical Weather Prediction, October 14–18, 1991, Denver, CO,* American Meteorological Society.

Sun, J. and Lilly, D. K. (1991). Initializing storm-scale NWP from single-Doppler radar data. *Preprints, Ninth Conference on Numerical Weather Prediction, October 14–18, 1991, Denver, CO,* American Meteorological Society, 563–566.

Liou, Y. C., Gal-Chen, T. and Lilly, D. K. (1991). Retrieval of wind, temperature and pressure from single Doppler radar and a numerical model. *Preprints, Twenty-Fifth International Conference on Radar Meteorology, June 24–28, 1991, Paris, France,* American Meteorological Society, 151–154.

Kogan, Z. N., Lilly, D. K. and Kogan, Y. L. (1992). The sensitivity of solar radiative properties to cloud microphysics in low cloud layer simulations. *Preprints, Eleventh International Conference on Clouds and Precipitation, Montreal, Canada,* International Commission on Clouds and Precipitation and International Association of Meteorology and Atmospheric Physics, 793–796.

Kogan, Y. L., Lilly, D. K., Kogan, Z. N. and Filyushkin, V. V. (1992). A 3-D large-eddy simulation model of a stratocumulus cloud layer with explicit formulation of microphysical and radiative processes. *Preprints, Eleventh International Conference on Clouds and Precipitation, Montreal, Canada.* International Commission on Clouds and Precipitation and International Association of Meteorology and Atmospheric Physics, 340–343.

Lilly, D. K., Kogan, Y. L., Kogan, Z. N. and Filyushkin, V. V. (1993). Simulations of stratocumulus clouds with emphasis on particle physics and aerosol distributions. *Proceedings, Fourth Conference on Global Change Studies, January 17–22, 1993, Anaheim, CA,* American Meteorological Society, 203–207.

Kogan, Z. N., Lilly, D. K., Kogan, Y. L. and Filyushkin, V. V. (1993). The sensitivity of optical and radiative properties to cloud microphysics in simulations of stratocumulus clouds. *International Symposium on High Latitude Optics. Conference on Atmospheric Radiation, June 27–July 2, 1993, Tromso, Norway*, International Society for Optical Engineering.

Kogan, Y. L., Kogan, Z. N., Lilly, D. K. and Khayroutdinov, M. F. (1994). On parameterization of optical depth in GCM and global climate models. *Proceedings, Fourth ARM Science Team Meeting, February 28–March 3, 1994, Charleston, SC*, US Department of Energy.

(1995). Evaluation of two radiative parameterizations using a 3-D LES microphysical model. *Fourth ARM Science Team Meeting, February 28–March 3, 1994, Charleston, SC*, US Department of Energy, 207–210.

Kogan, Z. N., Kogan, Y. L. and Lilly, D. K. (1995). Simulations of stratocumulus clouds with emphasis on aerosols as cooling influence on atmosphere. *Preprints, Conference on Clouds Physics, January 15–20, 1995, Dallas, TX*, American Meteorological Society, 171–176.

(1995). The effects of stratocumulus cloud layer microphysics on its optical depth. *Preprints, Conference on Clouds Physics, January 15–20, 1995, Dallas, TX*, American Meteorological Society, 74–79.

Liu, Q.-F., Kogan, Y. L. and Lilly, D. K. (1995). The effects of surface moisture on the stratocumulus cloud layer microphysical properties. *Preprints, Conference on Clouds Physics, January 15–20, 1995, Dallas, TX*, American Meteorological Society, (J12)4–(J12)7.

(1995). Reducing the numerical dispersion of the cloud droplet spectrum in condensation calculations. *Preprints, Conference on Clouds Physics, January 15–20, 1995, Dallas, TX*, American Meteorological Society, 112–117.

Kogan, Y. L. Khairoutdinov, M. F., Lilly, D. K., Kogan, Z. N. and Liu, Q.-F. (1995). Study of the effects of stratocumulus cloud layer dynamics on microphysics using a 3-D large-eddy simulation model with explicit microphysics. *Preprints, Conference on Clouds Physics, January 15–20, 1995, Dallas, TX*, American Meteorological Society, 165–170.

Khairoutdinov, M. F., Kogan, Y. L. and Lilly, D. K. (1995). Comparison of the explicit and parameterized formulation of microphysics using a large-eddy simulation model of cloud-topped boundary layer. *Preprints, Conference on Clouds Physics, January 15–20, 1995, Dallas, TX*, American Meteorological Society, 378–380.

Filyushkin, V., Lilly, D. K., Xue, M. and Stames, K. (1995). Adaptive grid implementation in LES model of Arctic stratus clouds. *Preprints, Conference on Clouds Physics, January 15–20, 1995, Dallas, TX*, American Meteorological Society.

Kogan, Z. N., Kogan, Y. L. and Lilly, D. K. (1995). Evaluating aerosol indirect effect through marine stratocumulus cloud. *Fifth ARM Science Team Meeting, March 20–23, 1995, San Diego, CA*. US Department of Energy, 147–150.

(1997). Seasonal variation of the sulfate aerosol indirect effect using observation-derived cloud climatology. *Sixth ARM Science Team Meeting, March 4–7, 1996, San Antonio, TX*. US Department of Energy, 133–138.

Droegemeier, K. K., Bassett, G. M. and Lilly, D. K. (1996). Does helicity really play a role in supercell longevity? *Preprints, Eighteenth Conference on Severe Local Storms, February 19–23, 1996, San Francisco, CA*, American Meteorological Society, 205–209.

Kanak, K. M. and Lilly, D. K. (1996). The linear stability and structure of convection in a mean circular shear. *Preprints, Eighteenth Conference on Severe Local Storms, February 19–23, 1996, San Francisco, CA*, American Meteorological Society, 708–712.

Lilly, D. K. (1996). Numerical simulation and prediction of atmospheric convection. *Proceedings, Les Houches Summer School on Computational Fluid Dynamics, June 28–July 30, 1993*, Elsevier, 325–374.

Zhang, Q., Filyushkin, V. Stamnes, K. and Lilly, D. K. (1997). Large-eddy simulations of the summertime cloudy boundary layer over Arctic ocean. *Proceedings, seventh ARM Science Team Meeting, March 3–7, 1997, San Antonio, TX*, US Department of Energy.

Zhang, Q., Stames, K. and Lilly, D. K. (1998). The influence of radiation and large scale vertical motion on the persistence of Arctic stratus clouds. *Proceedings, Eighth ARM Science Team Meeting, March 23–27, 1998, Tuscon, AZ*, US Department of Energy, 857–860.

Kanak, K. M., Lilly, D. K. and Snow, J.T. (1998). The formation and maintenance of microscale atmospheric vortices. *Preprints, Ninteenth Conference on Severe Local Storms, September 14–18, 1998*, Minneapolis, MN, American Meteorological Society, 76–79.

Lilly, D. K. (2000). The meteorological development of large eddy simulation. In *IUTAM Symposium on Developments in Geophysical Turbulence, June 16–19, 1998*, Robert M. Kerr and Yoshifumi Kimura, eds., Dordrecht, The Netherlands: Kluwer Academic Publishers, 5–18.

Index